Application of Ionic Liquids in Drug Delivery

Masahiro Goto · Muhammad Moniruzzaman
Editors

Application of Ionic Liquids in Drug Delivery

 Springer

Editors
Masahiro Goto
Department of Applied Chemistry
Kyushu University
Fukuoka, Japan

Muhammad Moniruzzaman
Department of Chemical Engineering
Universiti Teknologi PETRONAS
Perak, Malaysia

ISBN 978-981-16-4367-5   ISBN 978-981-16-4365-1   (eBook)
https://doi.org/10.1007/978-981-16-4365-1

This Springer imprint is published by the registered company Springer Nature Singapore Pte Ltd.
The registered company address is: 152 Beach Road, #21-01/04 Gateway East, Singapore 189721,
Singapore

Preface

The pharmaceutical industry has been experiencing a series of challenges with newly developed solid-state drugs because most of them are insoluble or poorly soluble in water or most of the pharmaceutically accepted organic solvents or agents. Other notable limitations with solid-state drugs are polymorphisms and their low bioavailability. As a potential alternative to conventional organic solvents/agents and water, ionic liquids have been used as solvents or materials in pharmaceutics and medicinal fields owing to their excellent properties, which include the combination of "green" properties and adjustable physicochemical and biological properties. Recent development of third-generation ionic liquids comprising biocompatible cations and anions creates innovative opportunities for the development of smart drug formulations and delivery systems.

Written by an international group of experts, this book summarizes the recent works that support the use of ionic liquids for various applications in pharmaceutics and medicine, with a particular emphasis on addressing their critical pharmaceutical challenges including the low solubility, polymorphism, and bioavailability of solid-state drugs. Readers will find diverse approaches to the application of ionic liquids in drug solubility, Active Pharmaceutical Ingredient (API) formulation, and drug delivery, such as topical, transdermal, and oral delivery, with a specific focus on the latest developments. This book also provides insights into the development of biologically functionalized ionic liquid-assisted biopolymers, surfactants, and nano/micro carriers for enhanced drug delivery systems/technologies.

The book contains valuable information and significant evidence on the potential use of biocompatible ionic liquids and/or ionic liquid-based materials/technologies in drug formulations and/or drug delivery systems. The broad coverage also provides a comprehensive resource for researchers and students from different disciplines, such as chemical engineering, chemistry, material science polymer science, and pharmaceuticals/medical fields, who are interested in both ionic liquids and their applications in pharmaceutics and medicine.

Design principles for ionic liquids in drug delivery systems are addressed in Chap. 1, which was written by Chowdhury et al. The authors focus on the design of the ionic liquids, rather than on their implementation in drug delivery systems, and particularly on their precursor selection, biocompatibility status for safety, and potentiality for targeted drug delivery to achieve the maximum therapeutic efficacy with the fewest undesirable side effects. Chapter 2 was written by Moshikur and Goto, and it reviews the recent developments in ionic liquid-based API ingredient strategies to address the solubility and polymorphism challenges of crystalline APIs and explore their possible advantages in pharmaceutics by eliminating or at least minimizing common problems that are associated with solid APIs. The role of ionic liquids in transdermal delivery of APIs is described in Chap. 3 by Berton and Shamshina, where the authors focus on the Choline Geranate (CAGE) ionic liquid. Its scale-up and medical applications are also discussed. Chapter 4 by Moshikur et al. highlights the importance and advantages of ionic liquids as a potential solvent/agent for dissolving sparingly soluble drugs and explores the possible mechanism by which ionic liquids increase solubility during the preparation of drug formulations.

Chapter 5 by Shamshina and Rogers focuses on recent advances in the three-dimensional (3D) printing of cellulose and chitin from ionic liquids for drug delivery. There is a special focus on inkjet 3D printing and extrusion-based 3D printing of biopolymers from an ionic liquid. Recent advances in ionic liquid-based oral drug delivery systems are summarized in Chap. 6 by Islam and Goto. The authors highlight the potentials and limitations of oral therapy, and then they address the possible use of ionic liquids to bypass oral route constraints and enhance oral administration of many medicinal products. Chapter 7 by Pedro et al. reviews the latest applications of ionic liquids toward the development of polymer-based drug delivery systems. The authors also highlight the fundamental knowledge that is required to design multi-responsive copolymers from ionic liquid monomers to enhance the delivery of various drugs. Design and selection of potential ionic liquids for pharmaceutical applications are described using the Conductor-Like Screening Model (COSMO) in Chap. 8 by Khan et al., who highlight the advantages of COSMO-RS as a predictive tool.

Chapter 9 by Ali et al. presents Surface-Active Ionic Liquids (SAILs) as a potent biocompatible alternative to conventional surfactants and explores their pharmaceutical applications as promising surfactants/co-surfactants that help to form a stable formulation in aqueous or non-aqueous media. The authors also highlight why SAIL-based drug formulations represent favorable target drug delivery systems for pharmaceutical applications. Ionic liquid-based transdermal vaccination of drug molecules and antigen peptides is summarized in Chap. 10 by Tahara with a focus on how ionic liquid acts to enhance skin penetration. Chapter 11 by Elgharbawy et al. reviews ionic liquid-based antibiotics for resistant microbial strains and drug polymorphism, focusing on the antibiotics' solubility and bioavailability. Recent advances in ionic liquid-based microemulsions to enhance drug solubility and delivery are summarized in Chap. 12 by Salabat. The mechanism of enhancing the solubility of an active pharmaceutical ingredient and transdermal drug delivery using ionic liquid-based microemulsion is also highlighted.

Fukuoka, Japan   Masahiro Goto
Perak, Malaysia   Muhammad Moniruzzaman

Contents

Chapter 1
Design Principles for Ionic Liquids in Drug Delivery Systems

Md. Raihan Chowdhury, Md Nurunnabi, and Masahiro Goto

Abstract Ionic Liquids (ILs) have been a topic of interest in many scientific areas since the mid-1990s and can be classified as potential "green solvents," especially for use in drug delivery systems (DDSs). Because of the "green" and "designer" properties of ILs, the use of ILs has escalated dramatically in pharmaceutics and medicine. Although virtually an unlimited number of ILs could be produced from diverse sources of organic and inorganic cations and anions, including IL precursors derived from biological sources, special consideration must be given for the implementation of ILs in DDSs. Herein, the focus is on the design of the ILs, rather than their implementation in DDSs, particularly the precursor selection, biocompatibility status for safety (biosafety), and potential for targeted drug delivery. It is possible to design an IL suitable for a DDS with synergistic benefits by achieving the maximum therapeutic efficacy of the drug with minimum undesirable side effects caused by the delivery systems or vehicles.

Keywords Ionic liquids · Drug delivery system · Biocompatibility · Designed principle

M. R. Chowdhury (✉)
Department of Radiology, The Laboratory for Minimally Invasive Tumor Therapies, Beth Israel Deaconess Medical Center, Harvard Medical School, 1 Deaconess Rd, WCC 308-B, Boston, MA 02215, USA
e-mail: rchowdhu@bidmc.harvard.edu

M. Nurunnabi
Department of Pharmaceutical Sciences, School of Pharmacy, University of Texas, El Paso, TX 79902, USA

M. Goto
Department of Applied Chemistry, Graduate School of Engineering, Kyushu University, 744 Motooka, Nishi-ku, Fukuoka 819-0395, Japan

Abbreviations

ABDS	Affinity-Based Delivery Systems
API	Active Pharmaceutical Ingredients
BDOA	Benzyldimethyloctyl Ammonium
CAGE	Choline and Geranic Acid
COSMO-RS	Conductor-Like Screening Model for Real Solvents
DDS	Drug Delivery System
DHA	Docosahexaenoic Acid
IL	Ionic Liquids
siRNA	Small Interfering Ribonucleic Acid

1.1 Introduction

A drug delivery system (DDS) is an optimized formulation or device that can transport a therapeutic agent to a target site of action by avoiding or overcoming the biological barriers associated with cells, organs, and tissues. Some DDSs are simply engineered technologies that use various biocompatible materials for the targeted delivery and/or sustained release of therapeutic agents to maintain a desired level in the body (Erdine and De Andrés 2006; Kwon et al. 2008). New technologies are continuously being developed for use in DDSs, and current DDSs include liposomes, microemulsions, nanoparticles, niosomes, polymers, implants, and transdermal drug delivery systems (Vega-Vásquez et al. 2020). The design of optimized DDSs for particular purpose is challenging. Therefore, particular attention must be paid to the selection of the materials and the design of the vehicles to develop improved DDSs. In the last two decades, a new class of materials, the ionic liquids (ILs), has become popular for use in DDSs (Adawiyah et al. 2016a; Cabuy 2015; Huang et al. 2020; Karande et al. 2005; Pedro et al. 2020). ILs have been a topic of interest in many scientific areas since the mid-1990s, and can be classified as potential "green solvents," especially for use in DDSs (Moshikur et al. 2020b). ILs are the molten organic salts of unsymmetrical organic cations and inorganic or organic anions that have melting points barely above 100 °C (Marrucho et al. 2014; Moniruzzaman and Goto 2011; Shamshina et al. 2013). ILs are known as "designer solvents" because their physicochemical properties can be easily altered by simply changing the combinations of the cations and anions (Earle and Seddon 2007; Huang et al. 2020; Moshikur et al. 2020a). The designer and green solvent properties of ILs make them extremely useful materials for DDSs. The use of ILs has escalated dramatically, especially in the fields of pharmaceutics and medicine, in last two decades (Huang et al. 2020; Moshikur et al. 2020b). DDSs using either ILs or IL-based technologies, such as IL form of active pharmaceutical ingredients (IL-APIs), IL-in-oil microemulsions, IL-drug complexed nanoparticles, IL-protein/peptide complexes, or drugs in IL/IL-water binary solvent systems, have been developed for use in synthetic and medicinal chemistry, especially in pharmaceutical formulations (Adawiyah et al. 2016a; Agatemor et al. 2018; Amaral et al. 2021; Applicable 2013; Chowdhury et al. 2019a). In DDSs, ILs have been mainly used as media, formulation components, polymerizable agents, and plasticizers. Virtually, an unlimited number of ILs could be produced from diverse sources of organic and inorganic cations and anions, including IL precursors derived from biological sources, but special consideration must be given for the implementation of ILs in DDSs (Adawiyah et al. 2016b; Anselmo et al. 2018; Gomes et al. 2019; Kumar et al. 2017; McQueen and Lai 2019; Omar 2016; Rodríguez et al. 2008; Uddin et al. 2020a). However, it is worth mentioning that not all ILs are suitable for DDSs. Currently, the design of ILs is more important than their implementation in DDSs (Huang et al. 2020; Karande et al. 2005; Nebgen et al. 2018). The design of a suitable IL for a DDS should particularly consider the following: (1) the selection of precursors; (2) the biological safety; and (3) the targeted delivery system. Figure 1.1 shows the exponential growth of publications related to the application of ILs or IL-related technologies in different facets of DDSs in the pharmaceutical field, as compiled by Pedro et al. (2020). The data clearly indicates that the use of ILs and IL-related technologies in DDSs is increasing year by year and almost reached a saturated level over the last decade. However, intensive investigations are still required before the regulatory approval of any ILs or IL-related technologies, particularly when used in a DDS, can be acquired. The design of ILs and IL-related pharmaceutics has already been approved by the FDA with a special consideration that must be acquired from the regulatory authorities before use in pharmaceutical applications (Moshikur et al. 2020b; Zandu et al. 2019).

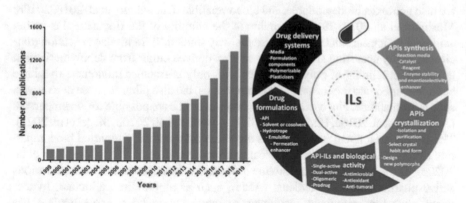

Fig. 1.1 Publications related to ILs or IL-related technologies used in DDSs over a 20-year period (left). An overview of the use of ILs in the pharmaceutical field (right). Reproduced from Pedro et al. (2020) with permission from the International Journal of Molecular Sciences and MDPI

1.2 Selection of Precursors

ILs consist of two precursor molecules, a cation and an anion, which can be derived from organic or inorganic sources (Welton 2018). According to the principal of green chemistry, an IL can be declared a "green" material/solvent/co-solvent only if it meets certain standards for toxicity and biodegradability (Gomes et al. 2019). An IL should be biocompatible if it is derived from biocompatible precursors. However, it has been demonstrated that the toxicity of ILs is an unpredictable parameter that does not follow any formal rules (Rodríguez et al. 2008). Several studies have investigated the relationship between the toxicity of an IL and its cation and anion precursors, but unfortunately no general correlations have been established yet (Chowdhury et al. 2018; Moshikur et al. 2020a). The accumulated toxicity and biodegradability data reported in the last few decades indicate that: (A) the ideal strategy for synthesizing biocompatible ILs is to use precursors from biocompatible sources; (B) the length of the alkyl side chain in the cations might have a strong influence on the toxicity; (C) the functional groups in the cations might contribute to the toxicity; (D) the nature of the cations and anions influences the toxicity; and (E) intermolecular interactions between the precursors also affect the toxicity (Egorova et al. 2017; Gomes et al. 2019; Marrucho et al. 2014; Omar 2016; Uddin et al. 2020b). As a result, third generation ILs have been designed with favorable biodegradable and biocompatible profiles especially for use in DDSs. The most common approach to design such ILs is to use biomolecules, such as amino acids, non-nutritive sweeteners, glucose, and carboxylic acids, as the precursors to minimize environmental, biochemical, and financial concerns (Egorova et al. 2017; Gomes et al. 2019; Rodríguez et al. 2008; Sivapragasam et al. 2020). Amino acids, which can be derived from protein hydrolysis and purification, can be considered as bio-renewable precursor molecules, and are also nontoxic, biodegradable, and biocompatible (Chowdhury et al. 2018, 2019b; Moshikur et al. 2018, 2019). To maintain the chirality of the ILs, natural α-amino acids and their ester salts have been extensively studied. To minimize the cation toxicity, naturally occurring sugars are an option, derived either from depolymerization or the direct refining of polysaccharides. Not only D-fructose monomers that have been converted into monosubstituted compounds, but also other fructose derivatives, such as isomannide, glucose, arabinose, and isosorbide are possible cation precursors (Adawiyah et al. 2016a; Huang et al. 2020; Moshikur et al. 2020b; Siegel et al. 2021). The cholinium ion and its derivatives are the most commonly reported biocompatible cations because of their biological source (Amaral et al. 2021; Moshikur et al. 2020b). The design of ILs with choline or its derivatives in combination with various active pharmaceutical ingredients (APIs), such as ampicillin, nalidixate, pyrazinoate, phenytoin, niflumate, picolinate, methotrexate, and 4-amino salicylate, has contributed to the improvement of pharmaceuticals, particularly pharmaceuticals used in DDSs (Amaral et al. 2021; Applicable 2013; Huang et al. 2020; Moshikur et al. 2020b; Pedro et al. 2020). In a recent study, according to the green chemistry metrics, a series of amino acid-derived surface active ILs with different head group precursors, including pyridium, imidazolium, and cholinium, has been investigated to determine the effect of the chain length on the biocompatibility and toxicity of ILs and the subsequent effects in DDSs (Pedro et al. 2020). Anions derived from natural organic acids; such as malic acid, succinic acid, and tartaric acid, are also options for the production of biocompatible ILs suitable for DDSs in combination with appropriate cationic precursors (Amaral et al. 2021; Moshikur et al. 2020b; Pedro et al. 2020). In a recent study, Tanner et al. outlined the design principles for ILs used for transdermal delivery; 16 ILs were used to investigate the dependence of the skin penetration of a drug on the chemical properties of the IL (Tanner et al. 2019). The study revealed that the ability of an IL to enhance the transdermal drug delivery was inversely correlated with the inter-ionic interactions. Finally, using knowledge of the effect of the ion stoichiometry of ILs on the skin penetration of drugs and knowledge of the inter-ionic interactions obtained by 2D NMR spectroscopy, an IL was designed that provided the highest delivery of the drug of all the 16 ILs investigated. This study provided a generalized framework for optimizing ILs for enhanced skin permeation using the appropriate precursor molecules. However, only the delivery potency of the ILs was described in this study, and it was strongly emphasized that additional investigations were required to evaluate the biological safety.

To design the best possible sustainable ILs for DDSs, the conductor-like screening model for real solvents (COSMO-RS) has become a very popular tool for the effective prediction of the toxicity and biocompatibility of ILs for implementation in DDSs (Lotfi et al. 2016). COSMO-RS is a quantum chemistry-based equilibrium thermodynamics method used for predicting the chemical potential (μ) in liquids by processing the screening charge density (σ) on the surface of molecules to calculate the chemical potential (μ) of each species in a solution by considering the temperature and pressure at 25 °C and 1 atm, respectively. The surface chemical potential, μ, with screening charge density, σ, in bulk, can be calculated using the following equation (Eq. 1.1):

$$\mu s(\sigma) = -\frac{RT}{Aeff} ln \left[Ps(\sigma\prime)exp\left\{ \frac{Aeff}{RT}[\mu s(\sigma\prime) - Emisfit(\sigma, \sigma\prime) \right. \right.$$
$$\left. \left. -EH.bonding(\sigma, \sigma\prime)]\right\}\right]d\sigma \tag{1.1}$$

where $Aeff$ is the effective contact area between two surfaces, $EH.bonding$ describes the energy share from H-bonding interactions, $Emisfit$ describes the electrostatic contact interaction energy, and $s()$ is the affinity of system S to the surface polarity σ measurement.

The ability of COSMO-RS to predict the solubility of chemicals in any pure or mixed solvents enables the program to be used to predict the solubilization of a drug in an IL, or mixture of ILs, as an effective pre-screening for ILs in DDSs. However, many other factors can have a detrimental effect on the toxicity and biocompatibility of ILs, and the exact mechanisms need to be investigated in detail in future ILs and IL-related research, especially for DDSs.

1.3 Biological Safety

Although ILs and IL-related techniques have tremendous potential for improving DDSs, as well as in drug development, their implementation is not always without problems. The main controversial issues with ILs and IL-related techniques in DDSs are the biological safety and biocompatibility, along with people's perceptions (Moshikur et al. 2020b). To develop a suitable IL or IL-related technique for DDSs, first, the biodegradability profile of the IL should be investigated using approved biodegradation methods to predict the minimum/maximum environmental effects upon direct/indirect interactions. However, the pharmaceutical application of ILs or IL-related techniques can only be approved upon satisfactory and detailed toxicological investigations, such as in vitro cytotoxicity assays using different human cell lines and in vivo cytotoxicity analysis using murine models (Petkovic et al. 2010a; b). In addition, the cumulative effect on biological systems needs to be investigated to determine the biocompatibility of an IL material. The 1st and 2nd generations of ILs were extensively studied regarding their cytotoxicity and environmental and microbial toxicity, which ultimately promoted the development of the 3rd generation of ILs that contain biocompatible precursors (Moshikur et al. 2020b). The 3rd generation ILs, which are most often nontoxic and readily biodegradable, could be appropriate materials for pharmaceutical applications, including DDSs. The addition of biological properties, along with the tunable physicochemical properties, is the main advantage of the 3rd generation of ILs that may direct them into biopharmaceutical applications, such as use as antibacterial, local anesthetic, anticholinergic, and antifungal agents. One of the best strategies to develop an IL-based biopharmaceutical is the use of an API or API precursor, either as the cation or anion or both, to achieve a pharmacologic effect (Egorova et al. 2017; Jordan and Gathergood 2015; Kumar et al. 2017). A few ILs are less toxic compared with their precursors, such as the IL from the venom of the fire ant, *Solenopsis* and formate from a different venom of its competitor *Nylanderia fulva*. Interestingly, the toxic effect of the precursor venom simply disappears in the synthesized IL (Moshikur et al. 2020b). In a study on the "design principals of ionic liquids for transdermal delivery," the IL based on choline and geranic acid (CAGE) was reported to have a lower dermal toxicity than that of its individual precursors (Tanner et al. 2019). However, it is not guaranteed that the final IL product will be safer compared with the precursors every time. Sometimes, a biocompatible IL or IL-drug formulation can be non-biocompatible at a higher concentration than the plasma isotonic level. The tolerable limit for plasma IV infusion is ~300–600 $mOsmL^{-1}$ (Moshikur et al. 2020b). Biological systems will treat an IL as a salt, as ILs consist of a cation and an anion, which can have considerable effects post-administration. A formulation of a biocompatible IL, cholinium dihydrogen phosphate, was considerably hypertonic when administered at 12%, w/v (~1520 $mOsmL^{-1}$) (Moshikur et al. 2020b). However, ILs can be designed based on their targeted route of administration, and any toxicological effects might be minimized by selecting the appropriate route of delivery. A database for the biological safety of ILs and IL-related techniques could be established by investigating the molecular aspects behind the toxicity/biocompatibility of the precursors, which would assist in the development of subsequent ILs for future pharmaceutical applications, especially for DDSs.

1.4 Targeted Delivery Systems

The therapeutic effects of many drugs are limited by their poor solubility in biological fluids, *in vivo* instability, insufficient plasma concentration of the drug or the time at the site of action, and, in particular, a lack of targeted delivery that can often cause undesirable side effects and high plasma fluctuations during therapeutic applications (Applicable 2013; Karande et al. 2005; Nebgen et al. 2018). In the last few decades, targeted DDSs have been explored extensively to overcome these problems to ensure the targeted and controlled delivery of a high payload of a drug, and also to maintain the desired level of the drug in the body within the therapeutic window (Deng et al. 2019; Hwu et al. 2009; Medi and Singh 2006; Wayne et al. 2019; Zhang et al. 2019). Figure 1.2 shows some examples of targeted DDS vehicles composed of different materials, which can provide an extended period of action for a drug, optimized drug concentrations at a target site, reduced drug degradation and thus the dosage required, less undesirable side effects, less fluctuations in the drug concentration in the plasma, enhanced bioavailability, and improved patient compliance (Amaral et al. 2021).

Currently, ILs and IL-related techniques are being explored extensively in targeted DDSs. The solubilization of increased amounts of a drug through the interactions between IL ions and specific groups present in the drug allows the delivery of higher

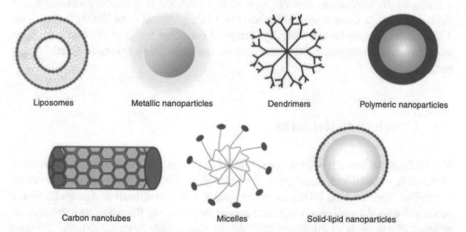

Fig. 1.2 Schematic illustration of targeted drug delivery systems for therapeutic applications. Reproduced from Amaral et al. (2021) with permission from Nanomedicine of Future Medicine Ltd

concentrations of the drug compared with the free drug (Islam et al. 2020a; b). The water solubility of paracetamol and diclofenac was reported to be enhanced significantly using N-acetyl amino acid N-alkyl cholinium-based ILs (Amaral et al. 2021). ILs based on choline and geranic acid (CAGE) successfully enabled the oral delivery of insulin with enhanced paracellular transport (Banerjee et al. 2018). The use of API-ILs offers enormous potential for promoting the dissolution of poorly soluble drugs. As a targeted delivery approach, the transdermal route has become very popular for delivering therapeutics, including small molecules and biologics (peptides and proteins), using ILs or IL-based techniques through and across the skin (Banerjee et al. 2017; Chowdhury et al. 2021; Tanner et al. 2018). The topical delivery and cellular internalization of RNAi (siRNA) were significantly enhanced using the IL, benzyldimethyloctyl ammonium-siRNA (Zakrewsky and Mitragotri 2016). ILs and IL-related techniques can deliver antigenic proteins and peptides to antigen presenting cells, such as Langerhans cells, in the dermis and epidermis by penetrating the lipid bilayers of the skin (Chowdhury et al. 2021; Tanner et al. 2019). It has been reported that the incorporation of ILs in transdermal and topical delivery systems could enhance the penetration and permeation of therapeutics into and across the skin in a non-invasive manner, upon the disruption of the lipid membranes (Anselmo et al. 2018). Nurunnabi et al. have shown that a CAGE IL could reduce body weight of treated mice by decreasing fat absorption through the intestine (Nurunnabi et al. 2019). The in vivo data showed 60–70% reduced absorption of the fat molecule, docosahexaenoic acid, in the intestine using CAGE. The mechanism for this effect could be that an interaction between CAGE and the fat molecules prevented the absorption of the fat by the intestinal tissue, which eventually provided a feeling of satiety. ILs have also been developed that incorporate metals to detect biomolecules, such as glucose in diabetic patients. Senthilkumar and coworkers have fabricated a metal-containing IL-based glucose sensor by immobilizing an IL containing Salophen on electrochemically reduced graphene oxide deposited onto a screen-printed carbon electrode (Amaral et al. 2021; Pedro et al. 2020). ILs can also be strategically designed for other targeted applications, such as anti-microbial agents, food preservatives, and stabilizing agents for proteins and enzymes.

1.5 Concluding Remarks

Much effort has already been devoted toward improving DDSs, but the incorporation of ILs or IL-related techniques provides a new dimension in the design of innovative, controlled, and targeted DDSs for therapeutics. It is possible to design an IL that is suitable for DDSs and has synergistic benefits by achieving the maximum therapeutic efficacy of a drug, with minimum undesirable side effects from the delivery systems or vehicles, by considering the biocompatibility of the precursors, and the safety and targeted delivery. The possible market for ILs in DDSs is large, provided these DDSs can gain regulatory approval. However, there is still a considerable amount of research required before the potential application of ILs in DDSs can be realized. We thank Edanz Group (https://en-author-services.edanz.com/ac) for editing a draft of this manuscript.

References

Adawiyah N, Moniruzzaman M, Hawatulaila S, Goto M (2016a) Ionic liquids as a potential tool for drug delivery systems. Medchemcomm 7:1881–1897. https://doi.org/10.1039/c6md00358c

Adawiyah N, Moniruzzaman M, Hawatulaila S, Goto M (2016b) Ionic liquids as a potential tool for drug delivery systems. The Royal Society of Chemistry, MedChemComm. https://doi.org/10.1039/c6md00358c

Agatemor C, Ibsen KN, Tanner EEL, Mitragotri S (2018) Ionic liquids for addressing unmet needs in healthcare. Bioeng Transl Med 3:7–25. https://doi.org/10.1002/btm2.10083

Amaral M, Pereiro AB, Gaspar MM, Reis CP (2021) Recent advances in ionic liquids and nanotechnology for drug delivery. Nanomedicine 16:63–80. https://doi.org/10.2217/nnm-2020-0340

Anselmo AC, Gokarn Y, Mitragotri S (2018) Non-invasive delivery strategies for biologics. Nat Rev Drug Discov 18:19–40. https://doi.org/10.1038/nrd.2018.183

Applicable P (2013) Drug delivery systems: advanced technologies potentially applicable in personalised treatment. Springer, US, Coimbra

Banerjee A, Ibsen K, Brown T, Chen R, Agatemor C, Mitragotri S (2018) Ionic liquids for oral insulin delivery. Proc Natl Acad Sci USA 115:7296–7301. https://doi.org/10.1073/pnas.1722338115

Banerjee A, Ibsen K, Iwao Y, Zakrewsky M, Mitragotri S (2017) Transdermal protein delivery using choline and geranate (CAGE) deep eutectic solvent. Adv Healthc Mater 6:1–11. https://doi.org/10.1002/adhm.201601411

Cabuy E (2015) Thermal ablation in cancer treatment. Therm Ablation Cancer Treat 2:1–68

Chowdhury MR, Moshikur RM, Wakabayashi R, Moniruzzaman M, Goto M (2021) Biocompatible ionic liquids assisted transdermal co-delivery of antigenic protein and adjuvant for cancer immunotherapy. Int J Pharm 601:120582. https://doi.org/10.1016/j.ijpharm.2021.120582

Chowdhury MR, Moshikur RM, Wakabayashi R, Tahara Y, Kamiya N, Moniruzzaman M, Goto M (2019a) Development of a novel ionic liquid-curcumin complex to enhance its solubility, stability, and activity. Chem Commun 55:7737–7740. https://doi.org/10.1039/c9cc02812a

Chowdhury MR, Moshikur RM, Wakabayashi R, Tahara Y, Kamiya N, Moniruzzaman M, Goto M (2019b) In vivo biocompatibility, pharmacokinetics, antitumor efficacy, and hypersensitivity evaluation of ionic liquid-mediated paclitaxel formulations. Int J Pharm 565:219–226. https://doi.org/10.1016/j.ijpharm.2019.05.020

Chowdhury MR, Moshikur RM, Wakabayashi R, Tahara Y, Kamiya N, Moniruzzaman M, Goto M (2018) Ionic-liquid-based paclitaxel preparation: a new potential formulation for cancer treatment. Mol Pharm 15:2484–2488. https://doi.org/10.1021/acs.molpharmaceut.8b00305

Deng C, Zhang Q, Jia M, Zhao J, Sun X, Gong T, Zhang Z (2019) Tumors and their microenvironment dual-targeting chemotherapy with local immune adjuvant therapy for effective antitumor immunity against breast cancer. Adv Sci 6:1801868. https://doi.org/10.1002/advs.201801868

Earle MJ, Seddon KR (2007) Ionic liquids. Green solvents for the future. Pure Appl Chem 72:1391–1398. https://doi.org/10.1351/pac200072071391

Egorova KS, Gordeev EG, Ananikov VP, Zelinsky ND (2017) Biological activity of ionic liquids and their application in pharmaceutics and medicine. Chem Rev 117:7132–7189. https://doi.org/10.1021/acs.chemrev.6b00562

Erdine S, De Andrés J (2006) Drug delivery systems. Pain Pract. https://doi.org/10.1111/j.1533-2500.2006.00059.x

Gomes JM, Silva SS, Reis RL (2019) Biocompatible ionic liquids: fundamental behaviours and applications. Chem Soc Rev 48:4317–4335. https://doi.org/10.1039/c9cs00016j

Huang W, Wu X, Qi J, Zhu Q, Wu W, Lu Y, Chen Z (2020) Ionic liquids: green and tailor-made solvents in drug delivery. Drug Discov Today 25:901–908. https://doi.org/10.1016/j.drudis.2019. 09.018

Hwu JR, Lin YS, Josephrajan T, Hsu MH, Cheng FY, Yeh CS, Su WC, Shieh DB (2009) Targeted paclitaxel by conjugation to iron oxide and gold nanoparticles. J Am Chem Soc 131:66–68. https://doi.org/10.1021/ja804947u

Islam MR, Chowdhury MR, Wakabayashi R, Kamiya N, Moniruzzaman M, Goto M (2020a) Ionic liquid-in-oil microemulsions prepared with biocompatible choline carboxylic acids for improving the transdermal delivery of a sparingly soluble drug. Pharmaceutics 12:1–18. https://doi.org/10. 3390/pharmaceutics12040392

Islam MR, , Chowdhury MR, Wakabayashi R, Tahara Y, Kamiya N, Moniruzzaman M, Goto M (2020b) Choline and amino acid based biocompatible ionic liquid mediated transdermal delivery of the sparingly soluble drug acyclovir. Int J Pharm 582. https://doi.org/10.1016/j.ijpharm.2020. 119335

Jordan A, Gathergood N (2015) Biodegradation of ionic liquids-a critical review. Chem Soc Rev 44:8200–8237. https://doi.org/10.1039/c5cs00444f

Karande PS, Jain AK, Mitragotri S (2005) Design principles of chemical permeation enhancers for transdermal drug delivery. AIChE Annu Meet Conf Proc 2005:8372

Kumar A, Bisht M, Venkatesu P (2017) Biocompatibility of ionic liquids towards protein stability: a comprehensive overview on the current understanding and their implications. Int J Biol Macromol 96:611–651. https://doi.org/10.1016/j.ijbiomac.2016.12.005

Kwon IK, Kim SW, Chaterji S, Vedantham K, Park K (2008) Drug delivery systems: smart drug delivery systems. In: Smart materials, pp 13-1–13-10. Elsevier. https://doi.org/10.1016/B978-0-12-816137-1.00078-7

Lotfi M, Moniruzzaman M, Abdul Mutalib MIB, Rajabi MS (2016) Predicting the solubility of pharmaceutical compound in ionic liquids using COSMo-RS model. ARPN J Eng Appl Sci 11:1618–1622

Marrucho IM, Branco LC, Rebelo LPN (2014) Ionic liquids in pharmaceutical applications. Annu Rev Chem Biomol Eng 5:527–546. https://doi.org/10.1146/annurev-chembioeng-060713-040024

McQueen L, Lai D (2019) Ionic liquid aqueous two-phase systems from a pharmaceutical perspective. Front Chem 7:1–9. https://doi.org/10.3389/fchem.2019.00135

Medi BM, Singh J (2006) Skin targeted DNA vaccine delivery using electroporation in rabbits II. Safety Int J Pharm 308:61–68. https://doi.org/10.1016/j.ijpharm.2005.10.035

Moniruzzaman M, Goto M (2011) Ionic liquids: future solvents and reagent for pharmaceuticals. J Chem Eng Japan 44:370–381

Moshikur RM, Chowdhury MR, Fujisawa H, Wakabayashi R, Moniruzzaman M, Goto M (2020a) Design and characterization of fatty acid based amino acid ester as a new "green" hydrophobic ionic liquid for drug delivery. ACS Sustain Chem Eng. https://doi.org/10.1021/acssuschemeng. 0c03419

Moshikur RM, Chowdhury MR, Moniruzzaman M, Goto M (2020b) Biocompatible ionic liquids and their applications in pharmaceutics. Green Chem 22:8116–8139. https://doi.org/10.1039/d0g c02387f

Moshikur RM, Chowdhury MR, Wakabayashi R, Tahara Y, Moniruzzaman M, Goto M (2019) Ionic liquids with methotrexate moieties as a potential anticancer prodrug: synthesis, characterization and solubility evaluation. J Mol Liq 278:226–233. https://doi.org/10.1016/j.molliq.2019.01.063

Moshikur RM, Chowdhury MR, Wakabayashi R, Tahara Y, Moniruzzaman M, Goto M (2018) Characterization and cytotoxicity evaluation of biocompatible amino acid esters used to convert salicylic acid into ionic liquids. Int J Pharm 546:31–38. https://doi.org/10.1016/j.ijpharm.2018. 05.021

Nebgen BT, Magurudeniya HD, Kwock KWC, Ringstrand BS, Ahmed T, Seifert S, Zhu JX, Tretiak S, Firestone MA (2018) Design principles from multiscale simulations to predict nanostructure in self-assembling ionic liquids. Faraday Discuss 206:159–181. https://doi.org/10.1039/c7fd00 154a

Nurunnabi M, Ibsen KN, Tanner EEL, Mitragotri S (2019) Oral ionic liquid for the treatment of diet-induced obesity. Proc Natl Acad Sci USA 116(50):25042–25047. https://doi.org/10.1073/pnas.1914426116

Omar M (2016) A review of ionic liquids for advance in drug delivery: theory and pharmaceutical implementation. UK J Pharm Biosci 4:41–44. https://doi.org/10.20510/ukjpb/4/i1/87844

Pedro SN, Freire CSR, Silvestre AJD, Freire MG (2020) The role of ionic liquids in the pharmaceutical field: an overview of relevant applications. Int J Mol Sci 21:1–50. https://doi.org/10.3390/ijms21218298

Petkovic M, Ferguson JL, Gunaratnc HN, Ferreira R, Leitao MC, Seddon KR, Rebelo LPN, Pereira CS (2010a) Novel biocompatible cholinium-based ionic liquids-toxicity and biodegradability. Green Chem 12:643–649

Petkovic M, Ferguson JL, Gunaratne HQN, Ferreira R, Leitão MC, Seddon KR, Rebelo LPN, Pereira CS (2010b) Novel biocompatible cholinium-based ionic liquids—toxicity and biodegradability. Green Chem 12:643. https://doi.org/10.1039/b922247b

Rodríguez H, Bica K, Rogers R (2008) Ionic liquid technology: a potential new platform for the pharmaceutical industry. Trop J Pharm Res 7:1011–1012

Shamshina JL, Barber PS, Rogers RD (2013) Ionic liquids in drug delivery. Expert Opin Drug Deliv 10:1367–1381. https://doi.org/10.1517/17425247.2013.808185

Siegel DJ, Anderson GI, Cyr N, Lambrecht DS, Zeller M, Hillesheim PC, Mirjafari A (2021) Molecular design principles of ionic liquids with a sulfonyl fluoride moiety. New J Chem 45:2443–2452. https://doi.org/10.1039/d0nj05603k

Sivapragasam M, Moniruzzaman M, Goto M (2020) An overview on the toxicological properties of ionic liquids toward microorganisms. Biotechnol J 15:9. https://doi.org/10.1002/biot.201900073

Tanner EEL, Curreri AM, Balkaran JPR, Selig-Wober NC, Yang AB, Kendig C, Fluhr MP, Kim N, Mitragotri S (2019) Design principles of ionic liquids for transdermal drug delivery. Adv Mater 31:1–10. https://doi.org/10.1002/adma.201901103

Tanner EEL, Ibsen KN, Mitragotri S (2018) Transdermal insulin delivery using choline-based ionic liquids (CAGE). J Control Release 286:137–144. https://doi.org/10.1016/j.jconrel.2018.07.029

Uddin MN, Basak D, Hopefl R, Minofar B (2020) Potential application of ionic liquids in pharmaceutical dosage forms for small molecule drug and vaccine delivery system. J Pharm Pharm Sci 23:158–176. https://doi.org/10.18433/jpps30965

Uddin S, Chowdhury MR, Wakabayashi R, Kamiya N, Moniruzzaman M, Goto M (2020b) Lipid based biocompatible ionic liquids: Synthesis, characterization and biocompatibility evaluation. Chem Commun 56:13756–13759. https://doi.org/10.1039/d0cc04491a

Vega-Vásquez P, Mosier NS, Irudayaraj J (2020) Nanoscale drug delivery systems: from medicine to agriculture. Front Bioeng Biotechnol. https://doi.org/10.3389/fbioe.2020.00079

Wayne EC, Long C, Haney MJ, Batrakova EV, Leisner TM, Parise LV, Kabanov AV (2019) Targeted delivery of siRNA lipoplexes to cancer cells using macrophage transient horizontal gene transfer. Adv Sci 6:1900582. https://doi.org/10.1002/advs.201900582

Welton T (2018) Ionic liquids: a brief history. Biophys Rev 10:691–706. https://doi.org/10.1007/s12551-018-0419-2

Zakrewsky M, Mitragotri S (2016) Therapeutic RNAi robed with ionic liquid moieties as a simple, scalable prodrug platform for treating skin disease. J Control Release 242:80–88. https://doi.org/10.1016/j.jconrel.2016.09.003

Zandu SK, Chopra H, Singh I (2019) Ionic liquids for therapeutic and drug delivery applications. Curr Drug Res Rev 12:26–41. https://doi.org/10.2174/2589977511666191125103338

Zhang L, Wu S, Qin Y, Fan F, Zhang Z, Huang C, Ji W, Lu L, Wang C, Sun H, Leng X, Kong D, Zhu D (2019) Targeted codelivery of an antigen and dual agonists by hybrid nanoparticles for enhanced cancer immunotherapy. Nano Lett 19:4237–4249. https://doi.org/10.1021/acs.nanolett.9b00030

Chapter 2
Ionic Liquids as Active Pharmaceutical Ingredients (APIs)

Rahman Md Moshikur⊙ and **Masahiro Goto**⊙

Abstract The development of effective drug delivery systems for poorly water-soluble drugs remains a significant challenge for the pharmaceutical industry because of their limited solubility, bioavailability, permeability and stability and their polymorphic conversion. A well-established approach to address these limitations is to convert the active compounds to salts; however, the challenges related to bioavailability, permeability and polymorphic transformation of crystalline drugs remain. The incorporation of active pharmaceutical ingredients (APIs) into ionic liquids (ILs) has been shown to be an attractive method for resolving these challenges and/or significantly increasing the pharmacokinetic and pharmacodynamic properties of drugs. To date, API-ILs have been designed to enhance the solubility of poorly water-soluble drugs in both water and simulated fluids, and to disrupt physiological barriers to deliver drugs to target sites. This chapter highlights the progress of ILs in API-related research. The discussion is focussed on the importance and advantages of the API-IL approach for the development of novel drugs, considering not only the physicochemical properties but also the pharmacological profiles of the API-ILs.

Keywords Active pharmaceutical ingredients · Ionic liquid · Dissolution · Drug development · Biomedical activity

R. Md Moshikur · M. Goto (✉)
Department of Applied Chemistry, Graduate School of Engineering, Kyushu University, 744 Motooka, Nishi-ku, Fukuoka 819-0395, Japan
e-mail: m-goto@mail.cstm.kyushu-u.ac.jp

M. Goto
Advanced Transdermal Drug Delivery System Centre, Kyushu University, 744 Motooka, Nishi-ku, Fukuoka 819-0395, Japan

Division of Biotechnology, Centre for Future Chemistry, Kyushu University, 744 Motooka, Nishi-ku, Fukuoka 819-0395, Japan

© The Author(s), under exclusive license to Springer Nature Singapore Pte Ltd. 2021
M. Goto and M. Moniruzzaman (eds.), *Application of Ionic Liquids in Drug Delivery*,
https://doi.org/10.1007/978-981-16-4365-1_2

Abbreviations

[C$_2$MIM]	1-Ethyl-3-methylimidazolium
[C$_{16}$Pyr]	1-Hexadecylpyridium
[N$_{2,2,2,2}$]	Tetraethylammonium
[N$_{4,4,4}$]	Tributylammonium
[N$_{4,4,4,4}$]	Tetrabutylammonium
[N$_{1,1,10,10}$]	Didecyldimethylammonium
[N$_{6,6,6,6}$]	Tetrahexylammonium
[P$_{4,4,4,4}$]	Tetrabutylphosphonium
[P$_{6,6,6,14}$]	Tributyl(tetradecyl)phosphonium
[C$_{16}$Pyr]	Cetylpyridinium
[HMPyr]	1-Methylpyrrolidinium
[AAE]	Amino acid ester
[AlaEt]	Alanine ethyl ester
[ProEt]	Proline ethyl ester
[AspEt]	Aspartic diethyl ester
[PheEt]	Phenylalanine ethyl ester
[C$_2$OHMIM]	1-(2-Hydroxyethyl)-3-methylimidazolium
[mPEG3N$_{444}$]	Triethylene glycol monomethyl ether tributylammonium
[(C$_{10}$)$_2$(C$_1$)$_2$ N]	Didecyldimethylammonium

2.1 Introduction

The pharmaceutical industry is facing unprecedented challenges related to the delivery of many solid active pharmaceutical ingredients (APIs) because of their limited solubility, insufficient bioavailability, polymorphism and poor stability, as well as formulation difficulties (Egorova et al. 2017; Ali et al. 2020). Approximately 40% of marketed drugs and up to 70% of drugs under development are poorly water-soluble, which leads to poor bioavailability and delivery difficulties, and thus causes them to fail in the later stages of development (Rodriguez-Aller et al. 2015; Moshikur et al. 2020b). Generally, the therapeutic efficacy of APIs is strongly influenced by their chemical structures, which contain various functional groups capable of forming strong inter/intramolecular interactions via hydrogen and/or halogen bonding, resulting in the formation of highly crystalline solids (Shamshina et al. 2015; Ali et al. 2021). These crystalline forms of APIs are accompanied by a suite of issues, such as lower aqueous solubility, irregular gastrointestinal absorption, pre-systemic metabolism, possible toxicity and side-effects of polymorphs and challenging particle size modification (Shamshina et al. 2015; Egorova et al. 2017). Large differences in bioavailability have been observed for different polymorphs of APIs, leading to toxic or potentially lethal ramifications if the wrong polymorph is administered (Ghielmetti et al. 1976). Difficulty controlling the particle size of solid APIs is another common problem that is often critical to the development of formulations with therapeutic efficacy. Based on the structural diversity of drug molecules (ionisable or non-ionisable), several approaches such as salt formation, prodrug conversion, nanoemulsion formation, micellisation and nanoparticle formation, are now widely used to improve their aqueous solubility and bioavailability (Rodriguez-Aller et al. 2015; Egorova et al. 2017). However, these techniques often use large quantities of organic solvents, leading to concerns for human health and ecosystems (Clarke et al. 2018). Green techniques that do not compromise the therapeutic efficacy of drugs are therefore required for effective delivery.

The use of API-ionic liquids (API-ILs, organic salts prepared by pairing an ionisable API with an appropriate IL-forming counterion that melt below body temperature) is a promising tool for addressing the polymorphism and aqueous solubility of solid drugs. ILs comprising APIs are expected to provide many advantageous physicochemical and biopharmaceutical properties over solid or crystalline APIs (Egorova et al. 2017; Moshikur et al. 2020b). The strategic design and appropriate choice of IL-forming counterions have the potential to not only stimulate the synergetic actions of API-ILs, but to allow tuning of physico-thermal properties such as solubility, polymorphism, hygroscopicity, permeability and thermal stability. In addition, the scientific and technical advantages of APIs in liquid form could allow facile formulation and delivery via various routes of administration compared with the solid or crystalline drug. Therefore, the API-IL strategy is a potential opportunity for pharmaceutical companies as one of their options in a competitive market.

The research activity related to the use of ILs in drug formulation and delivery has significantly expanded over the last two decades (Egorova et al. 2017; Moshikur et al. 2020b). Figure 2.1a shows the number of research publications that have reported the application of ILs/IL technologies in the pharmaceutical field since 2000. Notably, the API-ILs approach is the most studied area among the IL-based drug delivery

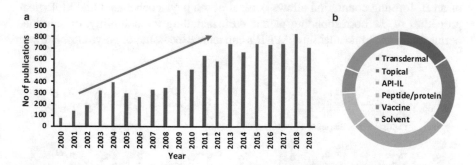

Fig. 2.1 a Publications related to ionic liquids (ILs)/IL technologies over the past two decades. **b** A comparison of the prevalence of API-IL technology and other commonly used IL-related technologies in the pharmaceutics literature, reproduced with permission from (Moshikur et al. 2020b)

strategies (Fig. 2.1b). Consequently, many companies that produce generic pharmaceuticals are increasingly motivated to design effective API-IL bearing drug delivery systems. However, the objective of this chapter is to summarise and further motivate biomedical application-driven exploration of API-ILs.

2.2 Design of API-ILs

The design of novel liquid forms of APIs is an appealing strategy for addressing the innate difficulties of many crystalline drugs. ILs derived from APIs may also provide new perspectives in terms of lowering the production costs or repurposing of classical drugs, and address the risk of toxicity arising from different undetected polymorphic phases that could cause harmful effects in patients. Converting solid drugs into API-ILs by pairing with tailor-made IL-forming counterions can lead to desirable physicochemical and biopharmaceutical properties. Their charged liquid state allows fine-tuning of their melting enthalpy barrier as well as the solubility/bioavailability. With effectively infinite options for design and flexibility, the use of the API-IL platform in drug delivery systems has already made advances through several approaches reported in the literature (Fig. 2.2). In this section, several implementation strategies for the API-IL platform developed over the last decade will be discussed.

2.2.1 Single-Active API-ILs

The combination of crystalline APIs with an appropriate IL-forming counterion is a promising technique for converting conventional pharmaceuticals into IL salts (Fig. 2.2a). These salts usually melt below body temperature and comprise one pharmaceutically active API and an IL-forming counterion. The appropriate selection of an IL-forming counterion allows control of the physicochemical and biological properties of the corresponding parent API, including the solubility, dissolution, permeability and bioavailability. API-ILs can reduce the issues of polymorphism and

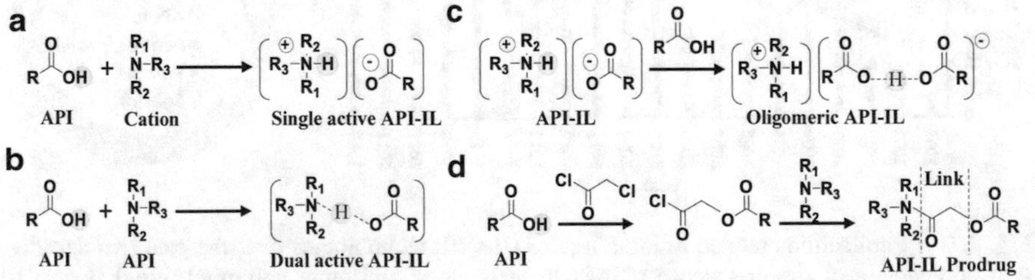

Fig. 2.2 Schematic examples of API-ILs in pharmaceutical approaches. **a** The preparation of a single-active API-IL, **b** dual-active API-IL, **c** oligomeric API-IL, and **d** API-IL prodrug

crystallinity that are related to the low aqueous solubility, poor therapeutic efficiency and thermal instability of drugs. A series of API-ILs with different pharmacological activities have been reported where solid APIs such as lidocaine, ibuprofen, sulfacetamide, sulfasalazine, indomethacin, procaine, aspirin, salicylic acid, methotrexate, piperacillin and penicillin are converted to the IL form by combining with IL-forming cations such as cholinium, amino acid ester, ammonium or phosphonium (Egorova et al. 2017; Moshikur et al. 2020b).

2.2.2 Dual-Active API-ILs

The dual-active API-IL strategy appears very attractive for the design of effective drug delivery systems owing to the dual-functional performances and possible synergistic effects beyond those of the parent APIs. Generally, any combination of two or more APIs is possible if both drugs form stable ions. Dual-active API-ILs are composed of an active cation and an active anion with different pharmaceutical activities (Fig. 2.2b). The role of the counterions is not only to influence the crystallinity of the drug molecules, but also to retain their own biomedical activity, resulting in dual-functional properties or providing new therapeutic properties not attainable with the two isolated APIs or known salt forms. Several notable examples of dual-active API-ILs have been reported to enhance their physicochemical and biomedical properties. A dual-functional API-IL was formed by the combination of acetylsalicylate with its main metabolite salicylic acid, resulting in the enhanced solubility of acetylsalicylate with minimal gastrointestinal distress (Endres 2010). Similar dual-active API-ILs were also formed from both salicylate and acetylsalicylate when paired with an analgesic tramadolium cation, antibacterial benzethonium cation, local anaesthetic lidocainium and procainium cations and an antiarrhythmic procainiumamide cation. Most of these API-ILs were liquids at room temperature and showed improved stability in air and moisture (Endres 2010). Antibacterial cations such as benzalkonium and didecyldimethylammonium were paired with the 'sweet' anions saccharinate (Sac) and acesulfame (Acesuf), which led to improved antimicrobial activity as well as good insect deterrent activity compared with the individual drugs. However, some of these API-ILs ($[(C_{10})_2(C_1)_2 \ N][Sac]$ and $[(C_{10})_2(C_1)_2 \ N][Acesulf])$ caused oral toxicity and skin irritation (Hough-Troutman et al. 2009). Ampicillin is a popular antibiotic that formed a dual-active API-IL when paired with the antiseptic cetylpyridinium ($C_{16}Pyr$), demonstrating significantly higher activity against several gram-positive and gram-negative bacterial strains compared with parent [Na][Amp] or $[C_{16}Pyr][Cl]$ (Ferraz et al. 2014, 2015). Notably, $[C_{16}Pyr][Amp]$ showed significantly higher growth inhibition in some tumour cell lines than [Na][Amp] (Ferraz et al. 2015). To improve the transdermal penetration, the nonsteroidal anti-inflammatory drugs, etodolac and ibuprofen, were combined with the local anaesthetic lidocaine, resulting in significant aqueous solubility as well as more efficient skin permeability. Although, the permeation of etodolac from lidocainium etodolac was significantly higher than that of parent etodolac, the opposite was observed for lidocaine (Miwa et al. 2016).

2.2.3 Oligomeric API-ILs

A nonstoichiometric approach for converting crystalline drugs into liquid forms is the formation of oligomeric API-ILs. Generally, oligomeric API-ILs are composed of hydrogen-bonded cations/anions and neutral non-ionised drug molecules, which share the delocalised protons between the deprotonated anion and protonated cation causing them to resist crystal formation (Fig. 2.2c). These API-ILs can be tuned by simply altering the stoichiometric ratio and/or complexity of the ions by introducing the free acids or bases of the conjugate bases or acids within the salt formulations. Use of the term 'ionic liquid' for this type of liquid salt formulation is debatable unless non-covalently bonded species behave as a single ion, because all species in API-ILs are partially to fully ionised liquids. Rogers' group first introduced this concept in 2010 by preparing 'oligomeric' tetrabutylphosphonium salicylates ($[P_{4,4,4,4}][H_n(Sal_{n+1})]$) (Bica and Rogers 2010). The formation of these oligomeric API-ILs can be explained by the formation of hydrogen-bonded dimer complexes between salicylic acid and the salicylate anion. All compositions in the range $[P_{4,4,4,4}][Sal]_{1.3-3}H_{0.7-2}$ were liquids. Lidocainium salicylate ($[HLid][Sal_{n+1}H_n]$ or $[HLid_{n+1}][Sal]$), a viscous liquid, is another example of an oligomeric API-IL, which was prepared by adding an excess of salicylic acid or lidocaine to lidocainium salicylate (Bica and Rogers 2010).

2.2.4 Prodrug API-ILs

The prodrug strategy is a promising approach for improving the therapeutic efficacy of APIs and extending their delivery to different routes of administration (Fig. 2.2d). A prodrug is a pharmacologically inactive molecule that is converted into an active substance through chemical and/or enzymatic transformation within the body (Pedro et al. 2020). Prodrugs are generally used to increase the solubility and site specificity, reduce rapid drug metabolism and cellular toxicity and achieve controlled drug release (Shamshina et al. 2013). However, prodrugs can exist as several polymorphs that suffer the same polymorphic problems as any solid API. For example, chloramphenicol palmitate—a prodrug of chloromycetin—exists in three polymorphic forms, A, B and C, but only the B polymorph is able to transfer into the blood plasma at the effective level (Aguiar et al. 1967). Combining the advantages of API-IL with a prodrug approach can be a potential mean of not only improving the drug efficacy, but also eliminating the unwanted toxicity of the solid drug. Prodrug API-ILs are prepared by adding hydrolysable functional groups, which can be easily biochemically cleaved, to neutral APIs, and then pairing them with IL-forming counterions to give new ILs (Cojocaru et al. 2013). The appropriate choice of IL-forming counterion would lead to a prodrug API-IL with the desired physicochemical and biopharmaceutical properties. A series of API-IL prodrugs prepared by pairing the acetaminophen bearing imidazolium, phosphonium, pyridinium and pyrrolidinium prodrugs with docusate, resulted in no melting point, lower aqueous solubilities and controlled release profiles in physiological fluids (Cojocaru et al. 2013).

2.3   Synthesis and Characterisation of API-ILs

2.3.1   Synthesis of API-ILs

The conversion of APIs to liquid forms is a promising strategy for boosting the efficacy and delivery of APIs. This can be effectively tuned by selecting biologically active ions and appropriate IL-forming counterions with a high degree of asymmetry and diffuse charge as well as a minimal number of potential H-bonds among the molecules (Balk et al. 2015a). The biologically active ions should also be on the list of substances 'generally regarded as safe' under sections 201(s) and 409 of the Federal Food, Drug, and Cosmetic Act (the Act), which is appraised and approved by the United States FDA. The initial development of API-ILs demanded an instinctual knowledge not only of the pharmacological activity of the ions, but also of their ability to form ILs. In general, API-ILs are synthesised by neutralisation of an acid with a base and/or metathesis reactions between the biologically active anion and the cation bearing salt (Moshikur et al. 2020b). In most cases, available salt forms of the cations and anions are dissolved in a solvent (e.g. water, methanol, ethanol, acetone, chloroform, tetrahydrofuran) and stirred at room temperature or a given temperature to afford the desired API-ILs by eliminating inorganic salts. For example, the first reported API-IL, lidocainium docusate, was synthesised by metathesis of lidocainium hydrochloride and sodium docusate in methanol. The efficiency of this method was also proven by synthesising a series of API-ILs including benzalkonium based penicillinate G, trans-cinnamate and thiomersalate. Alternative metathesis methods have also been used to synthesise API-ILs, for example, cationic salts were converted to the hydroxide forms in methanol using ion exchange or Amberlite resin, and then the basic solution was neutralised by adding the acidic API solutions. However, the use of large amounts of organic solvents during the preparation of API-ILs led to the formation of undesired impurities, which may be harmful to human health as well as to the environment. To address these issues, mechanochemical processing—a green alternative method for synthesising API-ILs—has attracted significant attention because it is based on a grinding technique and uses only small amounts of organic solvents, or no solvents at all (Fig. 2.3). Martins et al (2017) introduced this strategy to synthesise gabapentin and L-glutamic acid bearing mechnoAPI-ILs by simply grinding the API precursors with IL-forming counterions, suggesting a faster,

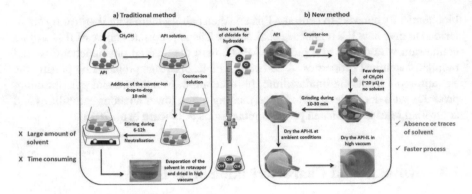

Fig. 2.3 Schematic diagrams of **a** the traditional method and **b** the mechanochemical method of API-IL preparation, reproduced with permission from Egorova et al. (2018)

solvent-free, reproducible, higher yield and greener synthetic approach for API-IL preparation. A series of ketoprofen-containing mechnoAPI-ILs were prepared with benzocainium, procainium or tetracainium using a mortar agate pestle. However, additional studies are required to clearly determine the mechanisms as well as to generalise these techniques for wider API-IL preparation.

2.3.2 Characterisation of API-ILs

Designing API-ILs for pharmaceutical applications requires careful assessment to ensure both appropriate physico-thermal properties and high therapeutic efficacy, with the minimal possible side-effects. Generally, the formation of API-ILs is confirmed by NMR spectroscopy through the qualitative characterisation of proton transfer—where the chemical shifts of ^{1}H, ^{13}C and ^{15}N are closely monitored to determine the degree of proton transfer from an acidic API to a basic IL-forming counterion, which indicates ionisation or H-bonded complex formation (Moreira et al. 2015). ^{1}H and ^{13}C NMR are used to monitor the reactions and to confirm the stoichiometry of the cation and anion of an API-IL that contains hydrogen and carbon atoms, from the change in the chemical shifts upon the proton transfer related to IL formation (Balk et al. 2015b). They are also used to evaluate the presence of by-products and degradation by simple integration of signals in the compound. Recently, nuclear overhauser enhancement experiments such as NOESY, ROESY and HOESY have been used to confirm the chemical structures of API-ILs. To eliminate the influence of the NMR solvents, several deuterated solvents such as methanol-d4, DMSO-d6 and chloroform-d have been used to measure the ^{1}H and ^{13}C NMR spectra of API-ILs by simply dissolving the ILs with the deuterated solvents in an NMR tube. In particular, ^{1}H NMR spectra of ILs can reflect the differences in intra- and intermolecular interaction between ions that vary in size and structure (Cremer et al. 2010). Infrared spectroscopy is also used to qualitatively assess the proton transfer, complete ionisation and impurities of API-IL moieties. X-ray diffraction of single crystals or powders is used to assess the crystallinity of API-ILs or the homogeneous distribution of APIs in the IL-forming counterionic matrix. Halide analyses are conducted to detect the presence of inorganic impurities using the metathesis reactions. Thermoanalytical methods such as differential scanning calorimetry, thermal gravimetric analysis or derivative thermal gravimetric analysis are carried out to assess the decomposition temperatures, melting point and phase transitions (glass transition temperature or solid–solid transitions) (Ali et al. 2019; Moshikur et al. 2020c). In addition, Karl Fischer titration is used to determine the water content, and hygroscopicity is assessed by dynamic vapour sorption during storage.

2.4 Physico-Thermal and Solubility Behaviour of API-ILs

The conversion of neutral APIs into IL forms is a promising approach for overcoming the inadequate physico-thermal properties and solubility challenges of APIs. The physicochemical and biomedical properties of API-ILs mainly depend on the appropriate choice of IL-forming cations. It has been suggested that using APIs in an IL form allows fine-tuning of their physicochemical and biological properties. Several notable studies have been conducted to emphasise the effect of counterions in controlling the various pharmaceutical cocktail properties of API-ILs, i.e. solubility, physico-thermal properties, stability, biological activity and bioavailability.

2.4.1 Physico-Thermal Properties

When crystalline solid APIs are converted into API-ILs, the thermal properties such as the glass transition temperature (T_g) and melting point (T_m) are significantly lower than those of the corresponding parent drugs. Cholinium-containing ibuprofen, ketoprofen and naproxen are liquid at room temperature (T_g in the range −90 to −70 °C), whereas the T_g values of cholinium-based acyclovir and methothexate were −50.2 and 27 °C, respectively (Shamshina et al. 2017; Chantereau et al. 2019; Moshikur et al. 2019). The T_g values of lidocaine ibuprofenate, ranitidinium ibuprofenate and ranitidinium sulfacetamide were −30, −12 and 25 °C, respectively; the melting points of these API-ILs were not detected in the range −80 to 300 °C. Similarly, docusate-containing diphenhydraminium, glycinium and ethylglycinium ILs showed no melting point, and their T_g values were observed below 0 °C (Araújo et al. 2014; Panić et al. 2020). Ampicillin- and penicillin-based ILs with cholinium, ammonium, pyridinium, phosphonium and imidazolium as cations, as well as cholinium-bearing nalidixic acid, pyrazinoic acid, niflumic acid, 4-aminosalicylic acid and other APIs as anions; showed significantly lower T_m than their corresponding starting APIs or conventional API salts (Balk et al. 2015c; Chantereau et al. 2019; Ferraz et al. 2020). A wide panel of ibuprofen-containing dual-active API-ILs with lidocaine, procainium, ranitidine, diphenhydramine and other counterions exhibited relatively low T_g and T_m, indicating acceptable thermal stability (Abednejad et al. 2019; Wu et al. 2019; Panić et al. 2020). The same was observed for dual-active API-ILs combining lidocaine with salicylate, diclofenate, naproxenate, etodolac and docusate (Miwa et al. 2016; Berton et al. 2017; Abednejad et al. 2019; Maneewattanapinyo et al. 2019; Panić et al. 2020). Methotrexate and acyclovir containing API-IL prodrugs with cholinium, ammonium, amino acid esters, phosphonium and/or imidazolium and docusate also demonstrated lower T_g and T_m values than the parent APIs. The obtained prodrug-ILs showed fast hydrolysis in both water and simulated body fluids (phosphate-buffered saline, simulated gastric and simulated intestinal fluids) (Shamshina et al. 2017; Moshikur et al. 2019). Lidocainium ibuprofenate was suggested for use as a stabiliser for the preparation of IL-mediated silver nanoparticles (20–30 nm), and was readily released from the nanoparticles (Jiang et al. 2018). An IL with the surface-active API 1-alkyl-3-methylimidazolium ibuprofenate, formed aggregate structures where the number of imidazolium cations in the micelles increased with the increasing alkyl chain length of the cations. API-ILs of mefenamic acid containing ammonium were suggested for use as an initiator for ring opening polymerisation of L-lactide, and the shape and size of API-IL-grafted–poly(L-lactide) was controlled by altering the number of hydroxyl groups in the cation (Halayqa et al. 2017). However, lidocaine salicylate showed higher density and viscosity with lower conductivity compared with lidocaine ibuprofenate, suggesting the smaller size of the salicylate anions facilitates dense packing of the lidocainium cations (Panić et al. 2020). A similar feature was found for diphenhydramine bearing ibuprofen and naproxen, where the diphenhydramine naproxenate showed higher viscosity with lower iconicity, diffusivity and ionic conductivity compared with diphenhydramine ibuprofenate (Wang et al. 2018).

2.4.2 Enhanced Solubility

The main anticipated advantage of the API-IL technique is the enhanced aqueous solubility and bioavailability compared with neutral APIs, which depend on the nature of the IL-forming counterions (Table 2.1). For example, the poorly water-soluble drugs acyclovir (ACV) and methotrexate (MTX) have been converted into IL forms using a series of IL-forming counterions. The solubility of MTX bearing cholinium and ammonium was at least 5000 times higher in both water and simulated body fluids than that of the neutral MTX, whereas for amino acid ester-based MTX-ILs, it was 4000 times higher (Moshikur et al. 2019). Similar variable solubility was also found in both water and simulated body fluids (up to 400 times higher than parent drug) when acyclovir was combined with IL-forming counterions such as ammonium, phosphonium, cholinium and docusate (Shamshina et al. 2017). The solubility of cholinium sulfasalazine in saline was improved almost 4000-fold, whereas that of cholinium niflumate in water was 56 000 times higher

Table 2.1 Pharmacological activity and main features of API-ILs

No.	APIs	API-ILs	Activity	Main features	References
1	Ibuprofen (IBU)	[IBU]: L-valine ethyl, propyl, isopropyl, butyl, pentyl and hexyl ester	Anti-inflammatory	– Improved solubility – lower partition coefficient – Enhanced the permeation	Janus et al. (2020)
2		[IBU]: N-methyl-2-pyrrolidone		Enhanced the drug retention in skin	Moshikur et al. (2020d)
3		[Cho]: [IBU]		– Improved aqueous solubility – Enhanced the rehydration ability	Chantereau et al. (2019)
4		[IBU]: [C$_4$MIM], [LID], [PRO], [RAN], [N$_{1,1,10,10}$], [N$_{6,6,6,6}$], [P$_{4,4,4,4}$], [C$_2$OHMIM], [P$_{6,6,6,14}$]		– Improved aqueous solubility – Enhanced the skin permeability	Wu et al. (2019)
5		[IBU][ProEt]		Enhanced the skin permeability	Furukawa et al. (2016)
6	Lidocaine (LID)	[LID]:[IBU], [Cl], [Doc]	Local anaesthetic Emollient	Rapidly absorbed the LID from [LID][IBU] than that of other ILs	Butt et al. (2018)
7		[LID]:[IBU], [SAL]		[LID][SAL]—improved the viscosity and density – decreased the conductivity	Panić et al. (2020)
8		[LID] [DIC]		– Lower the melting point – Improved thermal stability – Controlled release from IL-based transdermal patch	Maneewattanapinyo et al. (2019)

(continued)

Table 2.1 (continued)

No.	APIs	API-ILs	Activity	Main features	References
9		[LID]: [IBU], [NAP], [DIC]		– Improved aqueous solubility – Enhanced thermal stability – Enhanced drug release	Abednejad et al. (2019)
10	Donepezil (DON)	[DON]: [Doc], [IBU], [Ole], [Lin], [α-Lin], [DHA]	Neuro-protective agent	– Enhanced aqueous solubility – Lower partition coefficient – Enhanced the permeation	Wu et al. (2020)
11	Vitamin B	[Cho]: nicotinate, pantothenate, pyridoxylate [B6]	Supplement	– Improved aqueous solubility – Enhanced thermal stability – Enhanced the rehydration ability	Chantereau et al. (2020)
12	Amoxicillin (AMX) Penicillin (PEN)	[AMX], [PEN]: [Cho], [C_2MIM], [C_2OHMIM], [C_{16}Pyr] [$P_{6,6,6,14}$]	Antibacterial	– Lower the melting point – Enhanced the antibacterial activity	Ferraz et al. (2020)
13	Lumefan-trine	[Doc]: Lumefantrine	Antimalarial	– Improved the solubility – Enhanced the plasma exposure	Tay et al. (2020)
14	Methotrexate (MTX)	[MTX]: [ProEt], [AspEt], [PheEt], [C_2MIM], [$P_{4,4,4,4}$], [$N_{2,2,2,2}$], [Cho]	anticancer, anti-autoimmune disease agent	– Improved aqueous solubility – Enhanced thermal stability – Enhanced the antitumour activity	Moshikur et al. (2019)
15	Matrine (MAT)	[MAT]: hexanoate, octoate, decanoate, laurate, myristate	Anti-apoptotic, anti-proliferative	– Exhibited excellent physicochemical properties – Improved antibacterial and anticancer activities	Wang et al. (2019)

(continued)

Table 2.1 (continued)

No.	APIs	API-ILs	Activity	Main features	References
16	Ketoprofen (KET) Naproxen (NAP)	[Cho]: [KET], [NAP]	Anti-inflammatory	– Improved aqueous solubility – Enhanced physico-thermal stability – Enhanced the rehydration ability – Enhanced toxicity	Azevedo et al. (2017), Chantereau et al. (2019)
17	Diphenhydramine (DPH)	[DHP]: [IBU], [NAP]	Antihistamine	– Improved aqueous solubility – Enhanced thermal stability – Improved bioavailability	Wang et al. (2018)
18	Salicylic acid (SAL)	[CAP][SAL]	Anti-inflammatory	– Decreased the melting point – Excellent oral drug behaviour and non-toxic nature	Celik et al. (2020)
19		[Sal]: [AlaEt], [AspEt], [ProEt]		– Improved aqueous solubility – Enhanced thermal stability – Reduced the cytotoxicity – Enhanced skin permeation	Moshikur et al. (2018)
20		[SAL]:[Cho], [N$_{4,4,4}$], [HMPyr], [mPEG3N$_{4,4,4}$]		– Excellent thermal stability – Enhanced the skin permeability	Zavgorodnya et al. (2017)
21	Erlotinib (ERL) Gefitinib (GEF) Ceritinib (CER)	[Doc]:[ERL], [GEF], [CER], [Cabozantinib]	Anti-neoplastic agent	– Enhanced solubility in lipidic excipients – Enhanced oral absorption	Williams et al. (2018a)
22	Diclofenac	[DIC][N$_{4,4,4}$]	Anti-inflammatory	– Similar solubility and anti-inflammatory activity	Pore et al. (2017)

than that of the neutral API (Araújo et al. 2014; Shadid et al. 2015). API-ILs with cholinium containing nalidixic acid, 4-amino-salicylic acid, picolinic acid, pyrazinoic acid, naprox, ketoprofen, ibuprofen and betulinic acid, exhibited significantly improved aqueous solubility compared with the neutral APIs and sodium salts (Araújo et al. 2014; Azevedo et al. 2017; Chantereau et al. 2019). Amino acid esters—potent green IL-forming cations—were found to enhance the solubility of API-ILs. L-valine alkyl ester bearing ibuprofen ILs showed approximately 42 times better solubility than the parent API, whereas amino acid ester (alanine, proline and aspartic acid) bearing salicylate ILs were miscible (soluble) at any ratio with water (Moshikur et al. 2018; Janus et al. 2020). The API-ILs of indomethacin with tetramethylguanidine, 2-dimethylaminoethanol, 1,8-diazabicyclo[5.4.0]undec-7-ene and 1,4-diazabicyclo[2.2.2]octane showed very high aqueous solubility (5 000 000 times higher than that of the parent indomethacin) (Tojo 2018). Other IL-forming cations such as ammonium, phosphonium and docusate bearing API-ILs also improved the solubility in water compared with indomethacin. API-ILs of tetrabutylphosphonium bearing ibuprofen, ketoprofen, naproxen, diclofenac, sulfamethoxazole, sulfadiazine and tolbutamide showed enhanced solubility (1000 times higher than the parent drugs) in water (Balk et al. 2015c). Moreover, API-ILs bearing the hydrophobic IL-forming counterion docusate with glycinium, ethylglycinium, diphenhydraminium, ranitidinium sulfacetamide and lidocaine, exhibited similar aqueous solubility, whereas docusate-containing lumefantrine, erlotinib, gefitinib, ceritinib, cabozantinib, amlodipine and metformin ILs showed improved solubility in lipidic formulations (Frizzo et al. 2016; Williams et al. 2018a, b; Tay et al. 2020). Several dual-active API-ILs of lidocaine bearing ibuprofen, naproxen, diclofenac and etodolac showed approximately 470-fold higher aqueous solubility, whereas diphenhydramine containing ibuprofen and naproxen were considerably more water-soluble than the neutral APIs (Wang et al. 2018; Abednejad et al. 2019).

Although API-ILs are considered to have significant potential for application in the pharmaceutical industry, some drawbacks of the concept should be noted. In some cases, the polymorphism problem of solid APIs persists in their IL forms (Egorova et al. 2017). Ethambutol dibenzoate formed three polymorphs with different thermal stability. Procainium acetate was found to form a dihydrate salt that underwent irreversible crystallisation. Ibuprofen with 1-(2-hydroxyethyl)-3-methyl imidazolium showed high sensitivity to the water content. These findings suggest the importance of further investigation into the formation of polymorphs and hydrates in API-ILs.

2.5 Biomedical Activity

Notwithstanding numerous reports on the preparation and physicochemical properties of API-ILs, their biomedical activity has still only been evaluated using in vitro models. Many of the API-ILs showed improved aqueous solubility compared with the parent drugs, indicating the possibility of higher in vivo bioavailability. In addition, API-ILs addressed the polymorphism problems of many conventional solid

APIs, and the amorphous structure of API-ILs offered higher solubility with reduced polymorphism. However, the impact of converting the APIs into ILs on the biomedical activity is not always clear. Therefore, an ideal API-IL should be liquid with considerable physicochemical properties so polymorphs are not formed and should retain the level of biomedical activity exhibited by the parent API.

2.5.1 Enhanced Biological Activity

To date, only a few attempts have been made to investigate the biological activities of the numerous API-ILs that have been synthesised (Table 2.1). Imidazolium-containing API-ILs of amoxicillin and penicillin showed improved antibacterial activity compared with the conventional salts of the APIs. Ammonium, phosphonium and cetylpyridinium bearing API-ILs of ampicillin and penicillin exhibited improved antibacterial activity (Ferraz et al. 2020). Cholinium-bearing API-ILs of nalidixic, niflumic and pyrazinoic acids, showed similar cytotoxicity to that of the parent APIs, whereas the cholinium naproxenate and ketoprofenate API-ILs showed higher cytotoxicity (Araújo et al. 2014; Azevedo et al. 2017). N-methyl-2-pyrrolidone bearing ibprofenate exhibited lower toxicity than cholinium ibprofenate (Moshikur et al. 2020d). The introduction of amino acid esters (AAEs)—low toxicity cations—into the anticancer drug methotrexate (MTX) enhanced the antitumour activity, whereas the cholinium-bearing MTX showed similar antitumour activity compared with the parent drug. MTX-IL with IL-forming cations of different toxicity, showed different modes of antitumour activity (Moshikur et al. 2019). In addition, the low toxicity AAE-containing salicylate API-ILs showed reduced cytotoxicity compared with the parent salicylic acid drug, whereas the imidazolium-containing salicylate exhibited similar cytotoxicity to the parent drug (Egorova et al. 2015; Furukawa et al. 2016; Moshikur et al. 2018). The biological activity of API-ILs also depends on the structure of the ions. When antibacterial agents (benzalkonium, $[N_{1,1,10,10}]$, $[C_{16}Pyr]$ or 3-hydroxy-1-octyloxymethylpyridinium) were combined with an artificial sweetener (acesulfame or saccharinate), the antibacterial activity of the API-ILs was greatly influenced by the IL-forming cations (Hough-Troutman et al. 2009). Dual-active API-ILs of ibuprofenate-based ranitidinium and diphenhydraminium showed similar solubility and improved antibacterial activities compared with their precursors when tested against Candida. The same was observed for dual-active API-ILs combining docusate with glycinium, ethylglycinium and diphenhydraminium (Frizzo et al. 2016). Lidocainium docusate offered a longer and more pronounced analgesic effect than lidocainium hydrochloride, indicating a synergistic impact on the activity.

2.5.2 Enhanced Permeability

Many topical drugs have been found to be unable to penetrate the skin barrier and to crystallise prior to use (Moshikur et al. 2020a). For topical and transdermal delivery, the bioavailability and therapeutic efficacy are strongly related to the penetration of a drug through the skin. Therefore, a great deal of research is currently focussed on improving drug transportation through the skin barrier by employing a penetration enhancer (Table 2.1). As the salt forms of drugs have difficulty penetrating through the skin, API-ILs are a promising platform for skin delivery because of their skin enhancer properties. Several studies have been conducted to evaluate the role of ILs during permeation. In most cases, the in vitro permeability of API-ILs was evaluated using animal skins or artificial model membranes. For example, the Rogers group designed a series of API-ILs to investigate the permeability of a model membrane to various combinations of protic, acidic and basic APIs, demonstrating their more rapid penetration compared with that of the free APIs. They conducted a topical in vivo pharmacokinetic study on rats by dissolving the lidocaine-containing liquid co-crystal (Lid.Ibu) and salts ([Lid]Cl or [Lid][Doc]) in a carrier cream and observed faster transport/absorption of actives from the liquid co-crystals than from the salts. PEGylated salicylate IL exhibited faster transdermal transport (~2.5-fold) than PEG-free cations, and AAE-containing salicylate ILs enhanced the skin permeation compared with the parent drug (Zavgorodnya et al. 2017; Moshikur et al. 2018). Proline ethyl ester ibuprofenate exhibited tenfold better permeation than the parent API. In a study of ibuprofen ILs with L-valine alkyl esters, L-valine propyl and isopropyl ester bearing ibuprofenate exhibited higher rate of transport through the skin (Janus et al. 2020). Fatty acid-containing donepezil AIP-ILs improved the permeation through an artificial membrane by 1.9- and 1.55-fold for α-linolenic and docosahexaenoic acid based API-IL patches, respectively, compared with the corresponding free base patch (Wu et al. 2020). Compared with cholinium ibuprofenate, N-methyl-2-pyrrolidone-bearing ibuprofenate enhanced the skin retention 2.6 times, however, the skin penetration was 2.3-fold lower (Moshikur et al. 2020d).

2.6 Clinical Applications

Notwithstanding the progress in the preparation and characterisation of API-ILs, only one API-IL—lidocaine etodolac—has reached clinical trials to date (Egorova et al. 2017). In 2013, the Japanese company MEDRx published the phase I trial data for the MRX-7EAT lidocaine etodolac topical patch, which resulted in faster absorption of the active (etodolac) through the skin, with mild to moderate adverse effects. The Phase II and III trials have been conducted to evaluate the efficacy and safety in several types of pain treatment including ankle sprains (NCT01198834), acute tendonitis and bursitis (NCT01161615, NCT01506154), low back pain (NCT01968005) and shoulder pain (NCT01506154). At the end of November 2016, MEDRx decided to terminate the development of the MRX-7EAT topical patch owing to unsatisfactory trial results.

2.7 Conclusion and Future Directions

This chapter emphasised the importance and advantages of API-ILs as promising alternatives to solid crystalline APIs in pharmaceuticals. API-ILs have effectively addressed the solubility and polymorphism challenges observed with solid APIs, which were the primary barriers to the development of effective drug delivery systems. The possibility of rationally designing the API-ILs also allows the tuning of new drug forms with the desired physicochemical and biomedical properties. Advances in this area have revealed the ability of API-ILs to improve drug solubility in both water and simulated body fluids, as well as to enhance drug permeation through physiological skin barriers. The results reported have been very encouraging, with specific or dual pharmacological action and enhanced bioavailability and efficacy being the hallmarks of the IL-strategy. However, few studies have assessed the in vivo bioavailability of API-ILs, suggesting the necessity for pharmacokinetic and pharmacodynamic investigations to understand the pathways involved in their absorption, distribution, metabolism and elimination. In addition, experimental and fundamental methodologies are needed to synthesise new API-ILs and to better understand the physicochemical and biological properties of these systems. We believe that the current understanding of ILs provides an important starting point for the further exploration and development of API-ILs for pharmaceutical applications.

References

Abednejad A, Ghaee A, Morais ES, Sharma M, Neves BM, Freire MG, Nourmohammadi J, Mehrizi AA (2019) Polyvinylidene fluoride–hyaluronic acid wound dressing comprised of ionic liquids for controlled drug delivery and dual therapeutic behavior. Acta Biomater 100:142–157. https://doi.org/10.1016/j.actbio.2019.10.007

Aguiar AJ, Krc J, Kinkel AW, Samyn JC (1967) Effect of polymorphism on the absorption of chloramphenicol from chloramphenicol palmitate. J Pharm Sci 56:847–853. https://doi.org/10.1002/jps.2600560712

Ali MK, Moshikur RM, Wakabayashi R, Moniruzzaman M, Goto M (2021) Biocompatible ionic liquid-mediated micelles for enhanced transdermal delivery of paclitaxel. ACS Appl Mater Interface 13:19745–19755. https://doi.org/10.1021/acsami.1c03111

Ali MK, Moshikur RM, Wakabayashi R, Moniruzzaman M, Kamiya N, Goto M (2020) Biocompatible ionic liquid surfactant-based microemulsion as a potential carrier for sparingly soluble drugs. ACS Sustain Chem Eng. https://doi.org/10.1021/acssuschemeng.9b07773

Ali MK, Moshikur RM, Wakabayashi R, Moniruzzaman M, Kamiya N, Goto M (2019) Synthesis and characterization of choline–fatty-acid-based ionic liquids: a new biocompatible surfactant. J Colloid Interface Sci 551:72–80. https://doi.org/10.1016/j.jcis.2019.04.095

Araújo JMM, Florindo C, Pereiro AB, Vieira NSM, Matias AA, Duarte CMM, Rebelo LPN, Marrucho IM (2014) Cholinium-based ionic liquids with pharmaceutically active anions. RSC Adv 4:28126–28132. https://doi.org/10.1039/c3ra47615d

Azevedo AMO, Costa SPF, Dias AFV, Marques AHO, Pinto PCAG, Bica K, Ressmann AK, Passos MLC, Araújo ARTS, Reis S, Saraiva MLMFS (2017) Anti-inflammatory choline based ionic liquids: Insights into their lipophilicity, solubility and toxicity parameters. J Mol Liq 232:20–26. https://doi.org/10.1016/j.molliq.2017.02.027

Balk A, Holzgrabe U, Meinel L (2015a) Pro et contra' ionic liquid drugs—challenges and opportunities for pharmaceutical translation. Eur J Pharm Biopharm 94:291–304. https://doi.org/10.1016/j.ejpb.2015.05.027

Balk A, Widmer T, Wiest J, Bruhn H, Rybak JC, Matthes P, Müller-Buschbaum K, Sakalis A, Lühmann T, Berghausen J, Holzgrabe U, Galli B, Meinel L (2015b) Ionic liquid versus prodrug strategy to address formulation challenges. Pharm Res 32:2154–2167. https://doi.org/10.1007/s11095-014-1607-9

Balk A, Wiest J, Widmer T, Galli B, Holzgrabe U, Meinel L (2015c) Transformation of acidic poorly water soluble drugs into ionic liquids. Eur J Pharm Biopharm 94:73–82. https://doi.org/10.1016/j.ejpb.2015.04.034

Berton P, Di Bona KR, Yancey D, Rizvi SAA, Gray M, Gurau G, Shamshina JL, Rasco JF, Rogers RD (2017) Transdermal bioavailability in rats of lidocaine in the forms of ionic liquids, salts, and deep eutectic. ACS Med Chem Lett 8:498–503. https://doi.org/10.1021/acsmedchemlett.6b00504

Bica K, Rogers RD (2010) Confused ionic liquid ions—a "liquification" and dosage strategy for pharmaceutically active salts. Chem Commun 46:1215–1217. https://doi.org/10.1039/b925147b

Butt FI, Muhammad N, Hamid A, Moniruzzaman M, Sharif F (2018) Recent progress in the utilization of biosynthesized polyhydroxyalkanoates for biomedical applications—review. Int J Biol Macromol 120:1294–1305. https://doi.org/10.1016/j.ijbiomac.2018.09.002

Celik S, Albayrak AT, Akyuz S, Ozel AE (2020) Synthesis, molecular docking and ADMET study of ionic liquid as anticancer inhibitors of DNA and COX-2, TOPII enzymes. J Biomol Struct Dyn 38:1354–1364. https://doi.org/10.1080/07391102.2019.1604263

Chantereau G, Sharma M, Abednejad A, Vilela C, Costa EM, Veiga M, Antunes F, Pintado MM, Sèbe G, Coma V, Freire MG, Freire CSR, Silvestre AJD (2020) Bacterial nanocellulose membranes loaded with vitamin B-based ionic liquids for dermal care applications. J Mol Liq 302:112547. https://doi.org/10.1016/j.molliq.2020.112547

Chantereau G, Sharma M, Abednejad A, Neves BM, Sèbe G, Coma V, Freire MG, Freire CSR, Silvestre AJD (2019) Design of nonsteroidal anti-inflammatory drug-based ionic liquids with improved water solubility and drug delivery. ACS Sustain Chem Eng 7:14126–14134. https://doi.org/10.1021/acssuschemeng.9b02797

Clarke CJ, Tu WC, Levers O, Bröhl A, Hallett JP (2018) Green and sustainable solvents in chemical processes. Chem Rev 118:747–800. https://doi.org/10.1021/acs.chemrev.7b00571

Cojocaru OA, Bica K, Gurau G, Narita A, Shamshina MPD, JL, Barber PS, Rogers RD, (2013) Prodrug ionic liquids: functionalizing neutral active pharmaceutical ingredients to take advantage of the ionic liquid form. Medchemcomm 4:559–563. https://doi.org/10.1039/c3md20359j

Cremer T, Kolbeck C, Lovelock KRJ, Paape N, Wölfel R, Schulz PS, Wasserscheid P, Weber H, Thar J, Kirchner B, Maier F, Steinrück HP (2010) Towards a molecular understanding of cation-anion interactions-Probing the electronic structure of imidazolium ionic liquids by NMR spectroscopy, X-ray photoelectron spectroscopy and theoretical calculations. Chem Eur J 16:9018–9033. https://doi.org/10.1002/chem.201001032

Egorova KS, Gordeev EG, Ananikov VP (2018) Fundamental importance of ionic interactions in the liquid phase: a review of recent studies of ionic liquids in biomedical and pharmaceutical applications. J Mol Liq 272:271–300. https://doi.org/10.1016/j.molliq.2018.09.025

Egorova KS, Gordeev EG, Ananikov VP (2017) Biological activity of ionic liquids and their application in pharmaceutics and medicine. Chem Rev 117:7132–7189. https://doi.org/10.1021/acs.chemrev.6b00562

Egorova KS, Seitkalieva MM, Posvyatenko AV, Khrustalev VN, Ananikov VP (2015) Cytotoxic activity of salicylic acid-containing drug models with ionic and covalent binding. ACS Med Chem Lett 6:1099–1104. https://doi.org/10.1021/acsmedchemlett.5b00258

Endres F (2010) Physical chemistry of ionic liquids. Phys Chem Chem Phys 12:1648–1648. https://doi.org/10.1039/c001176m

Ferraz R, Costa-Rodrigues J, Fernandes MH, Santos MM, Marrucho IM, Rebelo LPN, Prudêncio C, Noronha JP, Petrovski Ž, Branco LC (2015) Antitumor activity of ionic liquids based on ampicillin. ChemMedChem 10:1480–1483. https://doi.org/10.1002/cmdc.201500142

Ferraz R, Silva D, Dias AR, Dias V, Santos MM, Pinheiro L, Prudêncio C, Noronha JP, Petrovski Ž, Branco LC (2020) Synthesis and antibacterial activity of ionic liquids and organic salts based on penicillin g and amoxicillin hydrolysate derivatives against resistant bacteria. Pharmaceutics 12:221. https://doi.org/10.3390/pharmaceutics12030221

Ferraz R, Teixeira V, Rodrigues D, Fernandes R, Prudêncio C, Noronha JP, Petrovski Ž, Branco LC (2014) Antibacterial activity of ionic liquids based on ampicillin against resistant bacteria. RSC Adv 4:4301–4307. https://doi.org/10.1039/c3ra44286a

Frizzo CP, Wust K, Tier AZ, Beck TS, Rodrigues LV, Vaucher RA, Bolzan LP, Terra S, Soares F, Martins MAP (2016) Novel ibuprofenate- and docusate-based ionic liquids: emergence of antimicrobial activity. RSC Adv 6:100476–100486. https://doi.org/10.1039/c6ra22237d

Furukawa S, Hattori G, Sakai S, Kamiya N (2016) Highly efficient and low toxic skin penetrants composed of amino acid ionic liquids. RSC Adv 6:87753–87755. https://doi.org/10.1039/C6R A16926K

Ghielmetti G, Bruzzese T, Bianchi C, Recusani F (1976) Relationship between acute toxicity in mice and polymorphic forms of polyene antibiotics. J Pharm Sci 65:905–907. https://doi.org/10.1002/jps.2600650625

Halayqa M, Zawadzki M, Domańska U, Plichta A (2017) API-ammonium ionic liquid—polymer compounds as a potential tool for delivery systems. J Mol Liq 248:972–980. https://doi.org/10.1016/j.molliq.2017.10.136

Hough-Troutman WL, Smiglak M, Griffin S, Matthew Reichert W, Mirska I, Jodynis-Liebert J, Adamska T, Nawrot J, Stasiewicz M, Rogers RD, Pernak J (2009) Ionic liquids with dual biological function: sweet and anti-microbial, hydrophobic quaternary ammonium-based salts. New J Chem 33:26–33. https://doi.org/10.1039/b813213p

Janus E, Ossowicz P, Klebeko J, Nowak A, Duchnik W, Kucharski Ł, Klimowicz A (2020) Enhancement of ibuprofen solubility and skin permeation by conjugation with l-valine alkyl esters. RSC Adv 10:7570–7584. https://doi.org/10.1039/d0ra00100g

Jiang Q, Yu S, Li X, Ma C, Li A (2018) Evaluation of local anesthetic effects of Lidocaine-Ibuprofen ionic liquid stabilized silver nanoparticles in Male Swiss mice. J Photochem Photobiol B Biol 178:367–370. https://doi.org/10.1016/j.jphotobiol.2017.11.028

Maneewattanapinyo P, Yeesamun A, Watthana F, Panrat K, Pichayakorn W, Suksaeree J (2019) Controlled release of lidocaine-diclofenac ionic liquid drug from freeze-thawed gelatin/poly(vinyl alcohol) transdermal patches. AAPS PharmSciTech 20:1–9. https://doi.org/10.1208/s12249-019-1545-2

Martins ICB, Oliveira MC, Diogo HP, Branco LC, Duarte MT (2017) MechanoAPI-ILs: pharmaceutical ionic liquids obtained through mechanochemical synthesis. Chemsuschem 10:1360–1363. https://doi.org/10.1002/cssc.201700153

Miwa Y, Hamamoto H, Ishida T (2016) Lidocaine self-sacrificially improves the skin permeation of the acidic and poorly water-soluble drug etodolac via its transformation into an ionic liquid. Eur J Pharm Biopharm 102:92–100. https://doi.org/10.1016/j.ejpb.2016.03.003

Moreira DN, Fresno N, Pérez-Fernández R, Frizzo CP, Goya P, Marco C, Martins MAP, Elguero J (2015) Brønsted acid-base pairs of drugs as dual ionic liquids: NMR ionicity studies. Tetrahedron 71:676–685. https://doi.org/10.1016/j.tet.2014.12.003

Moshikur RM, Chowdhury MR, Fujisawa H, Wakabayashi R, Moniruzzaman M, Goto M (2020a) Design and characterization of fatty acid-based amino acid ester as a new "green" hydrophobic ionic liquid for drug delivery. ACS Sustain Chem Eng 8:13660–13671. https://doi.org/10.1021/acssuschemeng.0c03419

Moshikur RM, Chowdhury MR, Moniruzzaman M, Goto M (2020b) Biocompatible ionic liquids and their applications in pharmaceutics. Green Chem 22:8116–8139. https://doi.org/10.1039/d0gc02387f

Moshikur RM, Ali MK, Wakabayashi R, Moniruzzaman M, Goto M (2020c) Formation and potential application of micelles composed of biocompatible N-lauroyl-amino acid ionic liquids surfactant. J Mol Liq 320:114424. https://doi.org/10.1016/j.molliq.2020.114424

Moshikur RM, Chowdhury MR, Wakabayashi R, Moniruzzaman M, Kamiya N, Goto M (2020d) Ionic liquids with N-methyl-2-pyrrolidonium cation as an enhancer for topical drug delivery: Synthesis, characterization, and skin-penetration evaluation. J Mol Liq 299:112166. https://doi.org/10.1016/j.molliq.2019.112166

Moshikur RM, Chowdhury MR, Wakabayashi R, Moniruzzaman M, Goto M (2019) Ionic liquids with methotrexate moieties as a potential anticancer prodrug: synthesis, characterization and solubility evaluation. J Mol Liq 278:226–233. https://doi.org/10.1016/j.molliq.2019.01.063

Moshikur RM, Chowdhury MR, Wakabayashi R, Moniruzzaman M, Goto M (2018) Characterization and cytotoxicity evaluation of biocompatible amino acid esters used to convert salicylic acid into ionic liquids. Int J Pharm 546:31–38. https://doi.org/10.1016/j.ijpharm.2018.05.021

Panić J, Tot A, Janković N, Drid P, Gadžurić S, Vraneš M (2020) Physicochemical and structural properties of lidocaine-based ionic liquids with anti-inflammatory anions. RSC Adv 10:14089–14098. https://doi.org/10.1039/c9ra08815f

Pedro SN, Freire CSR, Silvestre AJD, Freire MG (2020) The role of ionic liquids in the pharmaceutical field: an overview of relevant applications. Int J Mol Sci 21:1–50. https://doi.org/10.3390/ijms21218298

Pore Y, Mangrule V, Mane M, Chopade A (2017) Preparation, characterization and physicochemical studies of diclofenac ionic liquids. Asian J Pharm Pharmacol 3:208–214

Rodriguez-Aller M, Guillarme D, Veuthey JL, Gurny R (2015) Strategies for formulating and delivering poorly water-soluble drugs. J Drug Deliv Sci Technol 30:342–351. https://doi.org/10.1016/j.jddst.2015.05.009

Shadid M, Gurau G, Shamshina JL, Chuang BC, Hailu S, Guan E, Chowdhury SK, Wu JT, Rizvi SAA, Griffin RJ, Rogers RD (2015) Sulfasalazine in ionic liquid form with improved solubility and exposure. Medchemcomm 6:1837–1841. https://doi.org/10.1039/c5md00290g

Shamshina JL, Kelley SP, G, Gurau RDR, (2015) Develop ionic liquid drugs. Nature 528:188–189

Shamshina JL, Barber PS, Rogers RD (2013) Ionic liquids in drug delivery. Expert Opin Drug Deliv 10:1367–1381. https://doi.org/10.1517/17425247.2013.808185

Shamshina JL, Cojocaru OA, Kelley SP, Bica K, Wallace SP, Gurau G, Rogers RD (2017) Acyclovir as an ionic liquid cation or anion can improve aqueous solubility. ACS Omega 2:3483–3493. https://doi.org/10.1021/acsomega.7b00554

Tay E, Nguyen TH, Ford L, Williams HD, Benameur H, Scammells PJ, Porter CJH (2020) Ionic liquid forms of the antimalarial lumefantrine in combination with LFCS type IIIB lipid-based formulations preferentially increase lipid solubility, in vitro solubilization behavior and in vivo exposure. Pharmaceutics 12:17–24. https://doi.org/10.3390/pharmaceutics12010017

Tojo E (2018) New active pharmaceutical ingredient-ionic liquids (api-ils) derived from indomethacin and mebendazole. Proceedings 9:48. https://doi.org/10.3390/ecsoc-22-05781

Wang C, Chopade SA, Guo Y, Early JT, Tang B, Wang E, Hillmyer MA, Lodge TP, Sun CC (2018) Preparation, characterization, and formulation development of drug-drug protic ionic liquids of diphenhydramine with ibuprofen and naproxen. Mol Pharm 15:4190–4201. https://doi.org/10.1021/acs.molpharmaceut.8b00569

Wang Z, Zhang J, Lu B, Li Y, Liang Y, Yuan J, Zhao M, Wang B, Mai C, Zhang J (2019) Novel bio-renewable matrinium-based ionic liquids derived from Chinese herb medicine: synthesis, physicochemical properties and biological activity. J Mol Liq 296:111822. https://doi.org/10.1016/j.molliq.2019.111822

Williams HD, Ford L, Han S, Tangso KJ, Lim S, Shackleford DM, Vodak DT, Benameur H, Pouton CW, Scammells PJ, Porter CJH (2018a) Enhancing the oral absorption of kinase inhibitors using lipophilic salts and lipid-based formulations. Mol Pharm 15:5678–5696. https://doi.org/10.1021/acs.molpharmaceut.8b00858

Williams HD, Ford L, Lim S, Han S, Baumann J, Sullivan H, Vodak D, Igonin A, Benameur H, Pouton CW, Scammells PJ, Porter CJH (2018b) Transformation of biopharmaceutical classification system class I and III drugs into ionic liquids and lipophilic salts for enhanced developability using lipid formulations. J Pharm Sci 107:203–216. https://doi.org/10.1016/j.xphs.2017.05.019

Wu H, Deng Z, Zhou B, Qi M, Hong M, Ren G (2019) Improved transdermal permeability of ibuprofen by ionic liquid technology: correlation between counterion structure and the physicochemical and biological properties. J Mol Liq 283:399–409. https://doi.org/10.1016/j.molliq.2019.03.046

Wu H, Fang F, Zheng L, Ji W, Qi M, Hong M, Ren G (2020) Ionic liquid form of donepezil: preparation, characterization and formulation development. J Mol Liq 300:112308. https://doi.org/10.1016/j.molliq.2019.112308

Zavgorodnya O, Shamshina JL, Mittenthal M, McCrary PD, Rachiero GP, Titi HM, Rogers RD (2017) Polyethylene glycol derivatization of the non-active ion in active pharmaceutical ingredient ionic liquids enhances transdermal delivery. New J Chem 41:1499–1508. https://doi.org/10.1039/C6NJ03709G

Chapter 3
Ionic Liquids for Transdermal Drug Delivery: Choline Geranate System as a Case Study

Paula Berton and Julia L. Shamshina

Abstract Transdermal administration is of high interest for drug delivery due to its non-invasiveness and easy dosing. However, this route of drug delivery faces the challenges of low drug permeability across the skin. This chapter reviews the role of ionic liquids (ILs) for transdermal delivery of active pharmaceutical ingredients (APIs). In particular, it focuses on the case of the Choline Geranate (CAGE), including its synthesis, structural features, physical properties, and enhanced permeation across epithelial barriers as a transdermal drug carrier. Its scale-up and medical applications are also discussed. This system has potential to become a powerful excipient to improve the pharmacokinetic profiles of many pharmaceutical products in transdermal applications.

3.1 Introduction

Ionic liquids (ILs, loosely defined as salts with melting points below 100 °C (Welton 1999)) have attracted considerable attention in the field of pharmaceutical research due to their intrinsic tunable properties. In 1997, Chemical & Engineering News published their piece about ILs as "Designer Solvents" (Freemantle 1998) and effectively launched a renaissance in scientific and engineering interest in the field. In 2007, the concept of Active Pharmaceutical Ingredients ILs (API-ILs) was introduced (Hough et al. 2007), and soon thereafter IL-based systems have been proposed as solubilizers of poorly water-soluble active pharmaceutical ingredients (Jaitely et al. 2008; Mizuuchi et al. 2008), skin penetration enhancers (Sidat et al. 2019), and stabilizers for large molecules (Harada et al. 2018; Reslan and Kayser 2018).

P. Berton (✉)
Chemical and Petroleum Engineering Department, University of Calgary, Calgary, AB, Canada
e-mail: paula.berton@ucalgary.ca

J. L. Shamshina
Fiber and Biopolymer Research Institute, Department of Plant and Soil Science, Texas Tech University, Lubbock, TX, USA
e-mail: jshamshi@ttu.edu

Transdermal drug delivery is a non-parenteral administration technique that eliminates drug exposure to the gastrointestinal (GI) tract and first-pass metabolism. Instead, it delivers drugs locally to the skin to achieve systemic drug distribution and ensures the sustained and controlled release of drugs (Kalluri and Banga 2011). Although transdermal delivery offers advantages such as non-invasiveness and easy applicability, the properties of the skin barrier allow the permeation of only small (<500 Da) lipophilic molecules under normal conditions (Prausnitz and Langer 2008; Yang et al. 2017).

The rate-limiting barrier of the epidermis resides in the *stratum corneum* (SC). In other words, for the percutaneous absorption to occur, i.e., the penetration of substances into various layers of skin and permeation across the skin into the systemic circulation (Prausnitz et al. 2017; Brisson 1974) to take place, the drug must penetrate through the SC, the most superficial layer of the epidermis (Fig. 3.1). In general, once drug molecules cross the stratum corneal barrier, passage into deeper dermal layers and systemic uptake occurs relatively quickly and easily (Mehta 2004). Thus, permeability constant of water for SC is 1.1×10^{-7} cm/sec, while for dermis is 1.0×10^{-4} cm/sec, a 1000-fold faster (Brisson 1974). The SC has a thickness of 10–20 μm, composed of dead keratinocytes that are connected by glycoprotein, and produce the protein keratin which, together with the ceramide lipid component, form a dense structure (Prausnitz et al. 2017). Therefore, effective delivery of macromolecules across the skin, especially biologics which are becoming the major focus of pharmaceutical research, requires the use of perturbation methods to reduce skin barrier function.

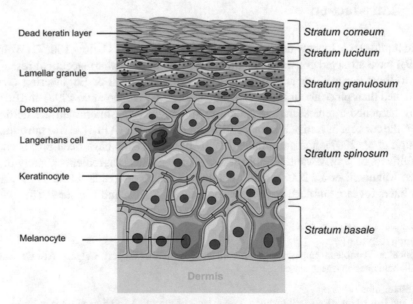

Fig. 3.1 Detailed representation of the epidermis layers of human skin. Reprinted with permission from Ramadon et al. Drug Delivery Transl Research, 2021 (Ramadon et al. 2021)

Several formulation-based approaches, comprising primarily of chemical perme-ation enhancers (CPEs) have also been developed (Karande et al. 2005,2004; Karande and Mitragotri 2009; Chen et al. 2014; Kim et al. 2008). The most common trans-dermal enhancers are ethanol, dimethylsulfoxide, Azone, terpenoids, and menthol ester derivatives. However, few of them are used clinically owing to their acute skin toxicity or irritation, as well as other safety issues that have not been clarified. Recently, ILs and deep eutectic solvents (DESs) have been introduced as a novel class of formulation-based approaches to enhance transdermal drug delivery for a wide range of molecules. The high tunability of these types of compounds allows achieving the desired physical, chemical, and biological properties of the CPE systems. This has generated immense research interests since the conventional CPEs have reached a plateau in terms of their discovery and efficacy.

The tunability of either ILs and/or DESs allows exploring different strategies, including their use as carriers (i.e., as solvents) solubilizing both hydrophilic and hydrophobic APIs of diverse classes, directly dissolving the API or in combination with oil, forming ionic liquid-in-oil microemulsion (Shamshina et al. 2018). A third option is the incorporation of the API into the IL composition, thus, changing the properties of the API itself (Shamshina et al. 2018). In this book chapter, we will review the latest attempts and strategies to use these systems as transdermal carriers, i.e., as potential replacements of CPEs.

3.2 Are These Systems ILs or DESs?

Here, we will not focus on ILs as a drug carrier *vs.* API-ILs, but the penetration of these systems through the skin. Usually, for APIs in the form of ILs (API-ILs), as well as for the IL as a drug carrier used in transdermal penetration, melting points below body temperature are targeted (Shamshina et al. 2018). The proper design with appropriate choice of counterions allows targeting the specific (<37 °C) melting point, and other targeted properties; here, we are interested mainly in dissolution capacity, permeability, and stability.

At the molecular level, the continuum of proton transfer *vs.* hydrogen bonding exists, and led to design rules for the "liquefaction" of higher melting salt forms, particularly through modulating ionicity of the ILs. As a result, liquefaction is achieved due to either complete proton transfer (classic IL systems) or due to the appearance of hydrogen bonding between two actives ("liquid co-crystals" or deep eutectics, classic DES systems) (Rebelo et al. 2007; Fei et al. 2006). One more strategy that results in liquefaction is an intentional change in the stoichiometry or complexity of the ions and/or complexes. Thus, ILs are defined as **salts** which melt below 100 °C, and DESs are **eutectic mixtures** of two or more distinct components which typically interact via strong hydrogen bonding, exist as mixtures of charged and neutral species, either in equimolar or non-stoichiometric ratios, and have a much lower melting point than the individual components (Zakrewsky et al. 2014a).

The first transdermally delivered IL was, indeed, one of the first API-ILs reported. It was reported by Rogers and coworkers (Hough et al. 2007) in 2007, and obtained by pairing lidocaine with sodium docusate, resulting in lidocaine docusate ([Lid][Doc]) IL, which was liquid at room temperature. The antinociceptive effect of this IL was superior to that of lidocaine as was evidenced by the improved and more extended therapeutic impact in a mouse model. Considering that fully ionized salts (e.g., metal salts, with each ion being solvated individually) normally diffuse poorly through lipid membranes, the result of the antinociceptive test was quite surprising. The next set of studies was reported by MacFarlane et al. (Stoimenovski and MacFarlane 2011). The study reported another example of transport of the pharmaceutically active ILs (protic ammonium salicylates) through a silicone membrane.

After a thorough spectroscopic investigation, it was found that even though it was initially suggested that API-ILs were composed entirely of dissociated ions, there is a wide range in the degree of ionicity in these liquid forms. ILs prepared by MacFarlane were even more permeable than corresponding inorganic salts, due to the formation of an *ion-pair complex*, and not separate solvated ions. Wang et al. expanded this approach showing that dual-active deep eutectics can form strong hydrogen-bonded complexes (i.e., DESs) without formal charges (Wang et al. 2014) in Lid·Ibu DES. When Lid·Ibu was dissolved in ethanol, the two actives were held together via strong hydrogen bond interactions, and simultaneously transported through a model silicone membrane, with almost identical transport rates of both. Park et al. hypothesized that such strongly hydrogen-bonded complexes might facilitate the transport of drugs through the skin, compared to commercial analogs (Park and Prauznitz 2010). These initial studies followed by extensive research and various degrees of ionicity were shown to exist in these systems, including deep eutectics, oligomeric ILs, etc., altogether termed liquid cocrystals (Bica and Rogers 2010; Kelley et al. 2013; Bica et al. 2011).

Indeed, it has been previously shown that by adding an excess of the neutral parent acid or base, it is possible to form materials that were termed "confused ionic liquids" (Bica and Rogers 2010). This included tetrabutylphosphonium salicylate ($[P_{4444}][Sal]$)/salicylic acid (HSal) salts $[P_{4444}][H_x(Sal)_{1+x}]$ with no melting point, lidocainium salicylate salts ($[HLid][H_x(Sal)_{1+x}]$), etc. With all these findings, Rogers has recently raised concern about the definition of the transdermally penetrating compound, CAGE (Choline And GEranate), that was presented in the literature as both an ionic liquid (IL) and deep eutectic solvent (DES), in a single phrase as "ionic liquid deep eutectic," and pointed out the inappropriateness of the use of both mutually exclusive definitions *simultaneously* (Rogers and Gurau 2018). In reply to Rogers' concern, the developers of CAGE, noted that the current definitions of either ILs or DESs are based on extreme behaviors and the reality is somewhere in the middle, and stated that *"Our choline and geranate (CAGE) compound fits the definition of an IL since it comprises largely ionic species, cholinium and geranate, and has a melting point below 100 °C. At the same time, CAGE is not a classical IL since it also contains neutral geranic acid* (Banerjee et al. 2018a)."

In our opinion, GAGE represents such an example of the formation of neither purely ionic nor neutral hydrogen-bonded oligomeric ions of parent acid and conjugate base, where the excess of geranic acid becomes tightly bound into the complex between choline and geranate, and the compound is still retaining the favorable properties of an IL; and suggest its proper structural formula would be [Cho][H(Ger)$_2$] rather than commonly used [Cho][Ger]$_2$, but we will use the ratio of the starting components to indicate the composition of the systems, to not confuse a reader even more.

The discussion of the nature of the components in solution is complex and sometimes not easy to elucidate. It is not limited to a discussion about proper nomenclature, but a critical and needed characterization, since it will determine the chemical and physical properties of the system and their potential applications (Kelley et al. 2013). As an example, the triethylammonium acetate system, although could be seen as a simple, protic ionic liquid, required extensive experimental studies to be classified as a complex system with oligomeric ions (Berton et al. 2018).

In the next sections, we will only focus on the effect of systems reported as transdermal carriers for active pharmaceutical ingredients, irrespective of the names assigned to the systems by the authors. For those systems still under discussion, such as CAGE, we will use the structural formula to avoid incorrect terminologies.

3.3 Using ILs as Transdermal Carriers

Recently, ILs and DESs (*the definitions and differences between ILs and DESs were given and discussed above, but henceforth we will use structural formula or ratio in "tricky" cases*) have demonstrated an excellent ability to enhance the permeation of macromolecules, acting as both enhancers and solvents in transdermal transport. An example of that is the study of Qi et al. who compared the performance of the 31 ammonium-based ILs with 44 conventional CPE systems for transdermal drug delivery, through examining the Attenuated Total Reflectance Fourier Transform Infrared (ATR-FTIR) spectra of skin SC exposed to both types of systems (Qi et al. 2020). Indeed, the ATR-FTIR allows to measure heterogeneity of SC and identifying lipid-rich and lipid-poor regions. The technique also allows determining the permeation and distribution of a drug (as well as solvents) into the SC.

When this analysis was conducted, a clear distinction was noted in the ATR-FTIR spectra after ILs and CPEs treatment, despite variations among each class: The former systems induced greater reductions in lipid peak areas but smaller reductions in the peaks of α-helixes of proteins (Qi et al. 2020). Karande et al. indicated that these two types of peaks have clear implications for transdermal drug delivery using CPEs, namely, that the extent of lipid extraction and fluidization was related to the potency of permeation enhancement, and the reduction in stable protein secondary structure correlated with the irritation potential (Karande et al. 2005). Based on the design principles that an ideal topical formulation should have strong permeation enhancement but low irritation potential, ILs exhibited a better performance than conventional CPEs.

Not only ammonium but choline ([Cho])-based systems were extensively studied. These systems were synthesized and tested to enhance skin penetration using mannitol as a model drug, chosen because of its hydrophobicity and thus low permeability into the skin (Zakrewsky et al. 2014b). The systems that were tested included choline urea (1:2 choline:urea ratio, Cho:Urea (1:2)), choline hexanoate (1:2 choline:hexanoic acid ratio, Cho:Hexanoic acid (1:2)), choline oleate (1:2 choline:oleic acid, Cho:Oleic acid (1:2)), choline geranate (1:2 choline:geranic acid, Cho:Geranic Acid (1:2), also abbreviated as CAGE), trihexyltetradecylphosphonium oleate ([P$_{66614}$][Oleate]), trihexyltetradecylphosphonium hexanoate ([P$_{66614}$][Hexanoate]), and trihexyltetradecylphosphonium geranate ([P$_{66614}$][Ger]).

The systems Cho:Urea (1:2) and Cho:Hexanoic Acid (1:2), reduced the permeation of mannitol, in comparison to the saline control, while Cho:Oleic Acid (1:2) and [P$_{66614}$][Oleate] showed no enhancement in skin permeation comparing with saline control. However, [P$_{66614}$][Oleate] enhanced transport of mannitol into deep skin as much as fivefold, indicating potential topic applications. The systems [P$_{66614}$][Hexanoate], Cho:Geranic Acid (1:2), and [P$_{66614}$][Ger] showed a significant transdermal enhancement. The highest enhancement was observed using Cho:Geranic Acid (1:2), with nearly fivefold enhancement of total delivery and more than 10 times more drug penetrating deep tissue layers. These findings showed that the depth of penetration could be controlled by the proper design of the systems: [P$_{66614}$][Oleate] resulted in enhanced delivery to *the superficial layer of the skin*, while [P$_{66614}$][Hexanoate] and Cho:Geranic Acid (1:2) exhibited delivery primarily *in the deep tissue layers*, i.e., [P$_{66614}$][Hexanoate] showing a preference for the epidermis and Cho:Geranic Acid (1:2) for the dermis.

The systems that showed enhancement of mannitol delivery were also tested for transdermal delivery of a model antibiotic, cefadroxil. All the tested systems delivered 15–20% of the applied dose, which translates to approximately fivefold enhancement of cefadroxil delivery into the skin. Again, Cho:Geranic Acid (1:2) led to the highest enhancement of cefadoxil delivery into the dermis (~16-fold compared with the aqueous solution), while other similar systems led to various degrees of enhanced transport (Zakrewsky et al. 2014b).

Since this initial publication in 2014 about its exceptional properties as a transdermal carrier (Zakrewsky et al. 2014b), Cho:Geranic Acid (1:2) arose as a leading solubilizing agent, with optimal antibacterial, cytotoxic, and skin-permeating properties, and has recently been patented (Zakrewsky et al. 2019). It is such an interesting case with huge potential that there is a company formed based on the platform this drug provides—CAGE Bio Inc., that uses structural heterogeneity of cations and anions to enhance membrane permeation (The ionic liquid platform 2020). It is worth spending a bit more time discussing these systems.

3.4 Choline:Geranic Acid Systems as Case Study

Synthesis

CAGE is typically synthesized by a salt metathesis reaction between choline bicarbonate and geranic acid (Fig. 3.2). The completion of the reaction is monitored through the evolution of CO_2 gas. The CAGE is a clear, colorless to yellow colored liquid, with a pH ~ 8.5, conductivity ~1.3 mS/cm, and a characteristic odor.

Toxicity

The starting components of Cho:Geranic Acid (1:2), i.e., choline and geranic acid, belong to the Generally Recognized as Safe (GRAS) list (Choline chloride 2020; Cohen et al. 2020). In vitro tests using normal human bronchial epithelial (NHBE) cells as a model nonskin cell to assess potential toxicity in the event of systemic absorption, showed no toxicity for Cho:Geranic Acid (1:2) (Zakrewsky et al. 2014b). ATR-FTIR (i.e., changes to α-helix content in the SC) and reconstructed human epidermal cultures (EpiDerm) were used to assess its irritation potential: It was found that while geranic acid and choline showed significant irritation potential, Cho:Geranic Acid (1:2) showed no significant changes with respect to the control (Zakrewsky et al. 2014b).

GLP dermal toxicity of CAGE was also studied in minipigs (Ko et al. 2020a), considered to be an excellent model to study the toxicity of topical skin formulations. Studies were conducted using Göttingen minipigs, by applying the Cho:Geranic Acid (1:2) as a 40 wt% gel formulated in common pharmaceutical excipients; 1 ml of this gel was applied over 10% of the pig body area once a day for 91 consecutive days. CAGE showed no signs of toxicity—clinical (ocular, electrocardiography, hematology, coagulation, clinical chemistry, urine analysis, and organ weights parameters) or non-clinical (body weight changes, food consumption).

Biological Activity

Besides mannitol and cefadroxil, Cho:Geranic Acid (1:2) has shown to induce an excellent solubility (450 times, from 0.016 mg/mL in water to 7.2 mg/mL in Cho:Geranic Acid (1:2)) of Nobiletin (NOB), a poorly water-soluble flavonoid of low oral bioavailability (Hattori et al. 2019). In vitro transdermal tests showed that Cho:Geranic Acid (1:2) was effective in enhancing transdermal absorption of NOB, compared to other penetration enhancers, such as ethanol-phosphate buffered saline (EtOH:PBS), and glycol monoethyl ether (DGME, Transcutol), following the order

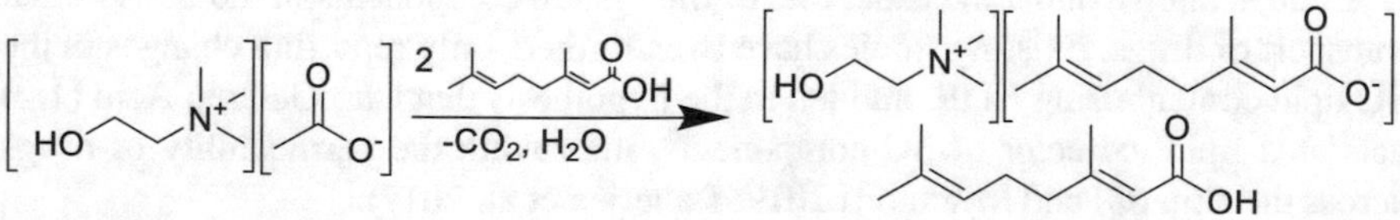

Fig. 3.2 Typical synthesis of Cho:Geranic Acid (1:2)

Cho:Geranic Acid (1:2) > EtOH:PBS > PBS > DGME. Subsequent in vivo tests were performed applying the NOB/Cho:Geranic Acid (1:2) to the shaved back of Sprague–Dawley rats. The NOB/Cho:Geranic Acid (1:2) transdermal formulation was 20-times more bioavailable than the oral solid form of NOB. Furthermore, NOB/Cho:Geranic Acid (1:2) drug formulation showed significant drops of the blood glucose level due to the hypoglycemic activity of NOB in the rats. These results indicated that NOB/Cho:Geranic Acid (1:2) formulation was able to penetrate the SC layer while PBS can only deliver NOB to the skin surface.

The use of Cho:Geranic Acid (1:2) was not limited to the solubilization of hydrophobic molecules of small size but was extended to transdermal delivery of macromolecules. An increase of permeation of dextran of different molecular weights (Mw 4, 20, 40, 70, and 150 kDa) was reported using Cho:Geranic Acid (1:2), with >40-fold increase observed for intermediate-size dextran (Mw 20 kDa) over control (PBS) (Qi and Mitragotri 2019), opening the possibility of delivering proteins, peptides, and even antibodies. Thus, without the use of any additional physical/chemical penetration enhancement techniques, Cho:Geranic Acid (1:2) was able to deliver proteins such as bovine serum albumin (BSA, Mw ~66 kDa), ovalbumin (OVA, Mw ~45 kDa), and insulin (Mw ~5.8 kDa) across porcine skin and significantly reduce blood glucose levels in rats when insulin-CAGE was topically administered (Banerjee et al. 2017).

The penetration of BSA, OVA, and insulin from Cho:Geranic Acid (1:2) was also compared with the commonly used CPEs such as DGME and ethanol. In all cases, penetration of BSA, OVA, and insulin was significantly higher in the presence of Cho:Geranic Acid (1:2) than in the presence of common CPEs. A significant 40% drop in blood glucose levels was observed 4 h after insulin–Cho:Geranic Acid (1:2) was administered to rat skin, compared to other formulation controls. The pharmacokinetic performance of the delivered insulin was studied, and interestingly, unlike efficacy of injected insulin that drops over time, the efficacy of insulin Cho:Geranic Acid (1:2) formulation indicated a sustained profile as serum insulin levels plateaued, suggesting that Cho:Geranic Acid (1:2) can be used to achieve long-term glycemic control, thereby eliminating the need for frequent insulin injections, and notably improving compliance among diabetic patients. The safety and efficacy of the system using insulin as a model drug hold significant promise as a platform technology to deliver injectables through the buccal route. In addition, insulin dispersed in Cho:Geranic Acid (1:2) was also tested for oral delivery, and was efficiently absorbed by the gastrointestinal tract after being orally administered as capsules to rats, indicating that Cho:Geranic Acid (1:2) was applicable as an oral delivery vehicle of insulin as well (Banerjee et al. 2018b).

Little is known about the exact role of the system components in the transdermal transport of drugs. Existing studies have been limited, only reporting changes in the SC lipid content using FTIR, and led to the hypothesis that Cho:Geranic Acid (1:2) acts as a lipid extractor of SC components, improving the permeability of drugs across the skin (Qi and Mitragotri 2019; Banerjee et al. 2017).

Influence of Water

Considering transdermal delivery applications, the water present in the epidermis and the dermis might interact with Cho:Geranic Acid (1:2) and influence its structure. Assuming that the SC depth is 20 μm, the epidermis depth is 200 μm, and the water content is 70% by mass (Bielfeldt et al. 2009; Warner et al. 1988), then the amount of water was calculated as 1.1 vol% water in the SC and 10.2 vol% water in the epidermis, in the case of the application of 500 μL of Cho:Geranic Acid (1:2) to 4 cm^2 skin (Takeda et al. 2021). Using Small-angle X-ray scattering (SAXS) and Nuclear Magnetic Resonance (NMR) analyses, it was shown that the addition of water up to 17% did not disturb the nanostructure of the CAGE substantially, that is, the structure of Cho:Geranic Acid (1:2) would be maintained within the skin when transdermally administered. When the amount of water exceeded 25 vol%, the viscosity of Cho:Geranic Acid (1:2) increased, and the IL exhibited low fluidity and formed micelles with an excess of water. This indicated that, even if the skin is highly hydrated, Cho:Geranic Acid (1:2) would form various structures such as lamellar and micellar phases. However, in a recent report, the authors indicated that with dilution, CAGE will gradually dissociate into its individual components (Ko et al. 2020b).

Structural Information

The important question that arises relates to ionic interactions that brought about such activity. To elucidate this, the ratio of Cho:Geranic Acid was explored by preparing not only the 1:2 Cho:Geranic Acid ratio, but also 1:1, 1:4, and 2:1 (Tanner et al. 2018). The new formulations were characterized using ^{1}H NMR spectroscopy, which suggested that the degree of hydrogen bonding in these systems is increased with the ratio of geranic acid to choline. The degree of hydrogen bonding influenced the physical and chemical properties of the system and its interactions with the skin and the dissolved drug. Properties such as viscosity, conductivity, and water miscibility were determined. Both the viscosity, conductivity, and miscibility in water decreased with the concentration of geranic acid. The 1:1 variant did not follow the linear trend, with the conductivity being lower and the viscosity being higher than expected, indicating an ion-pairing between choline and geranate. As the hydrophobicity increases with the geranic acid, those compositions with higher concentrations of this component showed higher lipid removal from the SC and enabled the dissolution of hydrophobic molecules, whilst a higher choline content increases the hydrophilicity of the IL, dissolving more hydrophilic molecules.

The degree to which the ions in each variant interact with each other will also influence their interactions with both drugs and skin. The 1:2 system was shown to be composed of choline, geranic acid, and the geranate anion in a 1:1:1 ratio ([Cho][H(Ger)$_2$]) which has been described as oligomeric (or confused) IL. ^{1}H NMR spectra of CAGE indicated a chemical shift of the protons compared to the position of these in starting materials, more so for the protons localized adjacent to the quaternary nitrogen in choline (Ko et al. 2020a). Analysis of the FTIR spectrum showed the carbonyl peak in the geranic acid to be located at 1645 cm^{-1}, while in CAGE an additional peak at 1549 cm^{-1} appears, not present in either precursor, indicating the formation of geranate. Magic-angle spinning (MAS) NMR, polarizing optical microscopy, small-angle X-ray scattering (SAXS), and mass spectrometry were used for the analysis of the closest analog of CAGE, GAGE-oct, where one equivalent of geranic acid was replaced with octanoic acid (Mann et al. 2020). Mass spectrometry and the ^{1}H NMR chemical shift of CAGE-oct indicated a dynamic system, with H-exchange occurring between hydrogen-bonded/ionic complexes in [Cho][H(Ger)(Oct)], and this provides an insight into the [Cho][H(Ger)$_2$] structural arrangement as well.

When the different systems were tested for in vitro delivery of insulin through the skin, it was observed that the systems containing 1:2 or 1:4 were effective in transporting the insulin through the SC, while neither the 2:1 nor the 1:1 samples— nor the starting materials—were able to transport the insulin into the skin (Tanner et al. 2018). These same systems were also tested for the transdermal delivery of the lipophilic drug ruxolitinib and the hydrophilic drug acarbose. In both cases, the delivery efficacy increased by approximately fourfold as the geranic acid content increased from 2:1 to 1:2 ratio, after which it plateaued, confirming that Cho:Geranic Acid (1:2) exhibited the highest delivery efficacy for hydrophilic as well as lipophilic drugs (Tanner et al. 2019).

The same model drugs, i.e., ruxolitinib and acarbose were used to compare the performance of Cho:Geranic Acid (1:2) with variants by replacing geranic acid with other organic acids in a 1:2 Cho:Organic Acid ratio, including citronellic acid, hexenoic acid, octanoic acid, octanoic acid, glutaric acid, decanoic acid, salicylic acid, and glycolic acid. As expected, the amount of drug delivered was dependent on the composition: In both cases, geranic acid outperformed other anions, followed by its close analog citronellic acid. Interestingly, octanoic acid, which possesses the same number of carbons as those in geranic acid backbone, performed well, ranked behind geranic acid and citronellic acid. Decanoic acid, on the other hand, which possesses an identical number of carbons as geranic acid, ranked very low and was poorly effective for both drugs. Octenoic acid ranked just below octanoic acid, like citronellic acid ranking below geranic acid, thus indicating that the enhancement is not simply a function of lipophilicity and that a single unsaturated bond makes an impact on delivery efficacy. Small anions including salicylic acid and glutaric acid were on the lower end of the ranking spectrum (Tanner et al. 2019). This is consistent with the expected mechanisms of action of ILs. Specifically, hydrophobic tails of anions and/or cations play a key role in the fluidization of the SC lipids. Since choline is highly hydrophilic and lacks long, aliphatic chains, the anions provide the primary mode of lipid disruption. Systematic modifications of the anion revealed that the ILs with the fewest interionic interactions were most successful at transdermal transport. Modifications of cations to further reduce the interionic interactions also improved delivery efficacy. The combination of the most effective cations and anions led to the selection of (2-hydroxyethyl)tributylammonium:citronellic acid (1:2 ratio) as the best delivery system, with 117.1 ± 12.7 μg cm^{-2} ruxolitinib measured in the dermis and acceptor after 24 h (Tanner et al. 2019).

Using the Cho:Geranic Acid (1:2) platform, similar ILs were synthesized at a stoichiometric ratio of 1:2 (cation:anion), and investigated to simultaneously stabilize small interfering RNAi (siRNA) and enhance its penetration into the skin following topical application (Mandal et al. 2020). The siRNA structure was found to be altered in the presence of Cho:Geranic Acid (1:2), Cho:Dimethylacrylic Acid (1:2), and Cho:Biphenyl-3-carboxylic Acid (1:2). On the other hand, Cho:Isovaleric Acid (1:2) and Cho:Phenylpropanoic Acid (1:2) retained the secondary structure of siRNA. The improved stability of siRNA in the presence of Cho:Phenylpropanoic Acid (1:2) suggested the possibility of synergistic effects between the ILs prepared from two structurally different anions. Consequently, the effect of IL mixtures on siRNA stability was assessed, to determine whether the compatibility of Cho:Phenylpropanoic Acid (1:2) with siRNA might offer additional protection against the adverse effects of Cho:Geranic Acid (1:2) and Cho:Biphenyl-3-carboxylic acid (1:2) on the siRNA structure. The combination of Cho:Geranic Acid (1:2) (25 vol%) and Cho:Phenylpropanoic Acid (1:2) (25 vol%) led to a prominent band indicative of retention of siRNA structure and generated the highest epidermal accumulation of siRNA, notably higher than any other individual ILs and/or combination. This formulation was then used in an imiquimod-induced psoriasis-like skin inflammation model that resembles plaque-type psoriasis in humans. Topical application of IL-siRNA for 4 consecutive days generated a substantial reduction in the levels of inflammatory cytokines and an array of psoriasis-related gene products (Mandal et al. 2020). Future work should focus on IL moiety optimization for improved cell internalization and biocompatibility, as well as applying this platform for the treatment of other skin diseases such as psoriasis, atopic dermatitis, and melasma (Zakrewsky and Mitragotri 2016).

Incorporating the Cho:Geranic Acid (1:2) Formulation to Solid Supports

These systems can also be used to dissolve drugs, and the mixture was incorporated into transdermal patches or devices as delivery systems. For example, Cho:Geranic Acid (with 1:1 or 1:2 molar ratio) were compared with Cho:Oleic Acid (with 1:1 or 1:2 molar ratio) to dissolve insulin, and the mixtures were incorporated into biofilms composed of xanthan gum and bacterial nanocellulose (Jorge et al. 2020). The films containing Cho:Geranic Acid (1:1) or (1:2) exhibited positive antimicrobial activity against *Escherichia coli* and *Staphylococcus aureus* but did not affect *Pseudomonas aeruginosa*. On the contrary, none of the Cho:Oleic Acid systems showed an effect upon E. coli and only a mild antimicrobial activity against S. aureus. On P. aeruginosa, only Cho:Oleic Acid (1:1) showed a positive mild antimicrobial activity. The Cho:Geranic Acid (1:2) that performed best in terms of lack of cytotoxicity, genotoxicity, and permeation enhancement was used to dissolve insulin and integrated into the biofilm, allowing an average permeated protein per skin area of 1.05 μg protein mm^{-2} skin.

In a recent publication, Silva and coworkers isolated lytic bacteriophages for *Staphylococcus intermedius* as a new antimicrobial treatment for pyoderma via cutaneous permeation of bacteriophage particles and used Cho:Geranic Acid (1:1) and (1:2) as permeation enhancers (Silva et al. 2021). The isolated phage particles, identified as Caudovirales, most likely belonging to the family Myoviridae, had particle sizes in the order of the nanometers (<100 nm). The phage particles were mixed with Cho:Geranic Acid (1:1) or (1:2), and the mixtures were integrated into hydroxyethyl cellulose (HEC) gel. When tested for transdermal delivery, the presence of Cho:Geranic Acid (1:2) led to increased average of permeated phage particles, as opposed to the permeation assays with Cho:Geranic Acid (1:1) or the absence of either Cho:Geranic Acid (1:1) or (1:2) in the HEC matrix. Combining phage particles and Cho:Geranic Acid (1:2) allowed the transdermal permeation of phage particles and will potentially allow the treatment of pyodermal infection. The same research group has developed a sodium alginate-based hydrogel integrating choline oleate as a carrier for a cocktail of lytic phage particles aiming at the treatment of deep skin infections in humans caused by *Acinetobacter baumannii* (Campos et al. 2020). Combining phage particles with choline-based ILs as carriers may be an excellent alternative to antibiotics to combat dermal infections.

Commercial Deployment

Scaling up the synthesis of CAGE and its medical applications are one of the greatest challenges for the translation of this compound into practice. The first report scaling up the synthesis of Cho:Geranic Acid (1:2), kg scale, was recently published in Ko et al. 2020a, where 40 wt% Cho:Geranic Acid (1:2) was incorporated as a gel using hydroxypropyl cellulose as a gelling agent; gel vehicle consisted of ~90% water along with some propylene glycol (cosolvent), a fragrance, and a gelling agent. This formulation, referred to as CGB400, was used for in vitro and in vivo tests. The resulting gel showed promising results, usingthe antimicrobial properties of Cho:Geranic Acid (1:2) for rosacea treatment. After proving for low toxicity and high antimicrobial effects, a 12-week open-label cosmetic study for rosacea was performed on humans. A marked improvement in symptoms was observed, with ~70% of the subjects reporting a greater reduction in lesions after treatment. Although this study did not include a drug and only explored the properties of Cho:Geranic Acid (1:2) previously shown in vitro and in mice as a broad-spectrum antimicrobial agent (Zakrewsky et al. 2016), it sets a precedent for the scaling up and low toxicity of the CAGE, now tested in humans.

Conclusions

Currently, CAGE Bio, Inc., a clinical-stage startup based in San Carlos, Calif. exploits CAGE and similar systems, to develop novel pharmaceutical products for dermatology, inflammation, immunology, etc. The startup has recently secured $7.2M in funding (CAGE Bio Inc 2021).

This is however not the first example of ILs and/or DESs with transdermal properties proposed for commercialization. The Lidocaine–Etodolac Patch (DES, containing the NSAID etodolac and the pain-reliever lidocaine) has shown to be effective in the treatment of postherpetic neuropathic (PHN) pain in placebo- and active-controlled comparisons and has been incorporated into a topical transdermal delivery system, Etodolac–Lidocaine Topical Patch (RX-7EAT or "Etoreat") (ClinicalTrials.gov 2018 ; Miwa et al. 2016). The etodolac products developed by a research and development Japanese company, MedRx, exhibited increased transdermal absorbability and improved efficacy of this IL. The new venture, IL Pharma

Inc. (MEDRx, Kagawa, Japan), owned 51% by Kaneka and 49% by MedRx, took over programs that MedRx was developing on its own. This patch was the first API-IL that has reached clinical trials for the treatment of ankle sprains, low back pain, and muscle soreness. MedRx Co. Ltd. has received approval to conduct human trials in the US and completed Phase I trials for acute pain in the USA (NCT02695381). Even though in vitro studies showed that this system significantly increased the skin permeation of etodolac (9.3-fold) compared to "free" etodolac (Miwa et al. 2016), due to the lack of statistically significant differences between Etoreat and placebo in humans, its development has been suspended.

Another system includes GSK2838232, the HIV-1 maturation inhibitor (MI) that is being developed for the treatment of HIV-1 infection in combination with other antiretroviral therapy, derived from the triterpene betulin natural product (Shamshina and Rogers 2020). It is currently in Phase IIa clinical studies, to evaluate its safety, pharmacokinetics, and antiviral activity in participants with HIV-1 virus (DeJesus et al. 2020).

Indeed, drug development is a long and expensive process, taking on average, 10–15 years, with ~14% of all drugs in clinical trials eventually getting approval from the FDA (Hale 2018). Based on this situation, IL (or DES)-based technology of producing transdermally active drugs, by selectively combining ionizable compounds from GRAS-list into a single IL (or DES) form, could take a shorter time to market. Owing to their easy tunability, ILs and DESs constitute a vast number of formulation candidates. The number of combinations is almost endless, by modifying functional groups, carbon-chain lengths, and ion ratios. However, ILs and DESs would require further characterization, and way more additional testing that is available now, and here, federal funding could play an important role by widening opportunities and providing more incentives to those investors who are willing to take risks.

References

Banerjee A, Ibsen K, Iwao Y, Zakrewsky M, Mitragotri S (2017) Transdermal protein delivery using choline and geranate (CAGE) deep eutectic solvent. Adv Healthcare Mater 6:1601411

Banerjee A, Ibsen K, Brown T, Chen R, Agatemor C, Mitragotri S (2018a) Reply to Rogers and Gurau: definitions of ionic liquids and deep eutectic solvents. Proc Natl Acad Sci USA 115:E11000–E11001

Banerjee A, Ibsen K, Brown T, Chen R, Agatemor C, Mitragotri S (2018b) Ionic liquids for oral insulin delivery. Proc Natl Acad Sci USA 115:7296–7301

Berton P, Kelley SP, Wang H, Rogers RD (2018) Elucidating the triethylammonium acetate system: Is it molecular or is it ionic? J Mol Liq 269:126–131

Bica K, Rogers RD (2010) Confused ionic liquid ions — A "liquification" and dosage strategy for pharmaceutically active salts. Chem Commun 46:1215–1217

Bica K, Shamshina JL, Hough WL, MacFarlane DR, Rogers RD (2011) Liquid forms of pharmaceutical co-crystals: exploring the boundaries of salt formation. Chem Commun 47:2267–2269

Bielfeldt S, Schoder V, Ely U, Van Der Pol A, De Sterke J, Wilhelm K-P (2009) Assessment of human stratum corneum thickness and its barrier properties by in-vivo confocal Raman spectroscopy. Int J Cosmet Sci 31:479–480

Brisson P (1974) Percutaneous absorption. Can Med Assoc J 110:1182–1185

CAGE Bio Inc (2021) CAGE Bio closes $7.2 Million round of series A funding. https://www.prn ewswire.com/news-releases/cage-bio-closes-7-2-million-round-of-series-a-funding-301229156. html. Last accessed 04–19–21

Campos WF, Silva EC, Oliveira TJ, Oliveira JM, Tubino M, Pereira C, Vila MMDC, Balcão VM (2020) Transdermal permeation of bacteriophage particles by choline oleate: potential for treatment of soft-tissue infections. Future Microbiol 15:881–896. https://doi.org/10.2217/fmb-2019-0290

Chen Y, Quan P, Liu X, Wang M, Fang L (2014) Novel chemical permeation enhancers for transdermal drug delivery. Asian J Pharm Sci 9:51–64

Choline chloride (April 1, 2020). FDA U.S. Food & Drug. https://www.accessdata.fda.gov/scripts/ cdrh/cfdocs/cfcfr/CFRSearch.cfm?fr=582.5252. Last accessed 04–19–21

ClinicalTrials.gov (2018) Identifier: NCT01198834, MRX-7EAT Etodolac-Lidocaine Topical Patch in the Treatment of Ankle Sprains. https://clinicaltrials.gov/ct2/show/NCT01198834. Last accessed 04–19–21.

Cohen SM, Eisenbrand G, Fukushima S, Gooderham NJ, Guengerich FP, Hecht SS, Rietjens IMCM, Harman C, Taylor SV (April 1, 2020) GRAS 28 Flavoring Substances. https://www.accessdata. fda.gov/scripts/cdrh/cfdocs/cfcfr/CFRSearch.cfm?fr=582.5252. Last accessed 04–19–21

DeJesus E, Harward S, Jewell RC, Johnson M, Dumont E, Wilches V, Halliday F, Talarico CL, Jeffrey J, Gan J, Xu J, Felizarta F, Scribner A, Ramgopal M, Benson P, Johns BA (2020) A phase IIa study evaluating safety, pharmacokinetics, and antiviral activity of GSK2838232, a novel, second-generation maturation inhibitor, in participants with human immunodeficiency virus type 1 infection. Clin Infect Dis 71:1255–1262

Fei Z, Geldbach TJ, Zhao D, Dyson PJ (2006) From dysfunction to bis-function: on the design and applications of functionalised ionic liquids. Chem Eur J 12:2122–2128

Freemantle M (1998) Designer Solvents. Ionic liquids may boost clean technology development. Chem Eng News 76:32–37

Hale C (2018) New MIT study puts clinical research success rate at 14 percent. https://www. centerwatch.com/articles/12702-new-mit-study-puts-clinical-research-success-rate-at-14-per cent#:~:text=Nearly%2014%20percent%20of%20all,MIT%20Sloan%20School%20of%20M anagement. Last accessed 04–19–21

Harada LK, Pereira JFB, Campos WF, Silva EC, Moutinho CG, Vila MMDC, Oliveira JM Jr, Teixeira JA, Balcão VM, Tubino M (2018) Insights into protein-ionic liquid interactions aiming at macromolecule delivery systems. J Braz Chem Soc 29:1983–1998

Hattori T, Tagawa H, Inai M, Kan T, Kimura S-I, Itai S, Mitragotri S, Iwao Y (2019) Transdermal delivery of nobiletin using ionic liquids. Sci Rep 9:20191. https://doi.org/10.1038/s41598-019-56731-1

Hough WL, Smiglak M, Rodríguez H, Swatloski RP, Spear SK, Daly DT, Pernak J, Grisel JE, Carliss RD, Soutullo MD, Davis JD Jr, Rogers RD (2007) The third evolution of ionic liquids: Active pharmaceutical ingredients. New J Chem 31:1429–1436

Jaitely V, Karatas A, Florence AT (2008) Water immiscible room temperature ionic liquids: some properties relevant to their pharmaceutical use. Int J Pharm 354:168–173

Jorge LR, Harada LK, Silva EC, Campos WF, Moreli FC, Shimamoto G, Pereira JFB, Oliveira Jr. JM, Tubino M, Vila MMDC, Balcão VM (2020) Non-invasive transdermal delivery of human insulin using ionic liquids: In vitro studies. Front Pharmacol 11, Article 243. https://doi.org/10. 3389/fphar.2020.00243

Kalluri H, Banga AK (2011) Transdermal delivery of proteins. AAPS PharmSciTech 12:431–441

Karande P, Mitragotri S (2009) Enhancement of transdermal drug delivery via synergistic action of chemicals. BBA-Biomem 1788:2362–2373

Karande P, Jain A, Mitragotri S (2004) Discovery of transdermal penetration enhancers by high-throughput screening. Nat Biotech 22:192–197

Karande P, Jain A, Ergun K, Kispersky V, Mitragotri S (2005) Design principles of chemical penetration enhancers for transdermal drug delivery. Proc Natl Acad Sci USA 102:4688–4693

Kelley SP, Narita A, Holbrey JD, Green KD, Reichert WM, Rogers RD (2013) Understanding the effects of ionicity in salts, solvates, co-crystals, ionic co-crystals, and ionic liquids, rather than nomenclature, is critical to understanding their behavior. Cryst Growth Des 13:965–975

Kim M-J, Doh H-J, Choi M-K, Chung S-J, Shim C-K, Kim D-D, Kim JS, Yong C-S, Choi H-G (2008) Skin permeation enhancement of diclofenac by fatty acids. Drug Delivery 15:373–379

Ko J, Mandal A, Dhawan S, Shevachman M, Mitragotri S, Joshi N (2020) Clinical translation of choline and geranic acid deep eutectic solvent. Bioeng Transl Med e10191. https://doi.org/10. 1002/btm2.10191

Ko J, Mandal A, Dhawan S, Shevachman M, Mitragotri S, Joshi N (2020) Clinical translation of choline and geranic acid deep eutectic solvent. Bioeng Transl Med e10191. Prepublication Peer Review and Author Responses. https://publons.com/publon/https://doi.org/10.1002/btm2.10191. Last accessed 04–19–21

Mandal A, Kumbhojkar N, Reilly C, Dharamdasani V, Ukidve A, Ingber DE, Mitragotri S (2020) Treatment of psoriasis with NFKBIZ siRNA using topical ionic liquid formulations. Sci Adv 6:eabb6049

Mann SK, Devgan MK, Franks WT, Huband S, Chan CL, Griffith J, Pugh D, Brooks NJ, Welton T, Pham TN, McQueen LL, Lewandowski JR, Brown SP (2020) MAS NMR Investigation of molecular order in ionic liquid crystal. J Phys Chem B 4975–4988

Mehta R (2004) Topical and transdermal drug delivery: what a pharmacist needs to know. InetCE. 1st, Arizona, 1–10

Miwa Y, Hamamoto H, Ishida T (2016) Lidocaine self-sacrificially improves the skin permeation of the acidic and poorly water-soluble drug etodolac via its transformation into an ionic liquid. Eur J Pharm Biopharm 102:92–100

Mizuuchi H, Jaitely V, Murdan S, Florence AT (2008) Room temperature ionic liquids and their mixtures: potential pharmaceutical solvents. Eur J Pharm Sci 33:326–331

Park HJ, Prauznitz MR (2010) Lidocaine-ibuprofen ionic liquid for dermal anesthesia. AIChE J 61:2732–2738

Prausnitz M, Langer R (2008) Transdermal drug delivery. Nat Biotechnol 26:1261–1268

Prausnitz MR, Elias PM, Franz TJ, Schmuth M, Tsai J-C, Menon GK, Holleran WM, Feingold KR (2017) Skin barrier and transdermal drug delivery. Chapter 124. In Bolognia J, Schaffer J, Cerroni L (eds) Dermatology: 2-volume set, 4th edn. Elsevier

Qi QM, Mitragotri S (2019) Mechanistic study of transdermal delivery of macromolecules assisted by ionic liquids. J Controlled Release 311–312:162–169

Qi QM, Duffy M, Curreri AM, Balkaran JPR, Tanner EEL, Mitragotri S (2020) Comparison of ionic liquids and chemical permeation enhancers for transdermal drug delivery. Adv Funct Mater 30:2004257

Ramadon D, McCrudden MTC, Courtenay AJ, Donnelly RF (2021) Enhancement strategies for transdermal drug delivery systems: current trends and applications. Drug Delivery Transl Res 1–34. https://doi.org/10.1007/s13346-021-00909-6

Rebelo LPN, Lopes JNC, Esperança JMSS, Guedes HJR, Lachwa J, Najdanovic-Visak V, Visak ZP (2007) Accounting for the unique, doubly dual nature of ionic liquids from a molecular thermodynamic and modeling standpoint. Acc Chem Res 40:1114–1121

Reslan M, Kayser V (2018) Ionic liquids as biocompatible stabilizers of proteins. Biophys Rev 10:781–793

Rogers RD, Gurau G (2018) Is "choline and geranate" an ionic liquid or deep eutectic solvent system? Proc Natl Acad Sci USA 115:E10999

Shamshina JL, Rogers RD (2020) Are myths and preconceptions preventing us from applying ionic liquid forms of antiviral medicines to the current health crisis? Int J Mol Sci 21:6002–6018

Shamshina JL, Berton P, Wang H, Zhou X, Gurau G, Rogers RD (2018) Ionic liquids in the pharmaceutical industry. In Zhang W, Cue B (eds) Green techniques for organic synthesis and medicinal chemistry, 2nd edn, Wiley, pp 541–577

Sidat Z, Marimuthu T, Kumar P, du Toit LC, Kondiah P, Choonara YE, Pillay V (2019) Ionic liquids as potential and synergistic permeation enhancers for transdermal drug delivery. Pharmaceutics 11:1–21

Silva EC, Oliveira TJ, Moreli FC, Harada LK, Vila MMDC, Balcão VM (2021) Newly isolated lytic bacteriophages for Staphylococcus intermedius, structurally and functionally stabilized in a hydroxyethylcellulose gel containing choline geranate: potential for transdermal permeation in veterinary phage therapy. Res Vet Sci 135:42–58

Stoimenovski J, MacFarlane DR (2011) Enhanced membrane transport of pharmaceutically active protic ionic liquids. Chem Commun 47:11429–11431

Takeda J, Iwao Y, Karashima M, Yamamoto K, Ikeda Y (2021) Structural evaluation of the choline and geranic acid/water complex by SAXS and NMR analyses. ACS Biomater Sci Eng 7:595–604

Tanner EEL, Ibsen KN, Mitragotri S (2018) Transdermal insulin delivery using choline-based ionic liquids (CAGE). J Control Release 286:137–144

Tanner E, Curreri A, Balkaran J, Selig-Wober N, Yang A, Kendig C, Fluhr M, Kim N, Mitragotri S (2019) Design principles of ionic liquids for transdermal drug delivery. Adv Mater 31:1901103

The ionic liquid platform (2020) https://cagebio.com/our-technology. Last accessed 04–19–21

Wang H, Gurau G, Shamshina J, Cojocaru OA, Janikowski J, MacFarlane DR, Davis JH Jr, Rogers RD (2014) Simultaneous membrane transport of two active pharmaceutical ingredients by charge assisted hydrogen bond complex formation. Chem Sci 5:3449–3456

Warner RR, Myers MC, Taylor DA (1988) Electron probe analysis of human skin: determination of the water concentration profile. J Invest Dermatol 90:218–224

Welton T (1999) Room-temperature ionic liquids. Solvents for synthesis and catalysis. Chem Rev 99:2071–2084

Yang R, Wei T, Goldberg H, Wang W, Cullion K, Kohane D (2017) Getting drugs across biological barriers. Adv Mater 29:1606596

Zakrewsky M, Mitragotri S (2016) Therapeutic RNAi robed with ionic liquid moieties as a simple, scalable prodrug platform for treating skin disease. J Controlled Release 242:80–88

Zakrewsky M, Lovejoy KS, Kern TL, Miller TE, Le V, Nagy A, Goumas AM, Iyer RS, Del Sesto RE, Koppisch AT, Fox DT, Mitragotri S (2014a) Biological applications of ionic liquids. Proc Natl Acad Sci USA 111:13313–13318

Zakrewsky M, Lovejoy KS, Kern TL, Miller TE, Le V, Nagy A, Goumas AM, Iyer RS, Del Sesto RE, Koppisch AT, Fox DT, Mitragotri S (2014b) Ionic liquids as a class of materials for transdermal delivery and pathogen neutralization. Proc Natl Acad Sci USA 111:13313–13318

Zakrewsky M, Banerjee A, Apte S, Kern TL, Jones MR, Del Sesto RE, Koppisch AT, Fox DT, Mitragotri S (2016) Choline and geranate deep eutectic solvent as a broad-spectrum antiseptic agent for preventive and therapeutic applications. Adv Healthcare Mater 5:1282–1289

Zakrewsky M, Mitragotri S, Fox DT, Koppisch A, Del Sesto RE, Lovejoy K (2019) Ionic liquids for transdermal drug delivery. US 10,449,254 B2, 22 October 2019

Chapter 4
Ionic Liquids for Increasing the Solubility of Sparingly Soluble Drug Molecules

Rahman Md Moshikur⑩, **Md. Korban Ali**⑩, **and Masahiro Goto**⑩

Abstract Ionic liquids (ILs) have attracted considerable attention as green alternatives to conventional toxic organic solvents. They have several advantageous properties that make them potentially useful in various areas of modern science, especially drug delivery. Owing to their intrinsic tunability and exceptional properties, ILs are undoubtedly capable of addressing the drawbacks associated with sparingly soluble drugs, such as polymorphism, limited solubility, and poor bioavailability. ILs can be used as replacements for conventional volatile organic solvents. Their task-specific optimization at the molecular level can be exploited to obtain effective drug formulations. The structural properties of ILs can be tailored by altering their constituent anions and cations to improve the pharmacokinetic and pharmacodynamic properties and biological activities of drugs. This chapter describes the progress of IL-related research in pharmaceutics. It comprises a discussion of the use of ILs, either as solubilizing agents or solvents/carriers for pharmaceutical drugs. Specifically, the chapter deals with the use of ILs as potential solvents/agents for dissolving sparingly soluble drugs, and with the exploitation of the possible mechanism by which ILs increase solubility. An understanding of the ionic interactions between drug molecules and ILs in an aqueous solution will also stimulate innovative IL-based technologies and the use of ILs in pharmaceuticals and medicine.

Keywords Sparingly soluble drug · Ionic liquid · Dissolution · Ionic interaction · Drug development

R. Md Moshikur · Md. K. Ali · M. Goto (✉)
Department of Applied Chemistry, Graduate School of Engineering, Kyushu University, 744 Motooka, Nishi-ku, Fukuoka 819-0395, Japan
e-mail: m-goto@mail.cstm.kyushu-u.ac.jp

Md. K. Ali
Department of Chemistry, Jashore University of Science and Technology, Jashore 7408, Bangladesh

M. Goto
Advanced Transdermal Drug Delivery System Centre, Kyushu University, 744 Motooka, Nishi-ku, Fukuoka 819-0395, Japan

Division of Biotechnology, Centre for Future Chemistry, Kyushu University, 744 Motooka, Nishi-ku, Fukuoka 819-0395, Japan

Abbreviations

[C_2MIM]	1-Ethyl-3-methylimidazolium
[C_4MIM]	1-Butyl-3-methylimidazolium
[C_6MIM]	1-Hexyl-3-methyl imidazolium
[C_8MIM]	1-Octyl-3-methylimidazolium
[C_{10}MIM]	1-Decyl 3-methylimidazolium
[$P_{6,6,6,14}$]	Trihexyl(tetradecyl)phosphonium
[C_4MPyr]	1-Butyl-1-methylpyrrolidinium
[C_8M_βPyr]	1-Octyl-3-methylpyridinium
[C_4NH$_3$]	*N*-Butylamine
[C_6NH$_3$]	*N*-Hexylamine
[C_8NH$_3$]	*N*-Octylamine
[C_2HOC$_{1,1}$EA]	*N*-Ethyl-2-hydroxy-*N*,*N*-dimethylethylammonium
[$N_{4,1,1,1}$]	Butyltrimethylammonium
[$N_{1,8,8,8}$]	Methyltrioctylammonium
[$N_{10,10,1,1}$]	Didecyldimethylammonium
[m-PEG350-NH$_3$]	Polyethylene glycol ammonium
[C_6HOCPyr]	1-Hexyl-3-hexyloxycarbonylpyridinium
[DC-7]	Dicationic cholinium
[BA]	(Benzyl)-dimethylalkylammonium
[AAE]	Amino acid ester
[Cl]	Chloride
[Br]	Bromide
[NO$_3$]	Nitrate
[OH]	Hydroxide
[MPn]	Methyl phosphonate
[EPn]	Ethyl phosphonate
[OAc]	Acetate
[TFSA]	Bis(trifluoromethylsulphonyl) amide
[OTf]	Trifluoromethane sulphonate
[BF$_4$]	Tetrafluoroborate
[Hex]	Hexanoate
(MeO)$_2$PO$_2$	Dimethyl phosphate
[C_5COO]	Pentanoate
[C_7COO]	Heptanoate
[C_9COO]	Nonanoate
[PF$_6$]	Hexafluoro phosphate
[C_6SO$_4$]	Hexyl sulphate
[CF$_3$O$_3$S]	Trifluoromethane sulphonate

4.1 Introduction

The development of effective drug delivery systems for sparingly soluble drugs remains a significant challenge for the pharmaceutical industry because such drugs are prone to polymorphism, have limited solubility, and are difficult to formulate (Egorova et al. 2017; Agatemor et al. 2018). The irregular gastrointestinal absorption, pre-systemic metabolism, and high levels of p-glycoprotein-mediated efflux associated with sparingly soluble drugs are also major barriers to maximizing therapeutic efficacy and minimizing side effects (Gavhane and Yadav 2012; Sohail et al. 2018). Currently, 40% of marketed drugs and up to 70% of drugs under development have poor solubility in aqueous systems and, consequently, are compromised in terms of bioavailability and therapeutic efficacy (Rodriguez-Aller et al. 2015; Moshikur et al. 2020b). These drugs have low or variable absorption rates when delivered orally and require high-dosage administration in therapeutic applications. As a result, long-term gastrointestinal tract (GIT) toxicity or potentially lethal dosages are possible if the wrong polymorph is administered (Pedro et al. 2020). However, a variety of pharmaceutical strategies—including lyophilization, micellization, and the use of solid dispersions, microemulsions, nanoparticles, and drug–carrier conjugates—have been designed to address the challenges posed by the formulation and delivery of these problematic drugs (Rodriguez-Aller et al. 2015; Egorova et al. 2017; Moshikur et al. 2020b). These techniques have already demonstrated strong potential for improving the solubility and bioavailability of sparingly soluble drugs. However, numerous organic solvents—such as methanol, ethanol, acetone, dimethyl sulfoxide (DMSO), chloroform, isopropanol, and tetrahydrofuran—are used to formulate these drugs, leading to concerns for human health and ecosystems (Clarke et al. 2018). Therefore, green techniques that do not compromise the therapeutic efficacy of sparingly soluble drugs are required for their effective delivery.

Ionic liquids (ILs) could be used to address the issues of polymorphism and aqueous solubility described above, simultaneously reducing the necessity for organic solvents (Box 4.1) (Egorova et al. 2017; Moshikur et al. 2019).

Box 4.1 Advantages of ionic liquids (ILs) in comparison with conventional organic solvents
- Scope to design tuneable ILs with desired physicochemical properties
- Green alternatives to conventional toxic organic solvents
- Easy to synthesize by neutralization
- Wide-ranging solubility and miscibility
- Lower melting points or glass transition temperatures, making them user friendly
- In most cases, improved physicothermal stability
- Enhanced ability to dissolve sparingly soluble drugs
- Improved antimicrobial activity over free drugs

Generally, ILs are molten organic salts comprising asymmetrical organic cations and inorganic or organic anions with melting points at or below 100 °C (Welton 1999; Moniruzzaman and Goto 2011; Uddin et al. 2020). They are synthesized by acid–base neutralization, metathesis in a suitable solvent, or solvent-free grinding or melting (Pedro et al. 2020). During neutralization, either stoichiometric or nonstoi-chiometric ratios of acids and bases can be used to prepare specific ILs, resulting in full or partial ionization. After synthesis, ILs do not readily form regular crystalline structures comprising cations or anions owing to consist the large dimensions of ions that disperse charge (Earle et al. 2006; Pedro et al. 2020). However, ILs have been used extensively as green alternatives to traditional toxic, hazardous, flammable, and volatile organic solvents. This is because they have unique physical and chemical properties, such as low volatility, nonflammability, high thermal stability, and the ability to dissolve various organic and inorganic compounds (Agatemor et al. 2018; Huang et al. 2020; Pedro et al. 2020). These properties can be easily altered by simply modifying the combination of cations and anions, leading to the concept of 'designer solvents' for particular applications. The potential to alter the combina-tion of cations and anions enables the formation of a vast range of ILs, allowing their intrinsic physicochemical properties—including density, viscosity, conduc-tivity, melting point, polarity, surface tension, hydrophilicity/hydrophobicity, thermal stability, and solvation power—to be customized (Adawiyah et al. 2016; Egorova et al. 2017). The tailor-made physicochemical and biological properties of ILs have qualified them for use as solvents or materials, or both, in a wide range of fields such as electrochemistry, organic synthesis, biocatalysis, biofuel production, extraction, and pharmaceutics (Egorova et al. 2017; Moshikur et al. 2020b). ILs are increasingly used in pharmaceutical formulations because they have promising physicochemical and biopharmaceutical advantages over crystalline or solid drugs. They can also be tailored to provide innovative solutions to many of the problems encountered in pharmaceutics and medicine (Egorova et al. 2017; Pedro et al. 2020). ILs have been designed to dissolve, stabilize, and deliver poorly soluble drugs—e.g., in oral, topical, or transdermal formulations. Because they have some unique properties that are not present in conventional organic solvents, ILs are used for drug delivery, or as reaction media used to synthesize drug compounds (Egorova et al. 2017; Ali et al. 2020). They also have potential antimicrobial properties and could be used to prevent or treat various infectious diseases.

The number of research publications related to the use of ILs in drug formulations and delivery has increased significantly in the last two decades (Pedro et al. 2020; Moshikur et al. 2020b). Figure 4.1a shows the number of publications related to the application of ILs/IL technologies in the pharmaceutical field over the past two decades. Interestingly, ILs have been used as green substitutes for commonly used pharmaceutical solvents including ethanol, DMSO, and propylene glycol (Fig. 4.1b). Consequently, many companies that produce generic pharmaceuticals are increas-ingly motivated to design effective IL-based drug delivery systems. However, this chapter presents a general overview of the use of ILs as solubilizing agents and emphasizes the development of ILs in pharmaceutics and medicine.

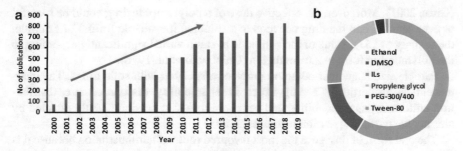

Fig. 4.1 a Publications related to ionic liquids (ILs)/IL technologies over the past two decades. **b** A comparison of the use of ILs and commonly used pharmaceutical solvents/co-solvents and surfactants/co-surfactants, reproduced with permission from Moshikur et al. (2020b)

4.2 Problems Associated with Conventional Drugs

The efficiency of any conventional drug in the human body mainly depends on its bioavailability, which can be ascribed to its solubility, permeability, and polymorphism (Gavhane and Yadav 2012; Egorova et al. 2017). The physiological accessibility of such drugs also depends on their ability to dissolve in bodily fluids and is directly related to the rate at which they are absorbed by the GIT. Limited aqueous solubility is a major challenge to the improvement of pharmacokinetic and pharmacodynamic properties in modern drug development, especially with regard to oral administration (Savjani et al. 2012). Generally, dissolution-limited absorption occurs when the solubility of a drug is less than 100 μg/mL of water or a physiological fluid (Savjani et al. 2012). In such cases, low permeability (a barrier to drug delivery), a narrow therapeutic window, and a short half-life with systemic side effects are observed in clinical applications (Ali et al. 2017; Egorova et al. 2017). As a result, higher doses of the drug are required to achieve an effective therapeutic response, which in the case of oral administration results in GIT toxicity. High-dose formulations are also problematic because they have a tendency to stick and exhibit poor powder flow during granulation and tabletting (Savjani et al. 2012). Furthermore, the overall production costs are high because large amounts of the drug are required to achieve an effective therapeutic dosage.

Polymorphism (derived from the Greek words 'polys' or multiple and 'morfe' or shape) is another major complication of solid drugs. It occurs when one crystalline chemical coexists as several polymorphs or pseudopolymorphs with different physicochemical and biopharmaceutical properties—such as melting point, solubility, stability, and bioavailability (Censi and Di Martino 2015; Egorova et al. 2017). These polymorphs can undergo unpredictable interchanges at any stage of manufacture or storage, and their formation is very difficult to control. The polymorphic phases may also be affected by manufacturing conditions (solvent, temperature, and pressure), and different forms—which can crystallize out or precipitate in solution or formulation—may coexist in the drug product, resulting in inadequate bioavailability

(Giron 2001). Moreover, the effective dose of a polymorphic drug could be harmful or even lethal when the drug converts to a soluble/bioavailable form. For example, the toxicity or LD_{50} value of the antibiotic nystatin varied significantly according to the polymorphic form in a formulation (Ghielmetti et al. 1976).

Particle size is another common problem associated with solid drugs. The pharmaceutical properties of a drug—such as its solubility, dissolution rate, distribution uniformity, suspendability, and penetrability—can be significantly affected by altering the particle size (Shamshina et al. 2013).

The oral form of dosage is the most favoured route of administration because it is simple, convenient, painless, flexible, and cost-effective. Moreover, it is associated with the highest rates of patient compliance and has the fewest constraints in terms of sterility compared with injection (Krishnaiah 2010). The biggest challenge for effective oral dosage is poor bioavailability, which can be attributed to limited aqueous solubility, poor intestinal permeability, pre-systemic and first-pass metabolism, and high levels of p-glycoprotein-mediated efflux (Gavhane and Yadav 2012; Sohail et al. 2018). These negative effects increase the cost and duration of manufacture and shift the burden to the patient. Therefore, it is important to manufacture solid drugs with controlled polymorphic form, controlled crystal size, and product solubility. The possible disadvantages of liquid formulations arising from issues with stability and palatability also need to be considered when designing effective delivery systems.

4.3   Importance of Drug Solubility in Pharmaceutics

The solubility of a drug is one of the key parameters that decisively influence its absorption and bioavailability. It also plays an important role in formulating the various forms of dosage, which include oral, parenteral, and transdermal (Savjani et al. 2012). Generally, the solubility of a drug indicates the degree to which it dissolves in a solvent to form a homogeneous solution, in which the atoms, molecules, or ions are uniformly distributed. Solubility mainly depends on both the chemical structure of the drug and the conditions of the solution, including pH, co-solvents, additives, ionic strength, temperature, and pressure. However, the limited solubility and dissolution rates of drugs in aqueous gastrointestinal fluids result in insufficient bioavailability and gastrointestinal mucosal toxicity. It has been reported that any drug can be absorbed in the GIT if it is present as an aqueous solution at the site of absorption (Egorova et al. 2017). Drugs with higher solubilities or dissolution rates have significantly improved absorption rates in the GIT, indicating the desired concentration in systemic circulation and the optimal pharmacological response (Egorova et al. 2017). Therefore, irrespective of the intended route of administration, the design of an effective therapeutic formulation depends on the drug's ability to dissolve in an aqueous medium.

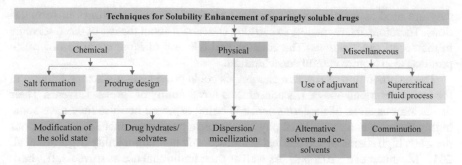

Fig. 4.2 Schematic representation of the various strategies used to increase the solubility of sparingly soluble drugs

4.4 Techniques for Solubility Enhancement

Technology is playing a crucial role in addressing the problem of the poor water solubility of drugs. A variety of pharmaceutical approaches have been considered for overcoming the limited aqueous solubilities and dissolution rates of sparingly soluble drugs (Fig. 4.2). These techniques are divided into several categories based on the nature of the modification—either physical or chemical—to the drug, which may be applied either individually or in combination. The use of prodrugs and salt formulations are popular chemical modification strategies. Crystallization and co-crystallization, solid dispersion, comminution (particle size reduction), micellization, hydration, solvation, and the use of alternative solvents and co-solvents are the common physical approaches to achieving therapeutic excellence or cornering a pharmaceutical market.

4.5 ILs as Novel Drug-Solubilizing Agents

The delivery of sparingly soluble drugs is a significant challenge owing to their poor solubility in conventional solvents and the difficulty associated with their formulation. Many valuable drugs—including conventional and nucleoside analogues—have limited solubility in conventional solvents and present delivery challenges (Egorova et al. 2017; Agatemor et al. 2018). To address these challenges, drugs with tailor-made properties have been modified chemically or biologically for effective drug delivery (Rodriguez-Aller et al. 2015). Several polar organic solvents—such as DMSO, pyridine, and *N,N*-dimethylformamide—have been used as excipients to address the issue of solubility in drug formulations. Currently, the use of these organic solubilizers for effective drug delivery is an undeniable challenge owing to their volatility, flammability, toxicity, and tendency to contaminate the final product (Agatemor et al. 2018; Pedro et al. 2020). It is also difficult to reuse or recover such solvents when they are used as reaction media for drug synthesis or as solubilizers in formulations. Therefore, the regulatory experts are concerned about the use of these solvents in therapeutic applications. The creation and disposal of large quantities of waste products also require careful consideration.

The incorporation of ILs/IL technologies could provide a satisfactory solution to the issues described above. ILs constitute a novel family of 'green' solvents. Their use in various areas of modern science is being explored because they have some highly tuneable and exceptional physicochemical properties, including low vapour pressure, high chemical and thermal stabilities, and nonflammability (Egorova et al. 2017; Moshikur et al. 2020b,d). As well as their traditional use as solvents, ILs have been considered for use in biomedical fields, both in catalytic media for drug synthesis and as potential components for effective biocompatible drug delivery (Egorova et al. 2017; Pedro et al. 2020). To date, several notable works have emphasized the advantages of ILs in drug development, especially with regard to the issue of drug solubility. However, the pharma industry has only focused on the salt or prodrug forms of solid drugs when attempting to overcome these difficulties and has not studied ILs in any great depth. Therefore, ILs are now being tailored as innovative solutions to many of the challenges associated with solid pharmaceuticals, including the dissolution and permeability of drugs, the disruption of physiological barriers to drug transport, the targeting of sites, and the development of new delivery options (Agatemor et al. 2018; Pedro et al. 2020). In addition to small chemical drugs, IL-based formulations can be used to deliver macromolecular drugs by significantly increasing their solubility and permeability.

4.5.1 Enhanced Solubility of Pharmaceutical Drugs

ILs are able to solvate sparingly soluble drugs by converting them into liquid form, thereby improving their absorption and bioavailability. It is notoriously difficult to design formulations comprising sparingly soluble drugs in simulated body fluids and/or US Food and Drug Administration (FDA)-approved organic solvents because some such drugs contain polar groups that impede their dissolution (Egorova et al. 2017; Moshikur et al. 2020b). ILs are promising alternatives to these solvents because they can be tailored to dissolve sparingly soluble drugs, producing formulations with improved shelf stability that can deliver drugs at high plasma concentrations. Generally, ILs can be used as solvents, anti-solvents, cosolvents, copolymers, and emulsifiers in formulations comprising sparingly soluble drugs. Several studies involving dissolution, micellization, and electrochemical and spectroscopic analyses have been carried out to investigate the behaviour of drugs in the presence of IL solutions and to compare it with the behaviour of those drugs in water or conventional organic solvents (Egorova et al. 2017; Pedro et al. 2020). The ionic interactions between ILs and drugs also dictate the solvation capabilities of the ILs, and consequently their ability to improve the solubility of the drugs and to form aggregates between themselves and the drugs. In contrast to DMSO—a universal and widely adopted solvent for preclinical pharmacokinetic studies—not all ILs improve solubility (Pedro et al. 2020). For example, amphotericin B was found to be seven times more soluble in DMSO than in fatty acid and polyethylene glycol (PEG)-based ILs (McCrary et al. 2013). ILs have been shown to increase the solubilities of 5-fluorouracil, methotrexate, acetaminophen, acyclovir, isoniazid, glibenclamide, and ibuprofen— which have limited solubilities in both water and organic solvents—by several to a dozen times (Huang et al. 2020).

The solvation power of an IL is highly dependent on the properties of its constituent cations and anions. The choice of an appropriate counterion has a significant impact on the rate at which the IL is able to solubilize drug molecules. The IL's component anions and cations may both influence its ability to dissolve drugs, although the effect of the anions remains ambiguous (Egorova et al. 2017). The cations usually have a significant effect on an IL's ability to dissolve drugs. Drug solubility also varies with the length of the alkyl groups of the IL-forming cations; variable drug solubilities have been observed as the alkyl groups of the cation increase in length (Table 4.1). The solubilities of penicillin, dexamethasone, progesterone, and dehydroepiandros-terone decrease as the lengths of the alkyl groups of imidazolium ILs increase from C_4 to C_8, whereas the solubility of albendazole increases as the alkyl groups increase in length from C_4 to C_8. The release rates of drugs slowly increase as the alkyl chain lengths of the IL-forming cations increase (by up to three times) (Jaitely et al. 2010). In one study, the solubility of isoniazid increased as the length of the alkyl chains of the imidazolium cation increased, because the acidity of the proton at the 2-position of the cation decreased (Forte et al. 2012). Forte et al. (2012) investigated the effect of anions on the solubility of isoniazid in imidazolium-based ILs comprising cations with various alkyl chain lengths ($[C_{2-10}MIM]^+$). The results demonstrated that the solubility of isoniazid in imidazolium ($[C_{2-10}MIM]^+$)-based ILs was significantly greater when the anion was [OTf] than when the anion was [TFSA], owing to the formation of a hydrogen bond (H-bond) between the NH of the carboxamide in the drug and the [OTf] anion (Forte et al. 2012). The solubilities of danazol and itra-conazole also increased by 3.6-fold and fivefold, respectively, when the hydrophobic [TFSA] anion was replaced with a hydrophilic $[N(CN)_2]$ anion (Williams et al. 2014). Hydrophilic ILs with [OAc] or $[(MeO)_2PO_2]$ anions also have higher solva-tion capabilities than ILs containing noncoordinating anions—such as $[BF_4]$, $[PF_6]$, and [TFSA]—owing to H-bonding and delocalization of charges between oxygen atoms, which explains the different solubilizing capabilities of ILs (McCrary et al. 2013; Huang et al. 2020).

Solubility is also correlated with the octanol–water partition coefficient of drugs. Mizuuchi et al. (2008) demonstrated that the solubilities of albendazole, danazol, acetaminophen, and caffeine could be significantly improved by increasing the hydrophobicity of the imidazolium cation ($[C_{4-8}MIM]^+$). In general, hydrophilic drugs are more soluble in hydrophilic ILs, and hydrophobic drugs are more soluble in hydrophobic ILs. For example, hydrophilic drugs—such as N-acetyl-L-cysteine and 4-hydroxycoumarin—are readily soluble in hydrophilic ILs with short alkyl side chains $[C_2MIM][OTf]$, whereas ibuprofen, coumarin, thymo-quinone, and 4-isobutylacetophenone are readily soluble in hydrophobic ILs, such as

Table 4.1 Solubilities of drugs in selected ionic liquids (ILs), organic solvents, and water

Drug	Solvent	Solubility[a]	References
Amphotericin B	Water	0.0002	(McCrary et al. 2013; Esson et al. 2020)
	DMSO	>33	
	$[C_4NH_3][OAc]$	30	
	$[C_6NH_3][OAc]$	30	
	$[C_8NH_3][OAc]$	20	
	$[C_2MIM][OAc]$	85	
	[DC-7][TFSA]	0.7	
	[Cho][Hex]	6	
4-Hydroxycoumarin	$[P_{6,6,6,14}][TFSA]$	0.0524^b	(dos Santos et al. 2013; Faria et al. 2015)
	$[C_2MIM][CF_3O_3S]$	0.1107^b	
	$[C_4MIM][CF_3O_3S]$	0.0907^b	
Acetylcysteine	$[C_2MIM][CF_3O_3S]$	0.171^b	(dos Santos et al. 2013)
	$[C_4MIM][CF_3O_3S]$	0.1088^b	
	$[C_4MIM][TFSA]$	0.0866^b	
	$[C_6MIM][TFSA]$	0.0635^b	
	$[C_{10}MIM][TFSA]$	0.0102^b	
Etodolac	$[C_4MIM][PF_6]$	374.33	(Goindi et al. 2015)
Fenofibrate	$[C_6HOCPyr][N(CN)_2]$	>125	(Williams et al. 2014)
	$[C_6HOCPyr][TFSA]$	>130	
Pyrazinecarboxamide	$[C_2HOC_{1,1}EA][TFSA]$	0.0165^b	(Lourenço et al. 2012; Melo et al. 2013; Faria et al. 2015)
	$[C_2MIM][TFSA]$	0.0048^b	
	$[C_4MIM][TFSA]$	0.0054^b	
	$[C_6MIM][TFSA]$	0.0050^b	
	$[C_8MIM][TFSA]$	0.0052^b	
	$[C_{10}MIM][TFSA]$	0.0046^b	
	$[C_4MIM][CF_3O_3S]$	0.0116^b	
	$[P_{6,6,6,14}][TFSA]$	0.0125^b	
Thymoquinone	$[P_{6,6,6,14}][TFSA]$	0.1105^b	(Faria et al. 2015)
Albendazole	Water	0.0005	(Jung et al. 1998; Mizuuchi et al. 2008)
	Methanol	0.25	
	Ethanol	0.35	
	DMSO	5.0	
	$[C_4MIM][BF_4]$	1.49^b	
	$[C_6MIM][BF_4]$	2.97^b	
	$[C_8MIM][BF_4]$	7.2^b	
	$[C_4MIM][PF_6]$	29^b	
	$[C_6MIM][PF_6]$	53^b	
	$[C_8MIM][PF_6]$	$>75^b$	
Danazol	Water	0.001	(Mizuuchi et al. 2008; McCrary et al. 2013; Williams et al. 2014)
	Soybean oil	4.8	
	$[C_8M_8Pyr][C_6SO_4]$	88.9	
	$[C_6HOCPyr][TFSA]$	26.3	
	$[C_6HOCPyr][N(CN)_2]$	>95	
	$[C_8MIM][BF_4]$	>19.9	

(continued)

Table 4.1 (continued)

Drug	Solvent	Solubility[a]	References
Itraconazole	Water	0.000001	(Mizuuchi et al. 2008;
	Ethanol	0.2	McCrary et al. 2013;
	[m-PEG350-NH$_3$][C$_5$COO]	5.4	Williams et al. 2014)
	[m-PEG350-NH$_3$][C$_7$COO]	5.3	
	[m-PEG350-NH$_3$][C$_9$COO]	5.4	
	[C$_6$HOCPyr][TFSA]	6.3	
	[C$_6$HOCPyr][N(CN)$_2$]	40.5	
Erythromycin	Water	0.00001[b]	(Wang et al. 2006; Manic
	Methanol	0.0143[b]	and Najdanovic-Visak 2012)
	Ethanol	0.0192[b]	
	Acetone	0.0131[b]	
	Chloroform	0.0054[b]	
	[C$_4$MIM][TFSA]	0.035[b]	
	[C$_{10}$MIM][TFSA]	0.072[b]	
	[N$_{4,1,1,1}$][TFSA]	0.053[b]	
	[N$_{1,8,8,8}$][TFSA]	0.015[b]	
	[C$_4$MPyr][TFSA]	0.017[b]	
	[P$_{6,6,6,14}$][Cl]	0.085[b]	
Coumarin	Water	0.00001[b]	(dos Santos et al. 2013)
	Octanol	0.059[b]	
	[C$_2$MIM][OTf]	0.24[b]	
	[C$_4$MIM][OTf]	0.37[b]	
	[C$_2$MIM][TFSA]	0.51[b]	
	[C$_4$MIM][TFSA]	0.51[b]	
	[C$_6$MIM][TFSA]	0.50[b]	
	[C$_{10}$MIM][TFSA]	0.58[b]	
Paclitaxel	Water	<0.004	(Chowdhury et al. 2018)
	[Cho][Gly]	22.34	
	[Cho][Ala]	18.52	
	[Cho][Pro]	16.16	
	[Cho][Phe]	14.15	
	[Cho][Ile]	9.39	
	[Cho][Ser]	7.32	
	[Cho][Leu]	6.61	
5-Fluorouracil	Water	12.21	(Goindi et al. 2014)
	Ethanol	31.19	
	Isopropyl myristate	0.36	
	[C$_4$MIM][Br]	31.19	

(continued)

[C$_{10}$MIM][TFSA], [(C$_6$)$_3$C$_{14}$P][Cl], and [(C$_6$)$_3$C$_{14}$P][TFSA] (Egorova et al. 2017; Huang et al. 2020; Pedro et al. 2020).

The use of surface-active ionic liquids (SAILs) with high critical micelle concentrations is an interesting alternative method of enhancing drug solubility (Egorova et al. 2017; Moshikur et al. 2020c). The hydrophilic–hydrophobic balance of a SAIL can be tailored by altering the functional groups and alkyl chain lengths of

R. Md Moshikur et al.

Table 4.1 (continued)

Drug	Solvent	Solubility[a]	References
Acetaminophen	Water	14.93	(Mizuuchi et al. 2008; Smith et al. 2011; Adawiyah et al. 2016)
	Chloroform	2.29	
	Isopropyl myristate	0.74	
	Dichloromethane	0.42	
	Toluene	0.32	
	[C_4MIM][BF_4]	> 20.0	
	[C_8MIM][BF_4]	19.05	
	[C_4MIM][PF_6]	7.86	
	[C_6MIM][PF_6]	13.21	
	[C_4MIM][Br]		
	[C_6MIM][Br]		
Methotrexate	Water	0.12	(Moniruzzaman et al. 2010a)
	Isopropyl myristate	0.02	
	[C_1MIM][$(MeO)_2PO_2$]	6.90	
Acyclovir	Water	0.05[c]	(Islam et al., 2020a,b; Moniruzzaman et al. 2010b)
	Isopropyl myristate	0.004[c]	
	[C_1MIM][$(MeO)_2PO_2$]	> 25[c]	
	[C_2MIM][$(MeO)_2PO_2$]	> 15[c]	
	[C_4MIM][Lac]	> 5[c]	
	[C_2MIM][OAc]	10-12[c]	
	[Cho][Gly]	250	
	[Cho][Ala]	210	
	[Cho][Ser]	135	
	[Cho][Lac]	203	
	[Cho][For]	208	
	[Cho][Prp]	278	
Isoniazid	Water	0.018[b]	(Forte et al. 2012; Melo et al. 2013)
	[$N_{10,10,1,1}$][NO_3]	0.045[b]	
	[BA][NO_3]	0.043[b]	
	[$C_2HOC_{1,1}EA$][TFSA]	0.024[b]	
	[C_2MIM][TFSA]	0.008[b]	
	[C_4MIM][TFSA]	0.004[b]	
	[C_6MIM][TFSA]	0.003[b]	
	[C_8MIM][TFSA]	0.003[b]	
	[$C_{10}MIM$][TFSA]	0.003[b]	
	[$C_{10}MIM$][OTf]	0.006[b]	
Glibenclamide	Water	0.024	(Alawi et al. 2015)
	Ethanol	5	
	DMSO	25	
	[Cho][Try]	> 160	
Ibuprofen	Water	0.12	(Garzón and Martínez 2004; Smith et al. 2011; Adawiyah et al. 2016)
	Chloroform	3.00[b]	
	Isopropyl myristate	0.01	
	Carbon tetrachloride	1.42	
	Octanol	1.91	
	[C_4MIM][PF_6]	12.18	
	[C_6MIM][PF_6]	26.38	

[a]mg/mL; [b]mole fraction (mol/mol); [c]weight percentage (g/g%)

its component cations and anions. Surfactant ILs are usually capable of increasing the solubility of a drug in an aqueous medium by reducing the interfacial tension between the surfactant solution and the drug molecules. De Faria et al. (2017) investigated the solubility of ursolic acid in aqueous solutions of various SAILs with long alkyl side chains $[C_{8-18}C_1MIM]X$ ($X = Cl$ and $[C_8H_{17}SO_4]$) and $[P_{444(14)}]Cl$. SAILs improve the solubility of ursolic acid in aqueous solution by eight orders of magnitude compared with its solubility in water. In a study by Cláudio et al. (2018), the solubility of oleanolic acid (a nutraceutical) was significantly improved in the presence of a SAIL $[C_{12}C_1MIM]Cl$ and water. As the concentration of the SAIL in the aqueous solution was increased up to 1000 mM, the solubility of oleanolic acid also increased to 21.10 mg/mL—hence the remarkable improvement (by up to 106-fold) in the solubility of the drug in water.

Despite their ability to significantly improve the solubility of drugs, the use of ILs with ammonium, phosphonium, imidazolium, pyridinium, morpholinium, or quinolinium cations is limited in pharmaceutical applications owing to their poor biocompatibility (Egorova et al. 2017; Moshikur et al. 2020b). ILs based on choline and organic acids (e.g. amino, fatty, or carboxylic acids), which have multi-functional properties, are now considered biocompatible and biodegradable owing to their biological origins (Moshikur et al. 2018; Ali et al. 2019, 2021). It has been reported that the solubility of drugs in biocompatible ILs is significantly greater than their solubility in water or a conventional organic solvent (Moshikur et al. 2020a,b). The solubilities of paclitaxel and acyclovir (ACV) in choline glycinate were improved by 5585- and 625-fold, respectively, compared with their solubilities in water (Chowdhury et al. 2018; Islam et al. 2020a). Biocompatible choline amino acid ILs also successfully addressed the issues of precipitation, stability, and hypersensitivity associated with the traditional paclitaxel formulation. Similarly, choline propionate improved the solubility of ACV (678-fold higher than in water). The solubilities of paracetamol and diclofenac were four times higher in aqueous N-acetyl amino acid N-alkyl choline-containing ILs than in an inorganic salt solution (Jesus et al. 2019). Recently, choline geranic acid (CAGE)—a biocompatible IL that acts as a broad-spectrum antimicrobial agent and a chemical enhancer of transdermal delivery—has been revealed as a potential platform for dissolving sparingly soluble hydrophobic drugs (e.g. sorafenib, mannitol, cefadroxil, and nobiletin) and macromolecules (e.g. bovine serum, albumin, ovalbumin, RNA, and insulin) (Agatemor et al. 2018; Moshikur et al. 2020b). Owing to its extraordinary solvation capability, CAGE can solubilize and/or formulate problematic drugs, ultimately boosting their therapeutic efficacy. The solubilities of nobiletin and sorafenib were 450- and 100 million-fold higher in CAGE than in water, respectively (Hattori et al. 2019; Shi et al. 2020). Similarly, the solubilities of dextrans, acarbose, ruxolitinib, insulin, peptides, and proteins are improved when CAGE is used as the solvent (Agatemor et al. 2018; Moshikur et al. 2020b). Therefore, ILs have better solvation capabilities than water and most pharmaceutically acceptable organic solvents.

Overall, the studies described above demonstrate that the solvation capability of an IL can be tailored by tuning its constituent cation and anion combination according to the nature of the drug and the molecular mechanisms involved. These variables make it difficult to establish heuristic rules and develop predictive models that could be used in a widespread manner.

4.5.2 Mechanisms of Solvation

Advances in the dissolution of sparingly soluble drugs in aqueous media play a significant role in the development of effective drug formulations (Rodriguez-Aller et al. 2015; Huang et al. 2020). The remarkable solvation capabilities of ILs are widely known and are being exploited to dissolve such drugs (Agatemor et al. 2018; Huang et al. 2020). The success of drug dissolution mainly depends on the ability of the IL to interact with the drug molecule to form IL–drug interactions that are more stable than the drug–drug and IL–IL interactions (Resende De Azevedo et al. 2014). The solubilizing mechanisms of ILs are still not completely understood, because the findings are incompatible with the dictum 'like dissolves like' (Huang et al. 2020). However, ILs can easily enhance the aqueous solubilities of drug molecules by forming aggregations between themselves and the drug molecules. The data from several experiments and molecular dynamics simulations suggest that the continuous polar network formed by the dissolution of an IL in water, which occurs with most pure ILs, is broken into smaller domains. Therefore, liquid–water matrices form within aqueous solutions of ILs through the incorporation of ionic filaments (Cláudio et al. 2015; Egorova et al. 2017). Multiple H-bonds form between the water molecules and the anions of these filaments, resulting in the excellent structural stability of the IL network within the aqueous solution (Fig. 4.3a). Cláudio et al. (2015) found that when vanillin (a model drug molecule) was added to an aqueous solution of [C$_4$MIM][N(CN)$_2$], the drug readily dissolved. This was because cation–vanillin clusters formed owing to dispersion forces and other intermolecular interactions— such as H-bonding and $\pi - \pi$ interactions—between the drug molecule and the IL (Fig. 4.3b–c).

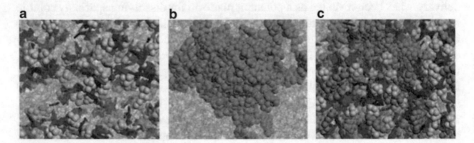

Fig. 4.3 The simulation snapshots of: **a** the ionic liquid (IL) [C$_4$MIM][N(CN)$_2$] in aqueous solution; **b** a vanillin–water mixture; and **c** vanillin in an aqueous solution of the IL. Figures reproduced with permission from Cláudio et al. (2015)

However, the dissolution of a drug in an IL depends on both the molecular structure of the drug and the structure of the constituent cation and anion of the IL. Moreover, it is induced by the multiple H-bonds between the drug molecules and the ILs. The solubility of LASSBio-294 (a cardiovascular drug) in an imidazolium-based IL has been investigated to determine the precise mechanism of solvation between the drug and the IL; both experimental and molecular dynamics simulation methods were used (Resende De Azevedo et al. 2014; Dasari and Mallik 2020). Imidazolium-based ILs are commonly used because they can act both as H-bond acceptors (the anion) and H-bond donors (the cation). Several interactions—including H-bonds, van der Waals forces, and $\pi - \pi$ interactions—between LASSBio-294 and the imidazolium-based ILs have been observed when the drug molecules are dissolved by the ILs. The improved solubility of LASSBio-294 in the ILs has been attributed to H-bonds between the carbonyl group oxygen atoms of the IL anions and the hydrogen atoms of the drug, or between the hydrogen atoms of the ILs and the carbonyl group oxygen atoms of the drug (Fig. 4.4) (Resende De Azevedo et al. 2014). Similarly, stacking interactions (the π–π interactions of the aromatic rings of the drug with the imidazolium cations of the IL) and van der Waals interactions (between the hydrogen atoms of the IL alkyl chains and the drug molecules) also help increase the solubility of the drug in the IL (Dasari and Mallik 2020). The solubility of nobiletin (NOB) increased significantly in CAGE owing to the multiple H-bond interactions between the NOB and the CAGE (Hattori et al. 2019). The oxygen atoms of the carbonyl and methoxy groups in NOB and the hydrogen atoms of the hydroxyl groups in CAGE may act as multipoint H-bond acceptors and donors, respectively. Similarly, multiple H-bond interactions may occur between the hydrogen atoms of the amino acid ester cations and the oxygen atoms of ibuprofen or those of the fatty acid anions

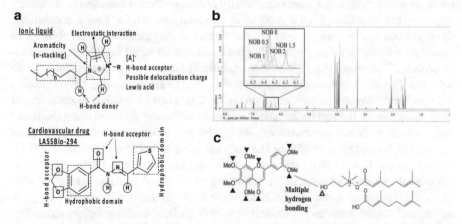

Fig. 4.4 a Schematic outline of hypothetical interactions between imidazolium-based ionic liquids (ILs) and LASSBio-294 (a cardiovascular drug); **b** superimposed proton nuclear magnetic resonance (^{1}H NMR) spectra of nobiletin–choline geranic acid (NOB–CAGE) complexes at each molar ratio; and **c** schematic diagram of the interactions between NOB and CAGE, reproduced with permission from (Hattori et al. 2019; Resende De Azevedo et al. 2014)

(Moshikur et al. 2020a). These multiple H-bond interactions between drugs and ILs, which form drug—IL complexes, are the driving force in the liquefaction of solid drugs.

Possible mechanisms for the solvation of sparingly soluble drugs in ILs may be investigated by nuclear magnetic resonance (NMR) analysis (Hattori et al. 2019; Moshikur et al. 2020a). To elucidate the mechanism by which ILs improve drug solubility, Hattori et al. (2019) prepared a series of drug–IL samples at different drug-to-IL molar ratios, then analyzed each sample using proton NMR (^{1}H NMR). When the NMR spectra of the NOB-containing CAGE samples were superimposed, the spectra differed between 6.1 and 6.5 ppm. This was attributed to the H-bond interactions between the NOB and the CAGE (Hattori et al. 2019). Similarly, Moshikur et al. (2020a) observed new characteristic peaks in the range 7.4−9.0 ppm of the NMR spectra of ibuprofen and an IL, indicating H-bond interactions between the drug and the IL (Moshikur et al. 2020a). Therefore, it is possible that multiple H-bond interactions between sparingly soluble drugs and ILs contribute to the enhancement of drug solubility in ILs.

4.6   Conclusion and Future Directions

This chapter has emphasized the advantages of ILs as potential solvents and/or agents for use in pharmaceutics over volatile organic solvents. The use of conventional organic solvents and co-solvents (such as ethanol, methanol, acetone, and DMSO), hydrotropes, and surface-active agents in pharmaceutical applications still raises concerns. They can contaminate the final product and may have an impact on the environment and on health. ILs have successfully overcome these issues and significantly improved the solubility of drugs in effective formulations. Studies have demonstrated the ability of ILs to improve—by several orders of magnitude—the solubilities of sparingly soluble and water-insoluble drugs comprising small molecules or macromolecules. Therefore, ILs represent a realistic alternative to conventional organic solvents. The physicochemical properties of ILs can be easily modified by exploiting the structural diversity of their component ions, and can be tailored to meet unmet biomedical needs. ILs can be designed to combat infectious diseases using the broad range of cation–anion combinations. As potent and broad-spectrum antimicrobial agents, the biological activities of ILs surpass those of many conventional antimicrobial agents. ILs are also emerging as potential ingredients and drug binders for effective drug formulations, and for enhancing the transport of drugs to targeted sites of action. However, there is significant uncertainty about the use of ILs in pharmaceutical applications owing to their potential intrinsic toxicity, biodegradability, and possible adverse impact on the environment. ILs with cations and anions of natural origin—such as choline, amino acids, and fatty acids—have some advantages over ILs based on imidazolium, ammonium, pyridinium, morpholinium, or phosphonium ions. In addition, the development of ILs for solubilizing sparingly soluble drugs requires a deep understanding of their structure–property relationships in the presence of water, because ultimately ILs contact water in biomedical applications.

References

Adawiyah N, Moniruzzaman M, Hawatulaila S, Goto M (2016) Ionic liquids as a potential tool for drug delivery systems. Med Chem Commun 7:1881–1897. https://doi.org/10.1039/C6MD00358C

Agatemor C, Ibsen KN, Tanner EEL, Mitragotri S (2018) Ionic liquids for addressing unmet needs in healthcare. Bioeng Transl Med 3:7–25. https://doi.org/10.1002/btm2.10083

Alawi MA, Hamdan II, AlsA S, Heshmeh NA (2015) Solubility enhancement of glibenclamide in choline-tryptophan ionic liquid: preparation, characterization and mechanism of solubilization. J Mol Liq 212:629–634. https://doi.org/10.1016/j.molliq.2015.10.006

Ali I, Lone MN, Alothman ZA, Alwarthan A (2017) Insights into the pharmacology of new heterocycles embedded with oxopyrrolidine rings: DNA binding, molecular docking, and anticancer studies. J Mol Liq 234:391–402. https://doi.org/10.1016/j.molliq.2017.03.112

Ali MK, Moshikur RM, Wakabayashi R, Moniruzzaman M, Goto M (2021) Biocompatible ionic liquid-mediated micelles for enhanced transdermal delivery of paclitaxel. ACS Appl Mater Interfaces 13:19745–19755. https://doi.org/10.1021/acsami.1c03111

Ali MK, Moshikur RM, Wakabayashi R, Moniruzzaman M, Kamiya N, Goto M (2020) Biocompatible ionic liquid surfactant-based microemulsion as a potential carrier for sparingly soluble drugs. ACS Sustain Chem Eng. https://doi.org/10.1021/acssuschemeng.9b07773

Ali MK, Moshikur RM, Wakabayashi R, Moniruzzaman M, Kamiya N, Goto M (2019) Synthesis and characterization of choline–fatty-acid-based ionic liquids: a new biocompatible surfactant. J Colloid Interface Sci 551:72–80. https://doi.org/10.1016/j.jcis.2019.04.095

Censi R, Di Martino P (2015) Polymorph impact on the bioavailability and stability of poorly soluble drugs. Molecules 20:18759–18776. https://doi.org/10.3390/molecules201018759

Chowdhury MR, Moshikur RM, Wakabayashi R, Moniruzzaman M, Kamiya N, Goto M (2018) Ionic-liquid-based paclitaxel preparation: a new potential formulation for cancer treatment. Mol Pharm 15:2484–2488. https://doi.org/10.1021/acs.molpharmaceut.8b00305

Clarke CJ, Tu WC, Levers O, Bröhl A, Hallett JP (2018) Green and sustainable solvents in chemical processes. Chem Rev 118:747–800. https://doi.org/10.1021/acs.chemrev.7b00571

Cláudio AFM, Cognigni A, de Faria ELP et al (2018) Valorization of olive tree leaves: extraction of oleanolic acid using aqueous solutions of surface-active ionic liquids. Sep Purif Technol 204:30–37. https://doi.org/10.1016/j.seppur.2018.04.042

Cláudio AFM, Neves MC, Shimizu K, Canongia Lopes JN, Freire MG, Coutinho JAP (2015) The magic of aqueous solutions of ionic liquids: ionic liquids as a powerful class of catanionic hydrotropes. Green Chem 17:3948–3963. https://doi.org/10.1039/c5gc00712g

Dasari S, Mallik BS (2020) Solubility and solvation free energy of a cardiovascular drug, LASSBio-294, in ionic liquids: a computational study. J Mol Liq 301:112449. https://doi.org/10.1016/j.molliq.2020.112449

De Faria ELP, Shabudin SV, Cláudio AFM et al (2017) Aqueous solutions of surface-active ionic liquids: remarkable alternative solvents to improve the solubility of triterpenic acids and their extraction from biomass. ACS Sustain Chem Eng 5:7344–7351. https://doi.org/10.1021/acssuschemeng.7b01616

dos Santos AD, Morais ARC, Melo C, Bogel-Łukasik R, Bogel-Łukasik E (2013) Solubility of pharmaceutical compounds in ionic liquids. Fluid Phase Equilib 356:18–29. https://doi.org/10.1016/j.fluid.2013.07.020

Earle MJ, Esperança JMSS, Gilea MA, Lopes JNC, Rebelo LPN, Magee JW, Seddon KR, Widegren JA (2006) The distillation and volatility of ionic liquids. Nature 439:831–834. https://doi.org/10.1038/nature04451

Egorova KS, Gordeev EG, Ananikov VP (2017) Biological activity of ionic liquids and their application in pharmaceutics and medicine. Chem Rev 117:7132–7189. https://doi.org/10.1021/acs.chemrev.6b00562

Esson MM, Mecozzi S, Mecozzi S (2020) Preparation, characterization, and formulation optimization of ionic-liquid-in-water nanoemulsions toward systemic delivery of amphotericin B. Mol Pharm 17:2221–2226. https://doi.org/10.1021/acs.molpharmaceut.9b00809

Faria RA, da Ponte MN, Bogel-Łukasik E (2015) Solubility studies on the system of trihexyl(tetradecyl)phosphonium bis[(trifluoromethyl)sulfonyl]amide) ionic liquid and pharmaceutical and bioactive compounds. Fluid Phase Equilib 385:1–9. https://doi.org/10.1016/j.fluid.2014.10.033

Forte A, Melo CI, Bogel-Łukasik R, Bogel-Łukasik E (2012) A favourable solubility of isoniazid, an antitubercular antibiotic drug, in alternative solvents. Fluid Phase Equilib 318:89–95. https://doi.org/10.1016/j.fluid.2012.01.022

Garzón LC, Martínez F (2004) Temperature dependence of solubility for ibuprofen in some organic and aqueous solvents. J Solution Chem 33:1379–1395. https://doi.org/10.1007/s10953-004-1051-2

Gavhane YN, Yadav AV (2012) Loss of orally administered drugs in GI tract. Saudi Pharm J 20:331–344. https://doi.org/10.1016/j.jsps.2012.03.005

Ghielmetti G, Bruzzese T, Bianchi C et al (1976) Relationship between acute toxicity in mice and polymorphic forms of polyene antibiotics. J Pharm Sci 65:905–907. https://doi.org/10.1002/jps.2600650625

Giron D (2001) Investigations of polymorphism and pseudo-polymorphism in pharmaceuticals by combined thermoanalytical techniques. J Therm Anal Calorim 64:37–60. https://doi.org/10.1023/A:1011572610005

Goindi S, Arora P, Kumar N, Puri A (2014) Development of novel ionic liquid-based microemulsion formulation for dermal delivery of 5-fluorouracil. AAPS PharmSciTech 15:810–821. https://doi.org/10.1208/s12249-014-0103-1

Goindi S, Kaur R, Kaur R (2015) An ionic liquid-in-water microemulsion as a potential carrier for topical delivery of poorly water soluble drug: development, ex-vivo and in-vivo evaluation. Int J Pharm 495:913–923. https://doi.org/10.1016/j.ijpharm.2015.09.066

Hattori T, Tagawa H, Inai M, Kan T, Kimura SI, Itai S, Mitragotri S, Iwao Y (2019) Transdermal delivery of nobiletin using ionic liquids. Sci Rep 9:1–11. https://doi.org/10.1038/s41598-019-56731-1

Huang W, Wu X, Qi J, Zhu Q, Wu W, Lu Y, Chen Z (2020) Ionic liquids: green and tailor-made solvents in drug delivery. Drug Discov Today 25:901–908. https://doi.org/10.1016/j.drudis.2019.09.018

Islam MR, Chowdhury MR, Wakabayashi R, Moniruzzaman M, Kamiya N, Goto M (2020a) Choline and amino acid based biocompatible ionic liquid mediated transdermal delivery of the sparingly soluble drug acyclovir. Int J Pharm 582:119335. https://doi.org/10.1016/j.ijpharm.2020.119335

Islam MR, Chowdhury MR, Wakabayashi R, Moniruzzaman M, Kamiya N, Goto M (2020b) Ionic liquid-in-oil microemulsions prepared with biocompatible choline carboxylic acids for improving the transdermal delivery of a sparingly soluble drug. Pharmaceutics 12:392. https://doi.org/10.3390/pharmaceutics12040392

Jaitely V, Mizuuchi H, Florence AT (2010) Current-stimulated release of solutes solubilized in water-immiscible room temperature ionic liquids (RTILs). J Drug Target 18:787–793. https://doi.org/10.3109/1061186X.2010.525653

Jesus AR, Soromenho MRC, Raposo LR, Esperança JMSS, Baptista PV, Fernandes AR, Reis PM (2019) Enhancement of water solubility of poorly water-soluble drugs by new biocompatible N-acetyl amino acid N-alkyl cholinium-based ionic liquids. Eur J Pharm Biopharm 137:227–232. https://doi.org/10.1016/j.ejpb.2019.03.004

Jung H, Medina L, García L, Fuentes I, Moreno-Esparza R (1998) Absorption studies of albendazole and some physicochemical properties of the drug and its metabolite albendazole sulphoxide. J Pharm Pharmacol 50:43–48. https://doi.org/10.1111/j.2042-7158.1998.tb03303.x

Krishnaiah YS (2010) Pharmaceutical technologies for enhancing oral bioavailability of poorly soluble drugs. J Bioequivalence Bioavailab 02:28–36. https://doi.org/10.4172/jbb.1000027

Lourenço C, Melo CI, Bogel-Łukasik R, Bogel-Łukasik E (2012) Solubility advantage of pyrazine-2-carboxamide: application of alternative solvents on the way to the future pharmaceutical development. J Chem Eng Data 57:1525–1533. https://doi.org/10.1021/je300044x

Manic MS, Najdanovic-Visak V (2012) Solubility of erythromycin in ionic liquids. J Chem Thermodyn 44:102–106. https://doi.org/10.1016/j.jct.2011.08.004

McCrary PD, Beasley PA, Gurau G, Narita A, Barber PS, Cojocaru OA, Rogers RD (2013) Drug specific, tuning of an ionic liquid's hydrophilic-lipophilic balance to improve water solubility of poorly soluble active pharmaceutical ingredients. New J Chem 37:2196–2202. https://doi.org/10.1039/c3nj00454f

Melo CI, Bogel-Łukasik R, Nunes da Ponte M, Bogel-Łukasik E (2013) Ammonium ionic liquids as green solvents for drugs. Fluid Phase Equilib 338:209–216. https://doi.org/10.1016/j.fluid.2012.11.029

Mizuuchi H, Jaitely V, Murdan S, Florence AT (2008) Room temperature ionic liquids and their mixtures: potential pharmaceutical solvents. Eur J Pharm Sci 33:326–331. https://doi.org/10.1016/j.ejps.2008.01.002

Moniruzzaman M, Goto M (2011) Ionic liquids: future solvents and reagents for pharmaceuticals. J Chem Eng Japan 44:370–381. https://doi.org/10.1252/jcej.11we015

Moniruzzaman M, Kamiya N, Goto M (2010a) Ionic liquid based microemulsion with pharmaceutically accepted components: formulation and potential applications. J Colloid Interface Sci 352:136–142. https://doi.org/10.1016/j.jcis.2010.08.035

Moniruzzaman M, Tahara Y, Tamura M et al (2010b) Ionic liquid-assisted transdermal delivery of sparingly soluble drugs. Chem Commun 46:1452–1454. https://doi.org/10.1039/b907462g

Moshikur RM, Chowdhury MR, Fujisawa H, Wakabayashi R, Moniruzzaman M, Goto M (2020a) Design and characterization of fatty acid-based amino acid ester as a new "green" hydrophobic ionic liquid for drug delivery. ACS Sustain Chem Eng 8:13660–13671. https://doi.org/10.1021/acssuschemeng.0c03419

Moshikur RM, Chowdhury MR, Moniruzzaman M, Goto M (2020b) Biocompatible ionic liquids and their applications in pharmaceutics. Green Chem 22:8116–8139. https://doi.org/10.1039/d0gc02387f

Moshikur RM, Ali MK, Wakabayashi R, Moniruzzaman M, Goto M (2020c) Formation and potential application of micelles composed of biocompatible N-lauroyl-amino acid ionic liquids surfactant. J Mol Liq 320:114424. https://doi.org/10.1016/j.molliq.2020.114424

Moshikur RM, Chowdhury MR, Wakabayashi R, Moniruzzaman M, Kamiya N, Goto M (2020d) Ionic liquids with N-methyl-2-pyrrolidonium cation as an enhancer for topical drug delivery: synthesis, characterization, and skin-penetration evaluation. J Mol Liq 299: 112166. https://doi.org/10.1016/j.molliq.2019.112166

Moshikur RM, Chowdhury MR, Wakabayashi R, Moniruzzaman M, Goto M (2019) Ionic liquids with methotrexate moieties as a potential anticancer prodrug: synthesis, characterization and solubility evaluation. J Mol Liq 278:226–233. https://doi.org/10.1016/j.molliq.2019.01.063

Moshikur RM, Chowdhury MR, Wakabayashi R, Moniruzzaman M, Goto M (2018) Characterization and cytotoxicity evaluation of biocompatible amino acid esters used to convert salicylic acid into ionic liquids. Int J Pharm 546:31–38. https://doi.org/10.1016/j.ijpharm.2018.05.021

Pedro SN, Freire CSR, Silvestre AJD, Freire MG (2020) The role of ionic liquids in the pharmaceutical field: an overview of relevant applications. Int J Mol Sci 21:1–50. https://doi.org/10.3390/ijms21218298

Resende De Azevedo J, Letourneau JJ, Espitalier F, Ré MI (2014) Solubility of a new cardioactive prototype drug in ionic liquids. J Chem Eng Data 59:1766–1773. https://doi.org/10.1021/je4009624

Rodriguez-Aller M, Guillarme D, Veuthey JL, Gurny R (2015) Strategies for formulating and delivering poorly water-soluble drugs. J Drug Deliv Sci Technol 30:342–351. https://doi.org/10.1016/j.jddst.2015.05.009

Savjani KT, Gajjar AK, Savjani JK (2012) Drug Solubility: importance and enhancement techniques. ISRN Pharm 2012:1–10. https://doi.org/10.5402/2012/195727

Shamshina JL, Barber PS, Rogers RD (2013) Ionic liquids in drug delivery. Expert Opin Drug Deliv 10:1367–1381. https://doi.org/10.1517/17425247.2013.808185

Shi Y, Zhao Z, Gao Y, Pan DC, Salinas AK, Tanner EEL, Guo J, Mitragotri S (2020) Oral delivery of sorafenib through spontaneous formation of ionic liquid nanocomplexes. J Control Release 322:602–609. https://doi.org/10.1016/j.jconrel.2020.03.018

Smith KB, Bridson RH, Leeke GA (2011) Solubilities of pharmaceutical compounds in ionic liquids. J Chem Eng Data 56:2039–2043. https://doi.org/10.1021/je101040p

Sohail MF, Rehman M, Sarwar HS, Naveed S, Salman O, Bukhari NI, Hussain I, Webster TJ, Shahnaz G (2018) Advancements in the oral delivery of docetaxel: challenges, current state-of-the-art and future trends. Int J Nanomed 13:3145–3161. https://doi.org/10.2147/IJN.S164518

Uddin S, Chowdhury MR, Wakabayashi R, Kamiya N, Moniruzzaman M, Goto M (2020) Lipid based biocompatible ionic liquids: synthesis, characterization and biocompatibility evaluation. Chem Commun 56:13756–13759. https://doi.org/10.1039/D0CC04491A

Wang Z, Wang J, Zhang M, Dang L (2006) Solubility of erythromycin a dihydrate in different pure solvents and acetone + water binary mixtures between 293 K and 323 K. J Chem Eng Data 51:1062–1065. https://doi.org/10.1002/cjoc.200690198

Welton T (1999) Room-temperature ionic liquids. solvents for synthesis and catalysis. Chem Rev 99:2071–2083. https://doi.org/10.1021/cr980032t

Williams HD, Sahbaz Y, Ford L, Nguyen TH, Scammells PJ, Porter CJH (2014) Ionic liquids provide unique opportunities for oral drug delivery: structure optimization and in vivo evidence of utility. Chem Commun 50:1688–1690. https://doi.org/10.1039/c3cc48650h

Chapter 5
3D Printing of Cellulose and Chitin from Ionic Liquids for Drug Delivery: A Mini-Review

Julia L. Shamshina and Robin D. Rogers

Abstract The largest group of chemicals that are currently used for the manufacturing of 3D-printed structures are synthetic polymers. Among those, the use of thermoplastics is very common, nearly exclusive, mainly because of the ease of their handling through melt processing. While in recent years, there has been a growing shift toward the use of biopolymers because of many attractive properties of those, traditional 3D printing methods from common solvents (such as volatile organic solvents, VOCs) or melts are not suitable for biopolymers. In fact, 3D printing of biopolymers is held back by the lack of suitable technology that would allow biopolymer liquefaction and layer-by-layer deposition in a liquid state, followed by solidification. The chapter focuses on recent advances in the 3D printing of cellulose and chitin realized by enabling technology based on a class of materials known as ionic liquids.

Abbreviations

ABDS	Affinity-Based Delivery Systems
API	Active Pharmaceutical Ingredients
DA	Degree of Acetylation
DMAc/LiCl	Dimethylacetamide/Lithium chloride
DMSO	Dimethyl sulfoxide
DP	Degree of Polymerization
ECM	Extracellular Matrix
FDA	Food and Drug Administration's
FDM	Fused Deposition Modeling

J. L. Shamshina (✉)
Department of Plant and Soil Science, Fiber and Biopolymer Research Institute, Texas Tech University, P.O. Box 45019, Lubbock, TX 79409, USA
e-mail: jshamshi@ttu.edu

J. L. Shamshina · R. D. Rogers
525 Solutions, Inc., P.O. Box 2206, Tuscaloosa, AL 35403, USA

© The Author(s), under exclusive license to Springer Nature Singapore Pte Ltd. 2021
M. Goto and M. Moniruzzaman (eds.), *Application of Ionic Liquids in Drug Delivery*,
https://doi.org/10.1007/978-981-16-4365-1_5

GMP	Good Manufacturing Practices
Ionic Liquids	ILs
[Amim][OAc]	1-Allyl-3-methylimidazolium acetate
[Amim]Cl	1-Allyl-3-methylimidazolium chloride
[C_4mim][OAc]	1-Butyl-3-methylimidazolium acetate
[C_4mim]Cl	1-Butyl-3-methylimidazolium chloride
[C_4mim][DCA]	1-Butyl-3-methylimidazolium dicyanamide
[C_1mim][$(CH_3)_2HPO_4$]	1,3-Dimethylimidazolium hydrogen phosphate
[C_2mim]Cl	1-Ethyl-3-methylimidazolium chloride
[C_2mim][OAc]	1-Ethyl-3-methylimidazolium acetate
[C_2mim]Cl	1-Ethyl-3-methylimidazolium chloride
[C_2mim][DCA]	1-Ethyl-3-methylimidazolium dicyanamide
[C_2mim][$(CH_3)_2PO_4$]	1-Ethyl-3-methylimidazolium dimethylphosphate
MCC	Microcrystalline Cellulose
MIC	Minimum Inhibition Concentration
MW	Molecular Weight
PAM	Pressure-Assisted Microsyringe
PET	Polyethylene Terephthalate
PLA	Poly-L-Lactic Acid
RBDS	Reservoir-Based Delivery Systems
USP	United States Pharmacopeia
VOC	Volatile Organic Solvent

5.1 Looking into the Future: Cellulose and Chitin as Plastic Replacement

The world's plastic pollution crisis (Ritchie 2020) has led to an inclination toward sustainable technologies. Corporations are taking major steps to improve environmental impact across their operations both sourcing renewable precursors and utilizing emerging sustainable practices. While major efforts are primarily focused on plastic packaging, biotech industries (healthcare, pharmaceutical, etc.) are as well looking for more environmentally responsible solutions (Fix and de Brower 2020). As a result, there is a window of opportunities in the area of biopolymeric materials for drug delivery (Yadav et al. 2015), which must meet specific criteria. These criteria are related to both raw material supply (economy of scale/long-term supply guarantee, production according to good manufacturing practices (GMPs)) and material requirements (e.g., biocompatibility, strength). Indeed, each end application has its own set of material requirements depending on the intended purpose (Fix and de Brower 2020).

Polymers that are used for encapsulation of active pharmaceutical ingredients (APIs) for drug delivery that have been used in the pharmaceutical industry include synthetic polymers such as poly(ethylene glycols), poly(N-vinyl pyrrolidone), poly(2-(diethylamino)ethyl methacrylate), 2-(dimethylamino)ethyl methacrylate), poly(2-aminoethyl methacrylate), N-(2-hydroxypropyl)methacrylamide, etc. (Rowe et al. 2005) for the manufacture of both affinity-based delivery systems (ABDSs) and reservoir-based delivery systems (RBDSs) (Willerth and Sakiyama-Elbert 2007; Yang et al. 2012) applied in situ (Madan et al. 2009) as components of an implant and drug delivery systems inside the human body (Major et al. 2020). In ABDSs, noncovalent interactions between biopolymeric device material and target drug exist to sustain and control its release from a polymeric matrix. Polymeric delivery matrix is, in this case, interacting with the matrix-immobilized active, through electrostatic, hydrogen bonding, hydrophobic interactions, etc., and control is achieved by changing the strength of the affinity. RBDSs, on the other hand, are porous materials in which the drug release rate is controlled by diffusion.

The polymeric matrices should be degradable, but more importantly, the properties of the polymeric matrix itself could be synergistic to those of the active. In this regard, the unique properties of abundant biopolymers, cellulose and chitin (Klemm 2004; Barikani et al. 2014) (Fig. 5.1) attracted significant attention for the development of drug delivery systems. The major advantages of using these polymers for biomaterial preparation are that they are biocompatible and bioactive (Garcia 2018), strong (Courtenay et al. 2018; Hickey and Pelling 2019), and biodegradable (Ji et al. 2014). They are, however, not suitable for thermal printing, because they do not melt but decompose at increased temperature due to the intrinsic lability of the glycosidic bond that connects monomeric sugar units. They are also not soluble in standard volatile organic solvents (VOCs), but in N,N-dimethylacetamide/lithium chloride (DMAc/LiCl) (Yusof et al. 2001; Austin 1977; Xie et al. 2017), 2,2,2-trifluoroethanol (Ji et al. 2014), hydroxide-urea (NaOH/urea) eutectic (Pillai et al. 2009; duan et al. 2015; Han et al. 2016), and other harsh chemicals that either degrade biopolymers (Wang et al. 2015) or are not suitable for manufacturing biomedical materials (Muzzarelli 1983; Seoudi and Nada 2007; Tamura et al. 2004).

It is also important to note that in order for these biopolymers to be used in medical applications, there must be a way to ensure they (or the materials made from them) meet requirements normally associated with GMP and "Medical Grade" polymers (Khor 2001), which also adds restrictions to the supply chain. Biopolymers should at least meet United States Pharmacopeia (USP) Class VI requirements, which refer to a series of biological tests performed on the materials to demonstrate their

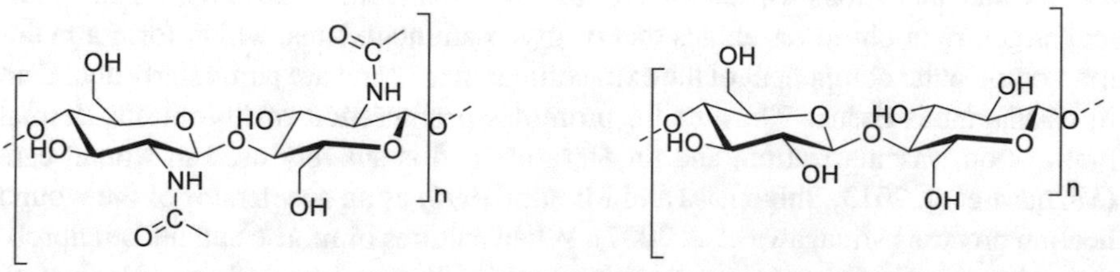

Fig. 5.1 Structure of chitin (left) and cellulose (right)

safety (Unites States Pharmacopeia 2021). Let's take a look at the properties of these polymers that make them attractive for health industries in more detail.

5.2 Useful Medical Properties of Cellulose and Chitin

Cellulose is a relatively inert polymer with a high hydroxyl group density (Clark and Gao 2002; Deppisch et al. 1998; Xu et al. 2009). It possesses high mechanical strength (especially nanocellulose), combined with suitable physical properties. The mechanical properties of nanocellulose, when dried, are of the same level as those for human bones, while the properties of a wet nanocellulose film resemble more of the extracellular matrix (ECM) (Lagerwall et al. 2014). In addition to the mechanical properties and high crystallinity, cellulose has good transport properties due to its ultrafine fiber network of high porosity, selectively permeable, and it is hydrophilic (highly water water-absorbent). It activates an alternative independent defense mechanism of the immune response, the so-called "complement pathway" (Caruso et al. 2020). Cellulose is also biocompatible.

Due to its high mechanical strength, microbial cellulose is used as an implant material (Narat et al. 1950), for bone tissue engineering as a matrix for obtaining different types of calcium carbonate crystals (Stoica-Guzun et al. 2012; Wan et al. 2009; Zhang et al. 2009), vascular grafts (Klemm et al. 2001; Backdahl et al. 2011), etc. As a result of its transport properties, cellulose is used in dialysis/hemodialysis membranes (Sindhu et al. 2015). Because of its ability to locally activate the coagulation cascade, cellulose is used as a hemostatic agent for controlling bleeding in trauma injuries (Khoshmohabat et al. 2019). Since cellulose-based materials as hydrogels, fibers, and nonwovens are able to absorb the excess of exudate, they maintain the proper moisture environment on/in the wound and are employed in wound care (Waston and Hodgkin 2005) and as skin tissue repair materials; often antimicrobial agents (copper, silver, and ZnO) are incorporated within the cellulosic matrix (Ruparelia et al. 2008).

Chitin, on the other hand, exhibits a myriad of useful biological properties (Elieh-Ali-Komi and Hamblin 2016): it is a hemostatic agent (Sundaram et al. 2016), an immunomodulator (Esteban et al. 2001), a stimulator of pathogen uptake by macrophages and dendritic cells via lectin binding (Elieh-Ali-Komi et al. 2018), and a controller of various aspects of cell function (Naseem et al. 2012). The chemical structure of chitin resembles that of glycosaminoglycans, which form a major macromolecular component of the extracellular matrix and are particularly abundant in cartilaginous tissues Thus, chitin promotes proliferation of fibroblasts, dermal granulation, vascularization, and regeneration and is actively used in wound care (Vázquez et al. 2013; Shigemasa and Minami 1996) as an accelerator of the wound healing process (Minagawa et al. 2007). When cultures of mouse and human fibroblast cells were seeded onto porous chitin matrix, they proliferated fast (Singh et al. 2008; Jayakumar et al. 2011), pointing out the likelihood of using chitin in cell transplant materials for tissue regeneration.

Chitin reduces inflammatory pain (Chow et al. 2001) and is somewhat antibacterial (although antibacterial activity depends on the bacterial strain, degree of acetylation (DA), and molecular weight of the polymer), with minimum inhibition concentration (MIC) ranging from 0.006% to 0.1% (Okamoto et al. 2002). In orthopedics, due to chitin's ability to enhance bone growth and achieve osseointegration when implanted in osseous sites, chitin composites with hydroxyapatites are used as bone healing agents for hard tissue substitute materials (Benhabiles et al. 2012), or temporary scaffolds/bone substitutes (Higashi et al. 1986). In dentistry, chitin is suitable for the manufacture of dental tissue materials (Singh et al. 2008).

In addition, chitin is biodegradable with the rate of degradation depending on its DA, repeat unit order, and molecular weight (MW); its reported degradation in the body (exemplified by implanted chitin fabric) is 12 weeks post-surgery (Shi et al. 2017). Chitin sutures are typically absorbed about 3–4 months in muscles, with no tissue reaction observed (Wan and Tai 2013). All of these properties make chitin a very important polymer suitable for the preparation of an implant or delivery system *inside the human body*.

5.3 3D Printing as Emerging Technology

3D printing is an emerging technology that uses computer-created 3D models to build solid materials in a "layer build up" fashion. The accuracy and effortless operation of 3D printing have established this technique as the method of choice in multiple areas, for the fast production of complex architectures. As the field grows and develops, new 3D-printing techniques (inkjet printing, extrusion-based, laser sintering, etc.) are applied for processing different types of materials (Nakajima et al. 1986; Gunasekera et al. 2016; Park et al. 2017; Shirazi et al. 2015) in various areas including broad application in healthcare as implants, drug delivery, or cell storage devices.

In the pharmaceutical industry, 3D printing has evolved rapidly since the Food and Drug Administration's (FDA) approval of the first 3D-printed medicine, SPRITAM®, a levetiracetam drug manufactured by Aprecia Pharmaceuticals Company (USA) (Spritam 2021; Ventola 2014). Aprecia's proprietary powder-based printing technology combined 3D printing and formulation science to produce rapidly disintegrating oral medicines for the treatment of seizures. It uses a proprietary 3D printing system/equipment assembly (Spritam) that allows spreading thin layers of powder containing levetiracetam onto a surface and bounding together selected regions of the powder by the controlled deposition ("printing") of a fluid.

Other and much more common examples of 3D printing are extrusion-based printing, namely, fused deposition modeling (FDM) and pressure-assisted micro-syringe (PAM). In FDM, the polymer is heated above its melting point, melted, and then solidified using synthetic polyethylene oxide systems, Kollidon® SR (a polyvinyl acetate and povidone mix), methacrylate-copolymer Eudragit EPO®, etc. (Yoo et al. 2014; Azad et al. 2020), while PAM 3D printing relies on the extrusion of a semi-solid polymer formulation.

PAM extrusion becomes possible from VOCs or water (as aqueous slurries) (Pattinson and Hart 2017; Li et al. 2017, 2018), in which the solidification of the print using VOCs is done through rapid solvent evaporation or a post-print freeze-drying step. Here, the printing of biopolymers in their native form using the same approach remains a challenge due to their insolubility in aqueous solutions and VOCs. Finally, inkjet printing that is used in the emerging field of bioprinting (Singh et al. 2002; Hinton et al. 2015) to fabricate biomedical parts that maximally imitate natural tissue characteristics presents a challenge of preparing biopolymer-containing inks as again, most biopolymers are not soluble in aqueous solutions and VOCs.

Hence, 3D printing of biopolymers is held back by the lack of suitable technology that would allow biopolymer dissolution, its layer-by-layer deposition in a liquid state, followed by solidification. In this, ionic liquids may be an enabling technology.

5.4 Ionic Liquids for 3D Printing as Enabling Technology

Ionic liquids (ILs, salts that are liquid below 100 °C) are known to be excellent solvents for biopolymers (and many synthetic polymers as well). The first report appeared in 2005 (Swatloski et al. 2002). Typical ILs for biopolymers dissolution include 1-butyl-3-methylimidazolium chloride ([C_4mim][Cl]) and acetate ([C_4mim][OAc]), 1-ethyl-3-methylimidazolium chloride ([C_2mim][Cl]) and acetate ([C_2mim][OAc]), 1-allyl-3-methylimidazolium chloride ([Amim][Cl]), and acetate ([Amim][OAc]) (Swatloski et al. 2002; Xie et al. 2006; Takegawa et al. 2010; Shamshina 2019) Less common ILs for biopolymers dissolution include 1,3-dimethylimidazolium and 1-ethyl-3 methylimidazolium phosphates and hydrogen phosphates ([C_1mim][(CH_3)$_2$HPO$_4$] and [C_2mim][(CH_3)$_2$HPO$_4$], respectively) (Wang et al. 2010) and "statistical mixtures" (two different 1,3-dialkylimidazoliums, with a basic anion, typically acetate or chloride) (Gurau et al. 2012). The representative examples of "statistical mixtures" include 2:1:1 mixture of 1-ethyl-3-methylimidazolium, 1,3-diethylimidazolium, and 1,3-dimethylimidazolium (or 1-butyl-3-methylimidazolium, 1,3-dibutylimidazolium, and 1,3-dimethylimidazolium) acetate or chloride (Gurau et al. 2012).

The first step for the preparation of different materials is the dissolution of biopolymers in the IL. Ionic liquids–biopolymer(s) solutions (prepared via thermal or microwave-assisted process) can then be used to prepare spun fibers, films, hydrogels, beads, electrospun mats, etc. using "solution processing" techniques (Qin et al. 2010; King et al. 2017; Shen et al. 2016; Kadokawa 2016; Shamshina et al. 2017, 2014; Zavgorodnya et al. 2017; Turner et al. 2004, 2005; Sun et al. 2008, 2011; Bagheri et al. 2008; Maxim et al. 2012; Takegawa et al. 2010; Singh et al. 2013; Mundsinger et al. 2015). Contrarily to solutions of synthetic polymers in VOCs that have low viscosities, IL solutions of biopolymers have viscosities several orders of magnitude higher. Such differences are due to the various types of biopolymer–IL interactions and they directly translate into distinctive differences in the processability of the biopolymer–IL solutions from that of the polymer–VOC solutions (Chen et al. 2011; Remsing et al. 2006; Moulthrop et al. 2005).

5.4.1   Inkjet Printing of Biopolymers from an Ionic Liquid

Inkjet printing utilizes inkjet technology and can be thermal (a bubble of vapor in the ink forms due to resistor-controlled pulse heating and the expanding bubble ejects a drop of ink from the nozzle) or piezoelectric (a pressure pulse generated by a bending cantilever or membrane ejects the drop by a combination of short forward and reverse pulses) (Cui et al. 2012), with continuous (continuous inkjet printing) or on-demand (drop-on-demand inkjet printing) (Gunasekera et al. 2016) processing.

There are few reports that utilized inkjet printing of monomers of synthetic polymers (acrylate and methacrylate) with further polymerization, from low viscosity dicyanamide (DCA) ILs ([C$_2$mim][DCA], [C$_4$mim][DCA]), either alone or mixed with polyethylene glycols (Löffelmann et al. 2012). While these are synthetic monomers, it's important to note this work as it was the first one that utilized ILs as printing inks.

The initial report on inkjet printing of *biopolymers* from ILs was an extension of a natural fiber welding technology (DeLong et al. 2012) (Table 5.1) in which individual fibers get swollen by an appropriate ionic liquid-based solvent system to form a congealed network. The process involved piezoelectric inkjet printing (Epson) of a [C$_2$mim][OAc]/acetonitrile (1:2 mol/mol) *onto* natural polymeric substrates, initially with no polymer dissolved (Haverhals et al. 2012), where the possibility to deposit controlled volumes of IL in a reproducible way was investigated and demonstrated in a proof-of-concept application. In this series of experiments, the IL solutions were controllably deposited onto *a cellulosic paper* (100% cotton) as a natural fiber substrate, under various heating conditions (temperature, time) and coagulation conditions. After printing, the IL was removed with water, and the cellulose paper sample was dried.

A significant part of the work was directed to figuring out the conditions suitable for inkjet printing of the IL. The very construction of inkjet printers prevents handling highly viscous materials. Inkjet printing requires the usage of very non-viscous liquids whose viscosity should fall within the "jetting window" (usually 4–20 mPa s) (Biswas et al. 2019), and the variations of viscosity have a high impact on the print outcome (Krainer et al. 2019). Inkjet printing is not suitable for the formation of "sturdy" 3D structures unlike those that involve the fusion of thermoplastics or solidified pastes because the low viscosity required by inkjet printers prohibits the formation of sturdy post-deposition architectures. The optimal range for IL-based ink viscosity for deposition was determined to be 3–8 mPa s, and this viscosity was controlled by the addition of co-solvent (acetonitrile, 1-butanol or DMSO in the amount of co-solvent:[C$_2$mim][OAc] = 2:1 mol/mol)). Two years later, the same group reported inkjet printing of a mixture of [C$_2$mim][OAc]/carbon black/microcrystalline cellulose (MCC) onto cotton paper, Table 5.1 (Sweely et al.

Table 5.1 Published examples of printing of biopolymers from an Ionic Liquid

Polymer	IL/cosolvent (if any)	Printer model	Printing	Concentration, wt%	Viscosity, mPa s	Process	Print conditions	References
None	[C$_2$mim][OAc]/acetonitrile 1:2 mol/mol	Epson	Inkjet	–	5	Piezoelectric inkjet printing followed by heating, rinsing, and air-drying	Print heat temperature 25 °C, droplet spacing not given	Haverhals et al. (2012)
Microcrystalline cellulose (w/carbon black)	[C$_4$mim][OAc]	Epson	Inkjet	1–4	3–8	Dissolving cellulose, suspending carbon, piezoelectric inkjet printing, coagulation in water, washing, and air-drying	Print heat temperature 25 °C, droplet spacing not given	Sweely et al. (2014)
Cellulose	[C$_4$mim][OAc]/DMSO (59/41 w/w)	DIMATIX inkjet printer (now part of Fujifim)	Inkjet	1.2	16	Dissolving cellulose, controlled inkjet printing followed by air drying	Print head temperature 55 °C, droplet spacing of 150 μm	Gunasekera et al. (2016)
Cellulose	[C$_4$mim][OAc]/DMSO (53/47 w/w)	DIMATIX inkjet printer (now part of Fujifim)	Inkjet	1.0	16	Dissolving cellulose, coagulation in water, washing, and again air-drying		Gunasekera et al. (2016)
Cellulose avicel PH-101 (Fluka–Sigma Aldrich) DP 150–300	[C$_2$mim][OAc]	Customized printer (MakerBot Replicator™ 2; MakerBot) where the heating element and the filament head were removed and a syringe holder was fitted	Extrusion	4	~3 × 10^3 (25 °C)	Not printed (viscosity-related reasons)	NA	Marksteat et al. (2016)
Pulp (Domsjö AB) DP 750	[C$_2$mim][OAc]			4	~150 × 10^3 (25 °C) ~40 × 10^3 (50 °C) ~6 × 10^3 (80 °C)	Dissolving cellulose, controlled multilayered dispensing, and coagulation via printing into agar gel	Needles diameter: >0.41 mm (22 G); T = 50 °C; Pump rate 10 μl/min	Marksteat et al. (2016)

(continued)

Table 5.1 (continued)

Polymer	IL/cosolvent (if any)	Printer model	Printing	Concentration, wt%	Viscosity, mPa s	Process	Print conditions	References
Bacterial nanocellulose (BNC), DP 2000–8000	[C$_2$mim][OAc]			4	~6,000 × 10^3 (25 °C) ~800 × 10^3 (50 °C) ~200 × 10^3 (80 °C)	Not printed (viscosity-related reasons)	NA	Marksteat et al. (2016)
IL-extracted chitin	[C$_2$mim][OAc]	Printrbot Simple Metal 3D printer equipped with heated paste extruder (plunger syringe cap made of Teflon)	Extrusion	3	~2,800 (40 °C) ~400 (80 °C)	Dissolving chitin, loading into syringe @80 °C, controlled multilayered dispensing at 50 °C, and coagulation via placing into water bath	Needles diameter: 0.41–1.60 mm (14–22 G); T = 50 °C, Print speed: 30 mm/s	Rogers et al. (2019)
IL-extracted chitin-PLA (1:1 w/w)	[C$_2$mim][OAc]		Extrusion	6 (3:3)	ND	Dissolving chitin, controlled multilayered dispensing, and coagulation via placing into water bath	Needles diameter: 0.41–1.60 mm (14–22 G); T = 50 °C, Print speed: 30 mm/s	Rogers et al. (2019)

2014) demonstrating the ability of IL-based inks to not only be deposited but also entrap biopolymers and micro-scale functional additives in the ink matrix.

Another usage of ILs with cosolvents was reported by Wildman's Group (Gunasekera et al. 2016) (Table 5.1) who exploited [C$_2$mim][OAc] and 1-butyl-3-methylimidazolium acetate ([C$_4$mim][OAc]) for inkjet printing of 2 and 5 wt% cellulose solutions onto polyethylene terephthalate (PET) and glass substrates. Adjustment of viscosity was attempted with 1-butanol and DMSO as rheological modifiers to ensure consistent printing. Among two ILs, [C$_4$mim][OAc] worked better for the purpose and DMSO cosolvent (41–47 wt%) allowed successful adjustment of the cellulose solution viscosity to a favorable range (12–18 mPa s), while the addition of butanol was expectedly not successful, resulting in precipitation of dissolved cellulose. The printing was conducted using a DIMATIX DMP-2831 piezoelectric inkjet printer equipped with a 16-nozzle cartridge, at 55 °C achieved by means of a built-in heater. Cellulose was regenerated by adding 0.1 mL of deionized (DI) water onto inkjet-printed samples for different time intervals (30 min–2 h) before drying for approximately 18 h at 50 °C. The surface profile images indicated that the height of the surface, corresponding to the amount of material deposited, on average stayed between 80 and 680 nm.

These works established desired rheological behavior of solutions *for inkjet printing* and demonstrated the potential of inkjet-printed IL inks. However, no printing of highly viscous solutions via extrusion-based printing in a continuous process was demonstrated at the time.

5.4.2 Extrusion-Based Printing of Biopolymers from an Ionic Liquid

5.4.2.1 3D Printing of Cellulose Structures from an Ionic Liquid

In extrusion-based printing, the polymer is either heated above its melting point or formulated into a paste, extruded, and then solidified, once printed. While some studies say that for polysaccharides, only dilute solutions are expected to be printable emphasizing high viscosity solutions being not suitable for 3D printing (Gunasekera et al. 2016), this is only true for inkjet printing.

Viscosity of the melt polymers or polymeric formulations lies in the range 100–10,000 Pa s (Markstedt et al. 2014), and this implies that IL-biopolymers solutions should be suitable for extrusion-based printing "as is", with no co-solvent. For instance, the viscosity of a 1 wt% solution of chitin in [C$_2$mim][OAc] is ~2500 mPa s (Zavgorodnya et al. 2017). Usually seen as a drawback, high viscosity IL-biopolymer solutions would be beneficial in extrusion 3D printing because this allows supporting the shape during and after the printing process. Besides, the viscosity is a function of polymer molecular weight (MW), which is responsible for polymer chain entanglement where the MW of the biopolymer, in turn, depends mostly on the method of biopolymer isolation (Wineinger et al. 2020) and could thus be modified when necessary.

Marksteat (Markstedt et al. 2014) (Table 5.1) was the first to realize multilayered prints of cellulose via extrusion printing. The group used three types of different origin/different MW celluloses, namely, Avicel PH–101 with a degree of polymerization (DP) of 150–300 (Fluka–Sigma Aldrich), cellulose pulp with DP 750 (Domsjö AB), and bacterial nanocellulose (BNC), with DP 2000–8000 dissolved in [C_2mim][OAc] in the amount of 1, 2, and 4 wt%; the dissolution was conducted at 85 °C from 3 h (Avicel) to overnight (pulp and BNC), to prevent cellulose degradation.

As 3D models, 25 mm height cylinders were chosen for printing in this work. The printing was conducted through initial preparation of 4 wt% cellulose-IL solution followed by controlled multilayered dispensing in layer-by-layer fashion. The initial results of the rheology analysis demonstrated that higher concentrations (i.e., 4 wt%) of cellulose and cellulose of higher molecular weights (BNC) gave more viscous solutions. These solutions were not Newtonian exhibiting shear-thinning behavior (viscosity decreases under high shear) (Marksteat et al. 2016), which was advantageous for extrusion type printing. Shear thinning was due to disruption of the hydrogen bonding network by the IL and subsequent reorientation of polymeric chains upon pressure. As a result, shear thinning eliminated the need for complicated printing nozzle structures, and a typical paste extruder equipped with a syringe pump was sufficient for printing the solution (Detsch et al. 2008). The viscosity decreased under shear and it became easier for the solution to pass through the nozzle; when shear was removed, viscosity quickly rose (Faddoul et al. 2012) and the printed pattern was able to retain its shape after the structure was printed, ensuring dimensional stability of the product.

Once the structure was printed from the IL-biopolymer solution, the coagulation, a process when the biopolymer is regenerated from the solution, was conducted using water as an antisolvent. Normally, when preparing different materials (e.g., fibers), the cellulose is regenerated and IL is washed out leaving behind stable structures. Here, three techniques were used for coagulation: (a) spraying water using an atomizing nozzle, (b) coagulation bath into which the print was dispensed, and (c) coagulating gel, however, the first two methods presented some difficulties.

Spraying water using an atomizing nozzle while printing allowed retention of decent print resolution but resulted in complications when multiple layers were extruded on top of each other because of poor adhesion between layers and layer separation. Alternatively, coagulation was attempted in the coagulation bath in a layer-by-layer manner, with antisolvent diffusion "from the bottom up". In this approach, water was added after the first few layers of solution were printed, and the amount of water was such that the top of the print was not covered with it and, therefore, not fully gelled, making it possible for the next layer to adhere. Then water diffused "from the bottom up" through the print so that every added layer came in contact with the nonsolvent. However, at approximately six deposited layers (~8 mm in height), water diffusion was too slow and the coagulation did not proceed at a sufficient speed causing the formation of a few layers that were not coagulated, sliding of the top of the print, and subsequent print top collapse.

The printing of IL solution was successful using a vertical agar gel support (prepared from agar and water), where a good interchange of IL with water anti-solvent was achieved. The solution printed onto the gel was instantly coagulated by the agar gel, followed by the addition of freshwater, enabling a coagulation process and forming a gel structure of the print.

5.4.2.2 3D Printing of Chitin Structures from an Ionic Liquid

The 3D printing of chitin solubilized in IL was investigated by Rogers' group, and thus far it's the only example of printing chitin from ILs. For the work discussed here chitin, extracted with the help of $[C_2mim][OAc]$ IL, with a MW two and a half times larger than that of commercially available pulped chitin (Wineinger et al. 2020), was used.

The printing was achieved using Printrbot Simple Metal 3D printer (Rogers et al. 2019) equipped with heated paste extruder in which the rubber plunger syringe cap was substituted with custom-made Teflon analog (Fig. 5.2). The freshly prepared 3 wt% chitin solution in $[C_2mim][OAc]$ was transferred into a 60 mL plastic syringe right after dissolution and placed into the extruder, which was preheated to 40 °C. Print parameters and accuracy (print speed rate) were controlled by Ultimaker Cura 1.5 software (Ultimaker Cura). Print speed was set to 30 mm/s and the extruder temperature at 40 °C (Table 5.1).

In this proof-of-concept study of a 3D model, 15 mm height rings with diameters of 20 and 40 mm and a cube with an edge of 15 mm were chosen for printing. The

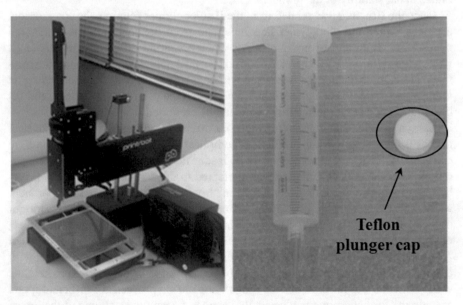

Fig. 5.2 Images of the 3D printer and example of a 10 mL syringe with Teflon plunger cap

Fig. 5.3 Chitin ring (40 mm diameter) printed on the solid support

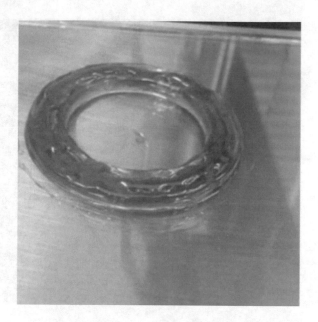

print shape was defined by the 3D model developed using Autodesk Fusion 360 Software (Fusion 360). The print layer thickness was controlled by using different sizes of blunt plastic needles (14 G–22 G). During the printing, the temperature of the extruder varied from 40 to 50 °C to achieve sufficient solution flow. Printed layers solidified on a glass plate at room temperature, but depending on the biopolymer concentration in IL, additional cooling could be beneficial. The print on a solid glass support is shown in Fig. 5.3.

After the 3D shapes were created and solidified (Fig. 5.3), the prints were coagulated in an aqueous coagulation bath (Fig. 5.4), followed by multiple washing steps to remove the IL. The materials after coagulation preserved the initial shape formed in layer-by-layer print fashion. The printed materials were freeze-dried from an aqueous solution (Fig. 5.5).

In a similar layer-by-layer fashion printing, the Rogers group conducted 3D-printing of chitin-poly-L-lactic acid (PLA) 1:1 (w/w) composites, with an overall polymeric load of 6 wt%. For 3D printing of chitin-PLA composites, chitin and PLA powder with a molecular weight of ~700,000 (6.5 dl/g) were simultaneously dissolved in [C_2mim][OAc] under constant stirring using thermal dissolution at 100 °C during 15 h. Similar to the work described above, the ring of 1.5 cm height and 20 mm diameter was chosen as a 3D model, in order to test the quality of layer adhesion. The prepared print that kept the shape after printing was coagulated in an aqueous bath, washed with water, and freeze-dried (Fig. 5.5). The 3D printing of chitin is the only example of extrusion printing of chitin from ILs.

Fig. 5.4 Chitin ring (40 mm diameter) coagulated in antisolvent (water)

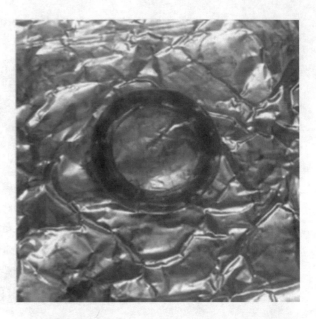

Fig. 5.5 Chitin ring (40 mm diameter) after freeze-drying

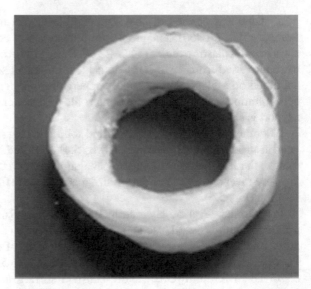

5.5 Outlook

This mini-review summarizes the 3D biopolymeric structures made through inkjet or extrusion-based 3D printing and studied in recently published articles. Indeed, biopolymers really offer great advantages as they are abundant yet degradable, strong, inexpensive, and are already utilized in delivery of pharmaceuticals, medical devices, and other biomedical applications, which would benefit from the kind of customization that 3D printing enables. 3D printing technology can be utilized for the fabrication of 3D objects with high resolution and multifunctionality in applications where a defined structure of the biopolymeric material is necessary. The applications go far beyond the delivery devices field since 3D bioprinting enables the formation of acellular scaffolds for tissue engineering, or even with cells to fabricate living tissues and organs.

Notes

Dr. Robin D. Rogers is a named inventor on related patents and applications and has partial ownership of 525 Solutions, Inc. J. L. Shamshina is an inventor on related patents and applications, former CSO and former employee of 525 Solutions, Inc., and former CSO of Mari Signum Mid-Atlantic, LLC.

References

Austin PR (1977) Chitin solution. US patent US4059457A

Azad MA, Olawuni D, Kimbell G, Badruddoza AZM, Hossain MS, Sultana T (2020) Polymers for extrusion-based 3D-printing of pharmaceuticals: a holistic materials–process perspective. Pharmaceutics 12:124–131. https://doi.org/10.3390/pharmaceutics12020124

Backdahl H, Risberg B, Gatenholm P (2011) Observations on bacterial cellulose tube formation for application as vascular graft. Mater Sci Eng C 31:14–21. https://doi.org/10.1016/j.msec.2010.07.010H

Bagheri M, Rodríguez H, Swatloski RP, Spear SK, Daly DT, Rogers RD (2008) Ionic liquid-based preparation of cellulose-dendrimer films as solid supports for enzyme immobilization. Biomacromol 9:381–387. https://doi.org/10.1021/bm701023w

Barikani M, Oliaei E, Seddiqi H, Honarkar H (2014) Preparation and application of chitin and its derivatives: a review. Iran Polym J 23:307–326. https://doi.org/10.1007/s13726-014-0225-z

Benhabiles MS, Salah R, Lounici H, Drouiche N, Goosen MFA, Mameri N (2012) Antibacterial activity of chitin, chitosan and its oligomers prepared from shrimp shell waste. Food Hydrocoll 29:48–56. https://doi.org/10.1016/j.foodhyd.2012.02.013

Biswas TT, Yu J, Nierstrasz VA (2019) Effects of ink characteristics and piezo-electric inkjetting parameters on lysozyme activity. Sci Rep 9:18252–18263. https://doi.org/10.1038/s41598-019-54723-9

Caruso A, Vollmer J, Machacek M, Kortvely E (2020) Modeling the activation of the alternative complement pathway and its effects on hemolysis in health and disease. PLoS Comput Biol 16(10):e1008139. https://doi.org/10.1371/journal.pcbi.1008139

Chen Q, Xu A, Li Z, Wang J, Zhang S (2011) Influence of anionic structure on the dissolution of chitosan in 1-butyl-3-methylimidazolium-based ionic liquids. Green Chem 13:3446–3452. https://doi.org/10.1039/C1GC15703E

Chow KS, Khor E, Wan AC (2001) Porous chitin matrices for tissue engineering: fabrication and in-vitro cytotoxic assessment. J Polym Res 8:27–35. https://doi.org/10.1007/s10965-006-0132-x

Clark WR (2002) Gao D (2002) Properties of membranes used for hemodialysis therapy. Semin Dial 15:191–195. https://doi.org/10.1046/j.1525-139x.2002.00019.x

Courtenay JC, Sharma RI, Scott JL (2018) Recent advances in modified cellulose for tissue culture applications. Molecules 23:E654. https://doi.org/10.3390/molecules23030654

Cui X, Boland T, D'Lima DD, Lotz MK (2012) Thermal inkjet printing in tissue engineering and regenerative medicine. Recent Pat Drug Deliv Formul 6(2):149–155

DeLong HC, Trulove PC, Haverhals LM, Reichert WM (2012) Natural fiber welding. US patent US8202379B1

Deppisch R, Storr M, Buck R, Göhl H (1998) Blood material interactions at the surfaces of membranes in medical applications. Sep Purif Technol 14:241–254

Detsch R, Uhl F, Deisinger U, Ziegler G (2008) 3D-cultivation of bone marrow stromal cells on hydroxyapatite scaffolds fabricated by dispense-plotting and negative mould technique. J Mater Sci Mater Med 19:1491–1496. https://doi.org/10.1007/s10856-007-3297-x

Duan B, Zheng X, Xia Z, Fan X, Guo L, Liu J, Wang Y, Ye Q, Zhang L (2015) Highly biocompatible nanofibrous microspheres self-assembled from chitin in NaOH/urea aqueous solution as cell carriers. Angew Chem Int Ed 54:5152–5156. https://doi.org/10.1002/anie.201412129

Elieh-Ali-Komi D, Hamblin MR (2016) Chitin and chitosan: production and application of versatile biomedical nanomaterials. Int J Adv Res (indore) 4(3):411–427

Elieh-Ali-Komi D, Sharma L, Dela Cruz CS (2018) Chitin and its effects on inflammatory and immune responses. Clin Rev Allergy Immunol 54(2):213–223. https://doi.org/10.1007/s12016-017-8600-0

Esteban MA, Cuesta A, Ortuño J, Meseguer J (2001) Immunomodulatory effects of dietary intake of chitin on gilthead seabream (Sparus aurata L.) innate immune system. Fish Shellfish Immunol 11(4):303–315. https://doi.org/10.1006/fsim.2000.0315

Faddoul R, Reverdy-Bruas N, Bourel J (2012) Silver content effect on rheological and electrical properties of silver pastes. J Mater Sci Mater Electron 23:1415–1426. https://doi.org/10.1007/s10854-011-0607-3

Fix A, de Brouwer H (2020) Sustainable polymers for healthcare help advance the circular economy. ONdrugDelivery 112:38–42

Fusion 360 software for 3D CAD modeling. https://www.autodesk.com. Accessed 16 Jul 2021

García MC (2018) Drug delivery systems based on nonimmunogenic biopolymers. In: Parambath A (ed) Engineering of biomaterials for drug delivery systems. Woodhead Publishing Series in Biomaterials (Elsevier), Cambridge, pp 317–344. https://doi.org/10.1016/B978-0-08-101750-0.00012-X

Gunasekera DHAT, Kuek S, Hasanaj D, He Y, Tuck C, Croft AK, Wildman RD (2016) Three-dimensional inkjet printing of biomaterials using ionic liquids and co-solvents. Faraday Discuss 190:509–523. https://doi.org/10.1039/C5FD00219B

Gurau G, Wang H, Qiao Y, Lu X, Zhang S, Rogers RD (2012) Chlorine-free alternatives to the synthesis of ionic liquids for biomass processing. Pure Appl Chem 84:745–754. https://doi.org/10.1351/PAC-CON-11-11-10

Han P, Shi J, Nie T, Zhang S, Wang X, Yang P, Wu H, Jiang Z (2016) Conferring natural-derived porous microspheres with surface multifunctionality through facile coordination-enabled self-assembly process. ACS Appl Mater Interfaces 8:8076–8085. https://doi.org/10.1021/acsami.6b00335

Haverhals LM, Brown EK, Foley MP, De Long HC, Trulove PC (2012) Formation of surface structures on biopolymer substrates through the inkjet printing of ionic liquids. ECS Trans 50:615–621

Hickey RJ, Pelling AE (2019) Cellulose biomaterials for tissue engineering. Front Bioeng Biotech 7:1–10. https://doi.org/10.3389/fbioe.2019.00045

Higashi S, Yamamuro T, Nakamura T, Ikada Y, Hyon SH, Jamshidi K (1986) Polymer-hydroxyapatite composites for biodegradable bone fillers. Biomaterials 7:183–187. https://doi.org/10.1016/0142-9612(86)90099-2

Hinton TJ, Jallerat Q, Palchesko RN, Park JH, Grodzicki MS, Shue HJ, Ramadan MH, Hudson AR, Feinberg AW (2015) Three-dimensional printing of complex biological structures by freeform reversible embedding of suspended hydrogels. Sci Adv 1:e1500758. https://doi.org/10.1126/sciadv.1500758

Jayakumar R, Chennazhi KP, Srinivasan S, Nair SV, Furuike T, Tamura H (2011) Chitin scaffolds in tissue engineering. Int J Mol Sci 12:1876–1887. https://doi.org/10.3390/ijms12031876

Ji Y, Liang K, Shen X, Bowlin G (2014) Electrospinning and characterization of chitin nanofibril/polycaprolactone nanocomposite fiber mats. Carbohydr Polym 101:68–74. https://doi.org/10.1016/j.carbpol.2013.09.012

Kadokawa J-i (2016) Dissolution, gelation, functionalization, and material preparation of chitin using ionic liquids. Pure Appl Chem 88:621–629. https://doi.org/10.1515/pac-2016-0503

Khor E (2001) Chapter 7: chitin and chitosan: making the grade. In: Khor E (ed) Chitin: fulfilling a biomaterials promise. Elsevier, Kidlington, Oxford, UK. https://doi.org/10.1016/B978-008044 018-7/50005-1

Khoshmohabat H, Paydar S, Makarem A, Karami MY, Dastgheib N, Zahraei SAH, Rezaei R, Mahmoudi Nezhad GS (2019) A review of the application of cellulose hemostatic agent on trauma injuries. Open Access Emerg Med 11:171–177. https://doi.org/10.2147/OAEM.S205006

King C, Shamshina JL, Gurau G, Berton P, Khan NSAF, Rogers RD (2017) A platform for more sustainable chitin films from an ionic liquid process. Green Chem 19:117–126. https://doi.org/10.1039/C6GC02201D

Klemm D (2004) Cellulose. In: De Baets S, Vandamme E, Steinbüchel AT (eds) Biopolymers in 10 volumes. Polysaccharides II: polysaccharides from eukaryotes, vol 6. Wiley-VCH, Germany

Klemm D, Schumann D, Udhardt U, Marsch S (2001) Bacterial synthesized cellulose: artificial blood vessels for microsurgery. Progress Polym Sci 26:1561–1603. https://doi.org/10.1016/S0079-6700(01)00021-1

Krainer S, Smith C, Hirn U (2019) The effect of viscosity and surface tension on inkjet printed picoliter dots. RSC Adv 9:31708–31719. https://doi.org/10.1039/C9RA04993B

Lagerwall J, Schütz C, Salajkova M, Noh J-H, Park JH, Scalia G, Bergström L (2014) Cellulose nanocrystal-based materials: from liquid crystal self-assembly and glass formation to multifunctional thin films. NPG Asia Mater 6:e80. https://doi.org/10.1038/am.2013.69

Li VCF, Dunn CK, Zhang Z, Deng Y, Qi HJ (2017) Direct ink write (DIW) 3D printed cellulose nanocrystal aerogel structures. Sci Rep 7:8018–8031. https://doi.org/10.1038/s41598-017-07771-y

Li VCF, Mulyadi A, Dunn CK, Deng Y, Qi HJ (2018) Direct ink write 3D printed cellulose nanofiber aerogel structures with highly deformable, shape recoverable, and functionalizable properties. ACS Sustain Chem Eng 6:2011–2022. https://doi.org/10.1021/acssuschemeng.7b03439

Löffelmann U, Mager D, Korvink JG (2012) Conductive and transparent gel microstructures fabricated by inkjet printing of ionic liquid based fluids. In: 2012 7th IEEE international conference on nano/micro engineered and molecular systems (NEMS). Kyoto, pp 312–317. https://doi.org/10.1109/NEMS.2012.6196782

Madan M, Bajaj A, Lewis S, Udupa N, Baig JA (2009) In situ forming polymeric drug delivery systems. Indian J Pharm Sci 71(3):242–251. https://doi.org/10.4103/0250-474X.56015

Major I, Lastakchi S, Dalton M, McConville CM (2020) Implantable drug delivery systems. In: Seyfoddin S, Dezfooli SM, Greene CA (eds) Engineering drug delivery systems. Woodhead Publishing (Elsevier), Cambridge, pp 111–146. https://doi.org/10.1016/B978-0-08-102548-2.00005-6

Markstedt K, Sundberg J, Gatenholm P (2014) 3D Bioprinting of cellulose structures from an ionic liquid. Addit Manuf 3:115–121. https://doi.org/10.1089/3DP.2014.0004

Maxim ML, White JF, Block LE, Gurau G, Rogers RD (2012) Advanced biopolymer composite materials from ionic liquid solutions in ionic liquids: science and applications. In: Visser AE, Bridges NJ, Rogers RD (eds) ACS symposium series. American Chemical Society, Washington, DC, pp 167–187

Minagawa T, Okamura Y, Shigemasa Y, Minami S, Okamoto Y (2007) Effects of molecular weight and deacetylation degree of chitin/chitosan on wound healing. Carbohydr Polym 67:640–644. https://doi.org/10.1016/j.carbpol.2006.07.007

Moulthrop JS, Swatloski RP, Moyna G, Rogers RD (2005) High-resolution ^{13}C NMR studies of cellulose and cellulose oligomers in ionic liquid solutions. Chem Commun 1557–1559. https://doi.org/10.1039/B417745B

Mundsinger K, Müller A, Beyera R, Hermanutz F, Buchmeiser MR (2015) Multifilament cellulose/chitin blend yarn spun from ionic liquids. Carbohydr Polym 131:34–40. https://doi.org/10.1016/j.carbpol.2015.05.065

Muzzarelli RAA (1983) Chitin and its derivatives: new trends of applied research. Carbohydr Polym 3:53–75. https://doi.org/10.1016/0144-8617(83)90012-7

Nakajima M, Atsumi K, Kifune K, Miura K, Kanamaru H (1986) Chitin is an effective material for sutures. Jpn J Surg 16:418–424

Narat JK, Cipolla AF, Cangelosi JP (1950) Fortisan® (regenerated cellulose yarn), a new suture material. Arch Surg 60(6):1218–1230. https://doi.org/10.1001/archsurg.1950.01250011243018

Naseem S, Parrino SM, Buenten DM, Konopka JB (2012) Novel roles for GlcNAc in cell signaling. Commun Integr Biol 5(2):156–159. https://doi.org/10.4161/cib.19034

Okamoto Y, Kawakami K, Miyatake K, Morimoto M, Shigemasa Y, Minami S (2002) Analgesic effects of chitin and chitosan. Carbohydr Polym 49:249–252. https://doi.org/10.1016/S0144-8617(01)00316-2

Park JS, Kim T, Kim WS (2017) Conductive cellulose composites with low percolation threshold for 3D-printed electronics. Sci Rep 7:3246–3256. https://doi.org/10.1038/s41598-017-03365-w

Pattinson SW, Hart AJ (2017) Additive manufacturing of cellulosic materials with robust mechanics and antimicrobial functionality. Adv Mater Techn 2(4):13. https://doi.org/10.1002/admt.201600084

Pillai CKS, Paul W, Sharma CP (2009) Chitin and chitosan polymers: chemistry, solubility and fiber formation. Prog Polym Sci 34:641–678. https://doi.org/10.1016/j.progpolymsci.2009.04.001

Qin Y, Lu X, Sun N, Rogers RD (2010) Dissolution or extraction of crustacean shells using ionic liquids to obtain high molecular weight purified chitin and direct production of chitin films and fibers. Green Chem 12:968–971. https://doi.org/10.1039/C003583A

Remsing RC, Swatloski RP, Rogers RD, Moyna G (2006) Mechanism of cellulose dissolution in the ionic liquid 1-n-butyl-3-methylimidazolium chloride: a ^{13}C and $^{35/37}$Cl NMR relaxation study on model systems. Chem Commun 1271−1273. https://doi.org/10.1039/B600586C

Ritchie H (2020) Plastic pollution. Available at OurWorldInData.org, 2018. https://ourworldindata.org/plastic-pollution. Last Accessed 15 Dec 2020

Rogers RD, Zavgorodnya O, Shamshina JL, Gurau G (2019) Printing of biopolymers from ionic liquid. US patent application 20190275199

Rowe RC, Sheskey PJ, Owen SC (2005) Handbook of pharmaceutical excipients, 5th edn. Pharmaceutical Press, Washington DC, p 850

Ruparelia JP, Chatterjee AK, Duttagupta SP, Mukherji S (2008) Strain specificity in antimicrobial activity of silver and copper nanoparticles. Acta Biomater 4(3):707–716. https://doi.org/10.1016/j.actbio.2007.11.006

Seoudi R, Nada AMA (2007) Molecular structure and dielectric properties studies of chitin and its treated by acid, base and hypochlorite. Carbohydr Polym 68:728–733. https://doi.org/10.1016/j.carbpol.2006.08.009

Shamshina JL (2019) Chitin in Ionic Liquids: historical insights on the polymer's dissolution and isolation. A review. Green Chem 21:3974–3993. https://doi.org/10.1039/C9GC01830A

Shamshina JL, Gurau G, Block LE, Hansen LK, Dingee C, Walters A, Rogers RD (2014) Chitin–calcium alginate composite fibers for wound care dressings spun from ionic liquid solution. J Mater Chem B 2:3924–3936. https://doi.org/10.1039/C4TB00329B

Shamshina JL, Zavgorodnya O, Bonner JR, Gurau G, Di Nardo T, Rogers RD (2017) "Practical" electrospinning of biopolymers in ionic liquids. ChemSusChem 10:106–111. https://doi.org/10.1002/cssc.201601372

Shen X, Shamshina JL, Berton P, Bandomir J, Wang H, Gurau G, Rogers RD (2016) Comparison of hydrogels prepared with ionic-liquid-isolated versus commercial chitin and cellulose. ACS Sustain Chem Eng 4:471–480. https://doi.org/10.1021/acssuschemeng.5b01400

Shi Q, Qian Z, Liu D, Liu H (2017) Surface modification of dental titanium implant by layer-by-layer electrostatic self-assembly. Front Physiol 8:574–581. https://doi.org/10.3389/fphys.2017.00574

Shigemasa Y, Minami S (1996) Applications of chitin and chitosan for biomaterials. Biotech Genetic Eng Rev 13:383–420. https://doi.org/10.1080/02648725.1996.10647935

Shirazi SFS, Gharehkhani S, Mehrali M, Yarmand H, Metselaar HSC, Kadri NA, Osman NAA (2015) A review on powder-based additive manufacturing for tissue engineering: selective laser sintering and inkjet 3D printing. Sci Technol Adv Mater 16:033502. https://doi.org/10.1088/1468-6996/16/3/033502

Sindhu KA, Prasanth R, Thakur VK (2015) Medical applications of cellulose and its derivatives: present and future. In: Thakur VK (ed) Nanocellulose polymer nanocomposites. Wiley, Scrivener Publishing LLC. https://doi.org/10.1002/9781118872246.ch16

Singh M, Haverinen HM, Dhagat P, Jabbour GE (2002) Inkjet printing—process and Its applications. Adv Materials 22:673–685. https://doi.org/10.1002/adma.200901141

Singh N, Koziol KKK, Chen J, Patil AJ, Gilman JW, Trulove PC, Kafienah W, Rahatekar SS (2013) Ionic liquids-based processing of electrically conducting chitin nanocomposite scaffolds for stem cell growth. Green Chem 15:1192–1202. https://doi.org/10.1039/C3GC37087A

Singh R, Chacharkar MP, Mathur AK (2008) Chitin membrane for wound dressing application–preparation, characterization and toxicological evaluation. Int Wound J 5:665–673. https://doi.org/10.1111/j.1742-481X.2008.00482.x

Spritam. https://www.spritam.com/#/patient. Last Accessed 31 Jan 2021

Stoica-Guzun A, Stroescu M, Jinga S, Jipa I, Dobre T, Dobre L (2012) Ultrasound influence upon calcium carbonate precipitation on bacterial cellulose membranes. Ultrasonics Sonochem 19:909–915. https://doi.org/10.1016/j.ultsonch.2011.12.002

Sun N, Li W, Stoner B, Jiang X, Lu X, Rogers RD (2011) Composite fibers spun directly from solutions of raw lignocellulosic biomass dissolved in ionic liquids. Green Chem 13:1158–1161. https://doi.org/10.1039/C1GC15033B

Sun N, Swatloski RP, Maxim ML, Rahman M, Harland AG, Haque A, Spear SK, Daly DT, Rogers RD (2008) Magnetite-embedded cellulose fibers prepared from ionic liquid. J Mater Chem 18:283–290. https://doi.org/10.1039/B713194A

Sundaram MNS, Mony U, Jayakumar, R (2016) Chitin and chitosan as hemostatic agents. Encycl Polym Sci Technol 1–12. https://doi.org/10.1002/0471440264.pst647

Swatloski RP, Spear SK, Holbrey JD, Rogers RD (2002) Dissolution of cellose with ionic liquids. J Am Chem Soc 124(18):4974–4975. https://doi.org/10.1021/ja025790m

Sweely KD, Fox ET, Brown EK, Haverhals LM, De Long HC, Trulove PC (2014) Inkjet printing ionic liquids for the fabrication of surface structures on biopolymer substrates. ECS Trans 64:575–582

Takegawa A, Murakami M-A, Kaneko Y, Kadokawa J-i (2010) Preparation of chitin/cellulose composite gels and films with ionic liquids. Carbohydr Polym 79:85–90. https://doi.org/10.1016/j.carbpol.2009.07.030

Tamura H, Hamaguchi T, Tokura S (2004) Destruction of rigid crystalline structure to prepare chitin solution. In: Boucher I, Jamieson K, Retnakaran A (eds) Advances in chitin science, Proceedings of the 9th international conference on chitin and chitosan. Montreal, pp 84–87

Turner MB, Spear SK, Holbrey JD, Daly DT, Rogers RD (2005) Ionic liquid-reconstituted cellulose composites as solid support matrices for biocatalyst immobilization. Biomacromol 6:2497–2502. https://doi.org/10.1021/bm050199d

Turner MB, Spear SK, Holbrey JD, Rogers RD (2004) Production of bioactive cellulose films reconstituted from ionic liquids. Biomacromol 5:1379–1384. https://doi.org/10.1021/bm049748q

Ultimaker cura: powerful, easy-to-use 3D printing software. https://ultimaker.com/en. Accessed 16 Jul 2021

Unites States pharmacopeia, biological reactivity tests. In: Vivo. http://www.pharmacopeia.cn/v29240/usp29nf24s0_c88.html. Last Accessed 1 Feb 2021

Vázquez JA, Rodríguez-Amado I, Montemayor MI, Fraguas J, González MDP, Murado MA (2013) Chondroitin sulfate, hyaluronic acid and chitin/chitosan production using marine waste sources: characteristics, applications and eco-friendly processes: a review. Marine Drugs 11(3):747–774. https://doi.org/10.3390/md11030747

Ventola CL (2014) Medical Applications for 3D Printing: current and projected uses. P. T 39:704–711

Wan ACA, Tai BCU (2013) Chitin: a promising biomaterial for tissue engineering and stem cell technology. Biotech Adv 31:1776–1785. https://doi.org/10.1016/j.biotechadv.2013.09.007

Wan YZ, Gao C, Luo HL, He F, Liang H, Li XL, Wang YL (2009) Early growth of nano-sized calcium phosphate on phosphorylated bacterial cellulose nanofibers. J Nanosci Nanotechnol 9:6494–6500. https://doi.org/10.1166/jnn.2009.1311

Wang WT, Zhu J, Wang XL (2010) Dissolution behavior of chitin in ionic liquids. J Macromol Sci B 49:528–541. https://doi.org/10.1016/j.carbpol.2009.07.030

Wang Y, Li Y, Liu S, Li B (2015) Fabrication of chitin microspheres and their multipurpose application as catalyst support and adsorbent. Carbohydr Polym 120:53–59. https://doi.org/10.1016/j.carbpol.2014.12.005

Waston NFC, Hodgkin W (2005) Wound dressings. Surgery (Oxford) 23:52–55

Willerth SM, Sakiyama-Elbert SE (2007) Approaches to neural tissue engineering using scaffolds for drug delivery. Adv Drug Delivery Rev 59(4–5):325–338. https://doi.org/10.1016/j.addr.2007.03.014

Wineinger HB, Shamshina JL, Kelly A, King C, Rogers RD (2020) A method for determining the uniquely high molecular weight of chitin extracted from raw shrimp shells using ionic liquids. Green Chem 22:3734–3741. https://doi.org/10.1039/D0GC00753F

Xie H, Zhang S, Li S (2006) Chitin and chitosan dissolved in ionic liquids as reversible sorbents of CO_2. Green Chem 8:630–633. https://doi.org/10.1039/B517297G

Xie M, Huan G, Xia W, Feng X, Chen L, Zhao Y (2017) Preparation and performance optimization of PVDF anti-fouling membrane modified by chitin. J Polym Eng 38:179–186. https://doi.org/10.1515/polyeng-2016-0372

Xu Z-K, Huang X-J, Wan l-S (2009) Membranes with glycosylated surface. In: Surface engineering of polymer membranes. Advanced topics in science and technology in China. Springer, Berlin.https://doi.org/10.1007/978-3-540-88413-2_7

Yadav P, Yadav H, Shah VG, Shah G, Dhaka G (2015) Biomedical biopolymers, their origin and evolution in biomedical sciences: a systematic review. J Clin Diagn Res 9(9):ZE21–ZE25. https://doi.org/10.7860/JCDR/2015/13907.6565

Yang S, Chen D, Li N, Mei X, Qi X, Li H, Xu Q, Lu J (2012) A facile preparation of targetable pH-sensitive polymeric nanocarriers with encapsulated magnetic nanoparticles for controlled drug release. J Mater Chem 22(48):25354–25361. https://doi.org/10.1039/c2jm34817a

Yoo J, Bradbury TJ, Bebb TJ, Iskra J, Surprenant HL, West TG (2014) Three-dimensional printing system and equipment assembly. US patent US8888480B2

Yusof NLBM, Lim LY, Khor E (2001) Preparation and characterization of chitin beads as a wound dressing precursor. J Biomed Mater Res 54:59–68. https://doi.org/10.1002/1097-4636(200101)54:1

Zavgorodnya O, Shamshina JL, Bonner JR, Rogers RD (2017) Electrospinning biopolymers from ionic liquids require control of different solution properties than volatile organic solvents. ACS Sustain Chem Eng 5:5512–5519. https://doi.org/10.1021/acssuschemeng.7b00863

Zhang S, He XGF, Huang Y, Wang Y, Wan Y (2009) Characterisation of hydroxyapatite/bacterial cellulose nanocomposite. Polym Polym Compos 17:353–358. https://doi.org/10.1177/096739110901700602

Chapter 6
Ionic Liquid-Based Oral Drug Delivery Systems

Md. Rafiqul Islam and Masahiro Goto

Abstract The oral route is the preferred mode of drug administration because it is non-invasive, self-administration is easy, and less physician supervision is required. However, the pharmaceutical industry faces several challenges in supplying many newly developed drugs by oral means because of their low water and other solvent solubility, formulation complexity, and limited bioavailability. In response to these constraints, tailor-made solvents known as ionic liquids (ILs) are increasingly being used in various pharmaceuticals sectors including drug delivery systems and active pharmaceutical ingredient formation because of their environmentally friendly and tunable physicochemical and biological properties. ILs and IL-based self-emulsifying drug delivery systems have been used for solubilizing poorly soluble drugs, protecting therapeutics from hostile gastrointestinal environments, improving permeability across the intestinal barrier, and enhancing absorption into the systemic circulation. This chapter highlights the potential of ILs as a tool for addressing key pharmaceutical issues in drug delivery/formulation, such as low solubility, stability, and bioavailability, with a specific focus on the latest developments.

Keywords Ionic liquid · Oral drug delivery systems · Bioavailability · Self-emulsifying delivery systems

Md. R. Islam · M. Goto (✉)
Department of Applied Chemistry, Graduate School of Engineering, Kyushu University, 744 Motooka, Nishi-ku, Fukuoka 819-0395, Japan
e-mail: m-goto@mail.cstm.kyushu-u.ac.jp

Md. R. Islam
e-mail: md.rafiqul.islam.764@m.kyushu-u.ac.jp

Md. R. Islam
Department of Applied Chemistry and Chemical Engineering, Noakhali Science and Technology University, Noakhali 3814, Bangladesh

M. Goto
Advanced Transdermal Drug Delivery System Center, Kyushu University, 744 Motooka, Nishi-ku, Fukuoka 819-0395, Japan

Division of Biotechnology, Center for Future Chemistry, Kyushu University, 744 Motooka, Nishi-ku, Fukuoka 819-0395, Japan

© The Author(s), under exclusive license to Springer Nature Singapore Pte Ltd. 2021
M. Goto and M. Moniruzzaman (eds.), *Application of Ionic Liquids in Drug Delivery*,
https://doi.org/10.1007/978-981-16-4365-1_6

Abbreviations

ADME	Absorption, delivery, metabolism, and excretion
API	Active pharmaceutical ingredient
Atvr	Atazanavir
[Atvr][NSA]	Atazanavir-2-naphthalene sulfonic acid
[Atvr][OSA2]	Atazanavir-dioctyl sulfosuccinic acid
BCS	Biopharmaceutics classification system
BGLs	Blood glucose levels
CAGE	Choline and geranate
Ciz	Cinnarizine
DDSs	Drug delivery systems
Dicf	Diclofenac
Dost	Docusate
FAs	Free acids
FB	Free basic
GI	Gastrointestinal
Haf	Halofantrine
ILs	Ionic liquids
[HHCPr][NTf2]	1-Hexyl-3-hexyloxycarbonylpyridinium bis(trifluoromethylsulfonyl)imide
[C8MβPr][C10SO4]	1-Octyl-3-methylpyridinium decyl sulfate
[C8MβPr][C18SO4]	1-Octyl-3-methylpyridinium octadecyl sulfate
Itrz	Itraconazole
LNF	Lipid-based non-aqueous formulation
MecA	Meclofenamic acid
N(CN)2	Dicyanamide
PMx	Physical mixtures
PVA	Poly(vinyl alcohol)
PWSDs	Poorly water-soluble drugs
RT	Room temperature
RTILs	Room temperature ionic liquids
Sb.C	Subcutaneous
SBO	Soybean oil
SEDSs	Self-emulsifying drug delivery systems
SRB	Sorafenib
Sulz	Sulfasalazine
[Sulz][Ch]	Sulz-choline
TolA	Tolfenamic acid
Trim	Triflimide

6.1 Introduction

Ionic liquids (ILs) are organic salts that are usually liquid at temperatures below 100 °C. Some ILs are liquid at room temperature and are, therefore, termed room temperature ILs (RTILs) (Adawiyah et al. 2016; Rogers and Seddon 2003). Reduced vapor pressure, excellent solubilizing ability, an extensive liquid limit, and high thermal stability are just a few of their distinctive potential features (Elgharbawy et al. 2018; Mirhoseini and Salabat 2015; Pernak et al. 2014). The potential for the features of ILs to be tailored via tunable ion matching allows application-specific designer organic components to be introduced (Adawiyah et al. 2016; Uddin et al. 2020). The pharmaceutical industry is faced with a series of challenges, which includes the need to administer rigid crystalline states of many medicines because of their poor aqueous solubility and bioavailability, and polymorphic transformation (Adawiyah et al. 2016; Shamshina et al. 2015). This issue becomes more pronounced when drugs are generally insoluble in commercially available pharmaceutical solvents. Fortunately, ILs are attracting interest as green solvents and chemicals for the formulation of pharmaceutical products with distinct properties that are not generally achievable with common solvents (Adawiyah et al. 2016; Ali et al. 2020; Chowdhury et al. 2019). ILs have gained popularity for a variety of pharmaceutical applications over the last decade, including the formation of active pharmaceutical ingredient-ILs (API-ILs) (Berton et al. 2017), as solvents and/or cosolvents for the synthesis and purification of pharmaceutical components (Freire et al. 2012; Sivapragasam et al. 2016; Ventura et al. 2017), for solubilization of insoluble and low aqueous solubility drugs (Moniruzzaman et al. 2010a, b), and for drug delivery systems (DDSs) (Rita Caparica et al. 2017; Santos de Almeida et al. 2017; Tahara et al. 2020). Initial reports of ILs highlighted their toxicity and antimicrobial activity (Docherty and Kulpa 2005; Sivapragasam et al. 2020). However, their use has, since, been expanded for biological applications such as enzyme breeding (Kragl et al. 2002), cellular biocatalysis (Yang and Pan 2005), protein solubility (Araki et al. 2015; Bisht et al. 2017), and protein stability (Adawiyah et al. 2016; Banerjee et al. 2018; Reslan and Kayser 2018).

IL-based DDSs can improve the efficiency of therapeutics delivered via different delivery routes. IL-assisted therapeutic administration via the oral route is the focus of this chapter. The oral route is considered the most widely recommended dosing technique as the administration is easy, it allows for greater patient compliance and less physician supervision, and it can be less expensive than injection and other prescription formulations (Adawiyah et al. 2016; Banerjee et al. 2018; Moshikur et al. 2020; Sahbaz et al. 2015; Williams et al. 2014). More than 70% of marketed medications are taken by mouth, typically in the form of solid dosage (e.g. tablets) or liquid dosage (e.g. solutions and dispersions) (Berardi and Bisharat 2017). Nonetheless, poor solubility and permeability, inadequate absorption into the gastrointestinal (GI) system, and a significant reduction of efficiency for certain drug molecules pose challenges in oral delivery (Moshikur et al. 2020). Orally dosed drugs are not immediately available in the systemic circulation to implement their therapeutic effects.

Before entering the bloodstream, they must first be consumed and migrate through the GI tract. Compared with parenteral administration, therapeutics administrated orally results in a delayed onset of action at best, or at worst can fail to reach the bloodstream entirely (Berardi and Bisharat 2017; York 2013). The chemical, physical, and biological interactions of a drug with the GI environment dictate whether or not the medication can enter the blood circulation and produce a therapeutic benefit as well as its effectiveness (Berardi and Bisharat 2017).

The potential and limitations of oral therapy are summarized in part I of this chapter. Part II of this chapter addresses the possible use of ILs to bypass constraints on oral routes and enhances the oral administration of many medicinal products. The special physiological interactions of ILs and IL-based self-emulsifying drug delivery systems (SEDSs) with various components of the GI tract are examined, highlighting how they can be beneficial for the effective oral supply of medicinal products (Berardi and Bisharat 2017). The main purpose of this chapter is to explain how the maximum benefit can be obtained by using ILs to distribute drugs via the oral route.

6.2 Physiological Impediments to Therapeutic Oral Administration

The degree of absorption of a drug into the systemic circulation is reflected in its oral bioavailability. Protection of macromolecules from degradation in the GI tract; permeation through the intestinal barrier; and absorption of molecules into the systemic circulation are all required for the successful oral delivery of drugs, proteins, and peptides (Banerjee et al. 2018). This section discusses the physiological aspects of the GI environment, physicochemical features of therapeutics, and issues relevant to the dosage form that influence drug bioavailability (Berardi and Bisharat 2017).

6.2.1 Physiological Factors of the GI Tract that Affect Bioavailability

A medicine can experience numerous biochemical obstacles to absorption as it passes across the GI tract. To begin, the medication must be able to tolerate the acidic pH of the stomach. In addition, enzymes secreted by the stomach and intestine can metabolize the substance (Banerjee et al. 2018; Berardi and Bisharat 2017). The oral distribution of drugs, proteins, and peptides is extremely difficult because of the aqueous nature of the GI tract operating environment and the release of digestive enzymes. Proteins are extremely sensitive to proteolytic enzymes that are found in various sites in the GI tract (Berardi and Bisharat 2017; Choonara et al. 2014). Owing to the limited surface area and non-absorptive properties of the epithelium, pepsin also induces chemical digestion of proteins in the stomach, resulting in reduced absorption (Berardi and Bisharat 2017; Choonara et al. 2014; Lee 2002). Moreover, the material must first move through the mucus layer that protects the epithelial cells of the GI tract until it can be absorbed from the lumen. At this point, the main barriers are the drug not diffusing through the membrane or getting trapped in the mucus (Berardi and Bisharat 2017; Cone 2009). Furthermore, the medicine can be exposed to first-pass metabolism by intestinal enzymes, which are present throughout the brush boundary and enterocyte cells of the intestine. This enzymatic metabolism is considered pre-systemic metabolism and is thought to be the primary cause of the low oral bioavailability of several medicines (Berardi and Bisharat 2017; Choonara et al. 2014; Thanki et al. 2013).

6.2.2   *Physicochemical Properties of Medications that Impair Bioavailability*

Fick's first law of diffusion simplifies drug absorption as: $J = PC$, where J is the flux of a transported drug across the GI tract, P is the permeability coefficient of the GI wall, and C is the concentration of the drug in the GI fluid. The concentration of drug depends not only on its solubility but also on the rate of dissolution and its stability against GI fluids (Choonara et al. 2014; Thanki et al. 2013). Passive transport, active transport, and endocytosis are all possible mechanisms for drug, protein, or peptide absorption from the GI tract (Artursson et al. 2007). The passive transport of therapeutic drugs across the intestinal membrane is facilitated by a concentration gradient and is dependent on the intrinsic physicochemical properties of the transported drugs and carrier. The low lipophilicity and high molecular mass of proteins and peptides limit their absorption via this mechanism (Artursson et al. 2007). In contrast, active transport is a chemically expensive process that is aided by a concentration gradient of the transported molecule across the GI epithelium, while endocytosis is an energy-consuming process (Artursson et al. 2007; Choonara et al. 2014). The Biopharmaceutics Classification System (BCS) separates drugs into four classes based on their solubility and permeability (Choonara et al. 2014; Lennern 1995), Fig. 6.1. According to the BCS, the low bioavailability of a drug is linked to its limited solubility, permeability, or both. Different strategies have been designed to improve the solubility, dissolution rate, and stability of drugs in the GI environment, including nanoparticles, SEDSs, and API-ILs transformation.

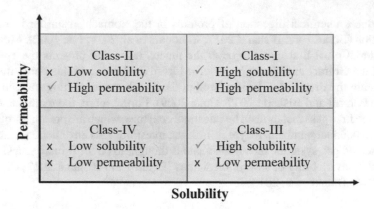

Fig. 6.1 BCS classification of drugs

6.2.3 Dosage Form-Related Factors that Influence Bioavailability

As stated earlier, oral bioavailability is influenced by both the physiological and physiochemical properties of medicines. In addition, the nature and properties of the dosage form can have a substantial impact on the bioavailability of a drug. For example, reformed release mechanisms can precisely modulate the volume of drug entering the bloodstream (Berardi and Bisharat 2017; Choonara et al. 2014). As previously discussed, delivery mechanisms to improve the oral bioavailability of sparingly soluble medications have been designed. The aim of this chapter is not to investigate the various dosage forms and methods used to increase the oral bioavailability of medication. Instead, the emphasis of this chapter is on using ILs to build delivery mechanisms that can increase oral bioavailability.

6.3 IL-Mediated Delivery Systems as Tools for Improving Oral Delivery

The physical and chemical barriers of the GI epithelium prevent drug, protein, and peptide absorption. The epithelial monolayer covering the large mucosal interface of the GI tract acts as an efficient barrier to protein and peptide absorption, preventing large molecules from entering the systemic circulation (Berardi and Bisharat 2017; Choonara et al. 2014). IL-based DDSs are promising as they can enhance the pharmacokinetic properties of both small and large molecules. Several IL-based DDSs have been used to address one or more of the challenges associated with the oral route (Banerjee et al. 2018; Moshikur et al. 2020). This section addresses recent developments in IL-based DDSs and their possible application to surmount oral transmission barriers.

6.3.1 ILs Enhance the Solubility of Poorly Soluble Drugs

It has been reported that water solubility, dissolution rate, and drug permeability all influence the bioavailability of drugs taken orally (Banerjee et al. 2018). The medication must first be dissolved in the gastrointestinal fluids before being absorbed (Banerjee et al. 2018; Berardi and Bisharat 2017). In aqueous environments, drugs with low solubility are hampered by several factors including reduced dissolution, irregular absorption, and limited oral bioavailability. Low aqueous solubility is, thus, a significant stumbling block preventing sufficient oral bioavailability for a large number of commercially available drugs (Choonara et al. 2014). Porter and co-workers are considered pioneers of IL-based oral drug delivery. They used 1-hexyl-3-hexyloxycarbonylpyridinium bis(trifluoromethylsulfonyl)imide ([HHCPr][NTf$_2$]) to enhance the solubility of three hydrophobic drugs; danazol, itraconazole (Itrz), and fenofibrate, and found 5.5, 100, and 1.7-fold improvements, respectively, compared with soybean oil (SBO). To improve the solubility further, they replaced the anion with hydrophilic dicyanamide (N(CN)$_2$) and found 20- and 500-fold improvements in the solubility of danazol and Itrz, respectively, compared with SBO (Williams et al. 2014), Fig. 6.2a.

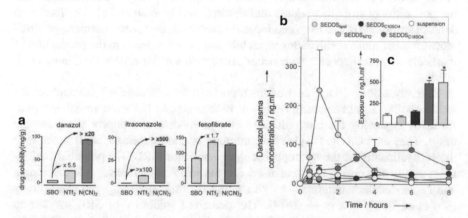

Fig. 6.2 a Comparative solubility study of danazol, Itrz, and fenofibrate in ILs consisting of [HHCPr]$^+$ cation and [NTf$_2$] or [N(CN)$_2$] anions, and SBO; (mean $\pm$ 1 SD, n = 3). **b** Danazol plasma concentrations in rats after oral danazol (25 mg/kg) administration via IL- or lipid-based SEDSs or a suspension formulation; (mean $\pm$ SEM, n $\geq$ 4). **c** Total danazol exposure over 8 h; (mean $\pm$ SEM, n $\geq$ 4, *p < 0.05 was significant compared with the suspension formulation. Reproduced with permission from ref. (Williams et al. 2014)

6.3.2 ILs Enhance the Oral Bioavailability of Hydrophobic Drugs

Low lipophilic therapeutic product oral bioavailability, which can be induced by poor aqueous miscibility, inadequate intestinal absorptivity, elevated P-glycoprotein efflux, and pre-systemic digestion, remains a significant obstacle (Rehman and Webster 2018; Shi et al. 2020). To address this issue, a variety of nanoparticle-based systems have been investigated, including nanosuspensions, nanoemulsions, solid lipid nanoparticles, and inorganic-based nanoparticles (Li et al. 2016; Shi et al. 2020). While these nanoparticle-based systems have specific advantages in terms of improving oral bioavailability, their performance to date has not been optimal (Shi et al. 2020).

Porter and colleagues used ILs ([HHCPr][NTf$_2$], 1-octyl-3-methylpyridinium decyl sulfate ([C$_8$M$_\beta$Pr][C$_{10}$SO$_4$]) and octadecyl sulfate ([C$_8$M$_\beta$Pr][C$_{18}$SO$_4$])) to boost the oral bioavailability of danazol by inserting it into formulations modeled on current lipid formulations that emulsify spontaneously in SEDSs when they come into contact with GI fluids. The oral bioavailability of SEDSs$_{C18SO4}$ was 4.3-fold higher than that of crystalline danazol suspension, and was comparable to that of the SEDSs$_{lipid}$, but with notable proof of persistent plasma concentrations and the possibility of sustained release and absorption (Fig. 6.2b and c). The increased danazol absorption of SEDSs$_{C18SO4}$ indicates that the IL mechanism, which permitted more efficient contact with endogenous bile salt micelles, leads to the production of markedly dispersed species with better absorptive surface access (Williams et al. 2014).

Recently, Mitragotri and co-workers reported a deep eutectic solvent made of RT stable choline and geranate (CAGE)—a biocompatible IL—as a promising platform for improving the bioavailability of hydrophobic and poorly water-soluble drugs. They used CAGE 1:2 (1:2 stoichiometry of choline:geranic acid) to improve the bioavailability of the hydrophobic drug sorafenib (SRB)—a potential multikinase antagonist presently being used for the treatment of renal cell carcinoma and hepatocellular carcinoma—by enhancing the drug solubility and absorptivity (Shi et al. 2020; Sun et al. 2017). The perceived solubility of SRB tosylate in CAGE was outstanding (>500 mg/mL), 100 million-fold higher than that in water (0.0044 μg/mL). Following oral administration, CAGE increased peak blood concentrations of SRB in rats 2.2-fold compared with the control SRB suspension formulation. The exclusion half-life of SRB was extended twofold, and the mean absorption period was 1.6-fold longer than that of the suspension formulation. In addition, SRB delivered by CAGE had a substantially more diverse biodistribution than control formulations. In particular, the concentrations of SRB in the lung and kidney were 4.4-fold and 6.2-fold higher, respectively, than those observed for SRB suspension (Shi et al. 2020). According to mechanistic studies, the SRB-CAGE solution self-assembled (427.41 nm) in the natural state, which may be the reason for the diverse biodistribution in vivo (Fig. 6.3) (Shi et al. 2020).

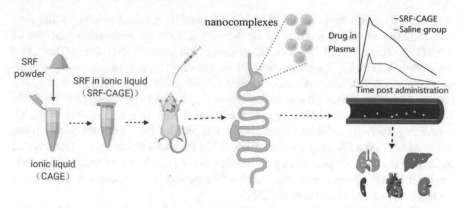

Fig. 6.3 Schematic illustration of SRB oral administration to rats though CAGE. CAGE increased SRB absorption by forming SRB-CAGE micelles in the GI fluids. Reproduced with permission from ref. (Shi et al. 2020)

6.3.3 ILs Enhance Solubility and Bioavailability by Transforming into Lipophilic API-ILs

It has been reported that owing to polymorphism, reduced solubility, and/or low bioavailability, approximately 40–70% of Food and Drug Administration's approved medications do not achieve their therapeutic potential (Moshikur et al. 2020). Several drug molecules that are poorly soluble in both aqueous and nonaqueous vehicles present a unique obstacle, and as a result, many struggles to advance from discovery to production and, eventually, to the pharmacy. High crystallinity and high intermolecular forces in the solid state—which are normally indicated by a high melting point—are the causes of these low solubility characteristics (Sahbaz et al. 2015). Developing an API-IL is a novel way to resolve this inherent problem for certain drugs (Moshikur et al. 2020).

In 2007, the Rogers group was the first to report an API-IL (Hough and Rogers 2007; Shadid et al. 2015). API-ILs of poorly water-soluble drugs (PWSDs) for oral delivery were first reported by Porter and colleagues in 2015 as a method of reducing crystallinity, boosting solubility, and facilitating drug absorption (Sahbaz et al. 2015). They treated three weakly basic PWSDs (Itrz, cinnarizine (Ciz), and halofantrine (Haf)) with lipophilic counter ions (triflimide (Trim), docusate (Dost)) to convert them into API-ILs. The resulting API-ILs were either miscible or highly soluble in lipid-based SEDSs. They found considerably higher plasma exposure for API-IL containing formulations (twofold for [Ciz][Trim] and 20-fold for [Itrz][Dost]) than for free basic (FB) drug suspension (or physical mixture) of the same dose. They also found significantly higher (two to threefold) plasma exposure for Itrz-IL than for a commercial Itrz formulation (Sporanox at the same dose). The findings show that the production of lipophiles can be used to increase the amount of dissolved medication in lipid-based formulations (Sahbaz et al. 2015).

They subsequently prepared API-ILs of weakly acidic PWSDs to improve the solubility in and the utility of lipid-based SEDSs by metathesis reaction of PWSDs (tolfenamic acid (TolA), meclofenamic acid (MecA), diclofenac (Dicf), and ibuprofen) with highly lipophilic cations based on alkyl amines or quaternary ammonium counterions. Compared with the corresponding free acids (FAs), the drug-ILs had lower melting points (in some cases below RT) and were noticeably more soluble in a model lipid-based non-aqueous formulation (LNF). In rats, low doses of LNF containing TolA-IL exhibited drug absorption profiles that were similar to those obtained when the FA drug was administered in the same formulation. However, at the higher doses made possible by conversion to the IL, altered plasma profiles (and some GI irritation) were apparent, and the time period of exposure was extended to up to 48 h postdose (Sahbaz et al. 2017).

In 2015, the Rogers group converted a BCS class-IV acidic drug, sulfasalazine (Sulz), to an API-IL (Shadid et al. 2015). They synthesized Sulz-choline API-IL ([Sulz][Ch]), by the metathesis reaction of Sulz drug with the generally recognized as safe choline cation (Islam et al. 2020a, b), to assess the absorption, delivery, metabolism, and excretion (ADME) properties in in vitro and in vivo models. They found 4000-fold improvement of the solubility of API-IL compared with free Sulz in saline. They also found that the absolute bioavailability of [Sulz][Ch] was 2.5-fold higher than that of Sulz-PMx (physical mixtures of 1:1 sulfasalazine and choline) (Sahbaz et al. 2015). This provided a new and versatile avenue for addressing the solubility challenges that are commonly encountered in drug development and in particular for intravenous administration (Shadid et al. 2015).

Later, Morgen and co-workers reported another SEDSs to improve the drug solubilizing capacity, drug loading capacity, and drug exposure profile. They converted a weakly basic PWSD, atazanavir (Atvr), into two lipophilic salts—Atvr-2-naphthalene sulfonic acid [Atvr][NSA] and Atvr-dioctyl sulfosuccinic acid [Atvr][OSA$_2$]—and found that [Atvr][NSA] improved solubility in SEDSs sixfold compared with IL-free drug. At high doses, the [Atvr][NSA] salt produced equivalent exposure to IL-free drug suspension, thereby minimizing inter-animal variability. Furthermore, lipophilic salts in lipid vehicles have elevated dissolved-drug levels that inhibit precipitation without the use of co-solvents (Morgen et al. 2017).

6.3.4 ILs Enhance the Oral Bioavailability of Insulin

Two decades have passed since the search for an oral insulin product commenced. Various technologies have been implemented to combat the gastrointestinal challenges of oral pharmacological absorption. Unfortunately, no formulation has successfully addressed all of the therapeutic barriers, meaning there are currently no commercially available oral insulin formulations. Furthermore, many products require multi-step formulation processes, assorted additives, or chemical protein alteration, all of which have their inherent drawbacks. With the worldwide diabetes epidemic on the rise, it is more important than ever to produce safe, efficient, and resilient oral insulin products (Banerjee et al. 2018). Mitragotri et al. formulated an oral insulin formulation based on ILs. They used a deep eutectic solvent CAGE that had previously shown exceptional effectiveness in the transdermal delivery of antibiotics (Zakrewsky et al. 2016) and insulin (Tanner et al. 2018). Single-step loaded insulin-CAGE was assessed in vitro and in vivo to measure its safety and effectiveness, as well as its storage stability. They found that CAGE improved paracellular insulin transport tenfold, elimination half-life twofold, and prolonged reduced blood glucose levels (BGLs) (up to 45%) Fig. 6.4a, b and c compared with subcutaneous injection (Banerjee et al. 2018). CAGE inhibits enzymatic degradation by forming a shield that prohibits intestinal enzymes from consuming primed insulin, thereby limiting their proteolytic function (Banerjee et al. 2018). In addition, CAGE helps to move insulin through the mucus membrane and facilitates paracellular permeation

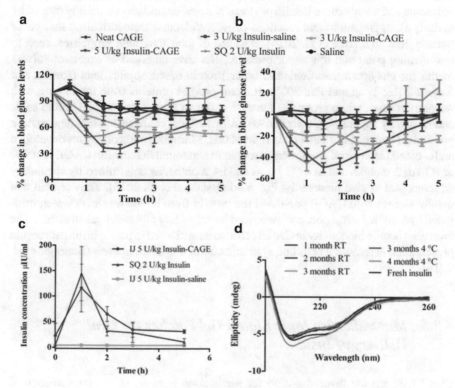

Fig. 6.4 a BGL dropping efficiency of insulin-loaded formulations compared with starting values; **b** Normalized BGL dropping efficiencies. Data were normalized by subtracting the BGL of saline groups from other groups and saline groups were considered the fasting group; (mean ± SD, n = 6, *P < 0.05 was significant compared with subcutaneously administered 2 U/kg insulin. **c** Insulin plasma concentration after oral administration to rats; (mean ± SD, n = 4). **d** Circular dichroism spectra after extracting insulin from CAGE at various time intervals. Reproduced with permission from ref. (Banerjee et al. 2018)

by opening tight connections. Because of the synergy of its ability to increase para-cellular absorption, suppress proteolytic enzymes, and penetrate through the mucus membrane, CAGE exhibited high oral provision efficiency (Banerjee et al. 2018).

6.3.5   ILs Enhance the Stability of Proteins and Peptides

Proteins are major biomacromolecules of clinically, biochemically, biotechnolog-ically, and genetically engineered live organisms. Weak interactions, including hydrogen bonding, hydrophobic, and ionic interactions are vital to preserve the three-dimensional structures of proteins (Banerjee et al. 2017). Changes in the protein microenvironment interaction initiate protein denaturation resulting in protein unfolding and inactivation. Insulin physicochemical degradation is mainly caused by hydrolysis, aggregation, and reactions of intermolecular transformation that cause potency loss (Banerjee et al. 2018). Because of their unrivaled properties, such as low melting point and low vapor pressure, ILs have emerged as excellent solvent media for enzymatic reactions and other protein-based applications (Reslan and Kayser 2018; Yang and Pan 2005). ILs can prevent proteins from interacting with water molecules, while protic ILs have been shown to stabilize several amino acids and insulin by suppressing their native conformational changes and self-aggregation proclivity (Banerjee et al. 2017). In light of this, Mitragotri and colleagues developed an IL-based insulin oral administration system and found that insulin CAGE is stable at RT for 2 months and at 4 °C for at least 4 months, as determined by secondary structure and in vivo bioactivity, Fig. 6.4d (Banerjee et al. 2018). They argued that insulin storage with CAGE prevented the protein from interacting with water, miti-gated hydrolytic interaction, and stabilized its secondary alpha-helical structure. The increased insulin bioactivity in CAGE can also be attributed in part to insulin molecule presentation in intestinal cells and systemic circulation as monomers (Banerjee et al. 2018).

6.3.6   Mucoadhesive Ionic Liquid Gel Patches for Oral Delivery of Insulin

Oral DDSs are the favored option for medication because of the convenience of use and resulting patient compliance (Banerjee et al. 2018; Moshikur et al. 2020). The acidic and enzymatic breakdown of the stomach; the active and "sticky" mucus layer of the epithelial layer; and the high barrier possession of tight junctions of the intestinal epithelium are all drug barriers (Peng et al. 2020). CAGE—an RT liquid salt—has previously been reported to be an excellent drug carrier and/or solvent for enhancing the oral absorption of hydrophobic small molecule drugs like SRB (Shi et al. 2020) and macromolecules like insulin (Banerjee et al. 2018). Insulin

CAGE administered orally to rats reduced BGLs significantly compared with insulin supplied in saline and increased bioavailability (Banerjee et al. 2018). The mechanism behind the effectiveness of oral administration of CAGE appears to be the temporary interruption of epithelial tight interconnections enabling paracellular uptake of drug. In CAGE, the efficacy of enzymes such as trypsin can be inhibited meaning insulin can be stored for longer periods (Banerjee et al. 2018; Peng et al. 2020).

In response to the problem of the poor liquid site in a GI setting, Mitragotri and collaborators developed a new IL gel that allows both the IL and drug to be released on a controlled and continuous basis. The "ionogel" CAGE-patches are a simple and effective way of encapsulating ILs and therapeutic drugs in a gel and comprise solid poly (vinyl alcohol) PVA and molten IL constituents (Fig. 6.5) (Peng et al. 2020). Insulin passage through epithelial layers was shown to be improved by CAGE-patches in vitro. The CAGE-patches provided significantly more insulin (<30%) than a CAGE solution with the same CAGE mass (0.3%). By focusing the generated CAGE into a specific area, CAGE-patches minimized the amount of CAGE needed for successful penetration improvement. Cellular uptake of insulin by Coca-2

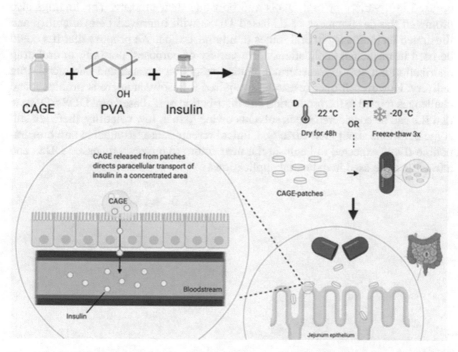

Fig. 6.5 Schematic illustration of CAGE-patches for an oral DDS. CAGE-patch ionogels were made up by mixing CAGE, PVA, and insulin. After being poured onto cell culture plates, the mixtures were either allowed to dry for 48 h or exposed to repetitive freeze–thaw cycles to produce dry-CAGE-patches or freeze-CAGE-patches, respectively. Reproduced with permission from ref. (Peng et al. 2020)

and HT29-MTX-E12 cells in the CAGE-patch groups was markedly lower than in the CAGE and saline control solutions (Peng et al. 2020).

The use of ILs for a number of purposes, including enhancing the solubility and oral bioavailability of poorly soluble materials, has been explored. A short overview of the various IL-based oral DDSs is presented in Table 6.1.

6.4 Conclusion

This chapter has illustrated the role and benefits of ILs as promising solvents, enhancers, and stabilizers in oral DDSs. Numerous sparingly soluble drugs have high solubility in ILs, allowing for modern, efficient distribution methods that are not feasible in traditional solvents. The issues of solid drug solubility and polymorphism that have been major roadblocks to developing innovative DDSs have been successfully overcome by transforming drugs into API-ILs. The structure and functional versatility of ILs support their inclusion in various media, enabling them to be used in a variety of oral DDSs and boosting their efficacy. The findings also prompted the development of IL-based DDSs with improved bioavailability and illustrated their efficacy in oral routes of administration. We propose that ILs could be used in medicinal applications for a variety of purposes, not only in oral drug distribution but also in transdermal, ocular, respiratory, nasal, and parenteral drug delivery. However, further research is required to resolve numerous problems and challenges related to the production and distribution of IL-based oral DDSs. Despite that ILs can be considered green solvents owing to their low volatility, there are still some questions about their toxicity. Clinical research into sustainable and biocompatible ILs is expected to begin in the near future to promote IL-based DDSs and allow ILs to be used in industrial applications.

Table 6.1 IL-assisted oral delivery systems

Number	Drugs	Role of ILs	Main findings	Remarks	References
1	Danazol, itraconazole and fenofibrate	a. [HHCPr][N(CN)$_2$] as a solubilizing agent b. [C$_8$M$_\beta$Pr][C$_{18}$SO$_4$] as a solubilizing and emulsifying agent	a. [HHCPr][N(CN)$_2$] improved solubility of danazol, itraconazole and fenofibrate by 20, 500 and, twofold, respectively, compared with soybean oil b. [C$_8$M$_\beta$Pr][C$_{18}$SO$_4$] improved oral bioavailability of danazol by 4.3-fold compared with crystalline danazol suspension c. Sustained plasma concentration compared with lipid formulation	High drug loading capacity and easy dispersion in GI fluids have been improved	Williams et al. (2014)

(continued)

Table 6.1 (continued)

Number	Drugs	Role of ILs	Main findings	Remarks	References
2	Cinnarizine, halofantrine and itraconazole	a. [Ciz][Trim] as an API-IL b. [Haf][Trim] as an API-IL c. [Itrz][Dost] as an API-IL	a. [Ciz][Trim] improved solubility by sevenfold compared with Ciz-FB drug and bioavailability by twofold compared with Ciz-FB suspension b. [Haf][Trim] improved solubility by fivefold compared with Haf-FB drug c. [Itrz][Dost] improved solubility by 50-fold compared with Itrz-FB drug and bioavailability by 20-fold compared with Itrz-FB suspension	Oral gavage administration and SEDSs	Sahbaz et al. (2015)

(continued)

Table 6.1 (continued)

Number	Drugs	Role of ILs	Main findings	Remarks	References
3	Tolfenamic acid, meclofenamic acid and diclofenac	a. [TolA][Decl$_2$Me$_2$] as an API-IL b. [MecA][Decl$_2$Me$_2$] as an API-IL c. [Dicf][Decl$_2$Me$_2$] as an API-IL	a. [TolA][Decl$_2$Me$_2$] improved solubility by 5.5-fold compared with TolA-FA drug b. [MecA][Decl$_2$Me$_2$] improved solubility by 10.1-fold compared with MecA-FA drug c. [Dicf][Decl$_2$Me$_2$] improved solubility by 3.8-fold compared with Dicf-FA drug d. Similar plasma concentration but sustained exposure profile compared with FA drug aqueous suspension	SEDSs with high drug loading capacity	Sahbaz et al. (2017)
4	Sulfasalazine	[Sulz][Ch] as an API-IL	a. Improved solubility by 4000-fold compared with IL free drug b. Improved bioavailability by 2.5-fold compared with Sulz-PMx suspension	Oral gavage administration to rat	Shadid et al. (2015)

(continued)

Table 6.1 (continued)

Number	Drugs	Role of ILs	Main findings	Remarks	References
5	Atazanavir	[Atvr][NSA] and (Atvr)[OSA$_2$] as API-IL	a. [Atvr][NSA] improved solubility by sixfold compared with IL free drug b. [Atvr][NSA] produced equivalent exposure to IL free drug suspension	SEDSs with high drug loading capacity	Morgen et al. (2017)
6	Insulin	CAGE as a solubilizing agent	a. Prolonged decreased of BGLs compared with Sb.C injection (up to 45%) b. Improved paracellular insulin transport by tenfold compared with control c. Improved elimination half-life by twofold compared with Sb.C. injection	High biocompatibility and biochemical activity were demonstrated	Banerjee et al. (2018)
7	Sorafenib	CAGE as a solubilizing agent	a. Improved solubility by 100 million-fold in CAGE, compared with water b. Enhanced drug elimination half-life, mean absorption time and peak blood concentration by twofold, 1.6-fold and 2.2-fold, respectively, and drug accumulation in lungs by 4.4-fold and kidneys by 6.2-fold over control suspension formulations	Spontaneous self-assembled structure of drug-CAGE	Shi et al. (2020)

(continued)

Table 6.1 (continued)

Number	Drugs	Role of ILs	Main findings	Remarks	References
8	Insulin	CAGE-gel patch	a. Improved in-vitro transport of insulin (>30%) through Coca-2 and HT29-MTX-E12 cells compared with control saline solution b. Improved biocompatibility compared with control (ethanol)	A novel concept to clinical localization in the GI tract	Peng et al. (2020)

References

Adawiyah N, Moniruzzaman M, Hawatulaila S, Goto M (2016) Ionic liquids as a potential tool for drug delivery systems. Medchemcomm 7:1881–1897. https://doi.org/10.1039/c6md00358c

Ali MK, Moshikur RM, Wakabayashi R, Moniruzzaman M, Kamiya N, Goto M (2020) Biocompatible ionic liquid surfactant-based microemulsion as a potential carrier for sparingly soluble drugs. ACS Sustain Chem Eng 8:6263–6272. https://doi.org/10.1021/acssuschemeng.9b07773

Araki S, Wakabayashi R, Moniruzzaman M, Kamiya N, Goto M (2015) Ionic liquid-mediated transcutaneous protein delivery with solid-in-oil nanodispersions. Medchemcomm 6:2124–2128. https://doi.org/10.1039/c5md00378d

Artursson P, Neuhoff S, Matsson P (2007) Passive permeability and active transport models for the prediction of oral absorption. Compr Med Chem II 5:59–278. https://doi.org/10.1016/B0-08-045 044-X/00126-7

Banerjee A, Ibsen K, Brown T, Chen R, Agatemor C, Mitragotri S (2018) Ionic liquids for oral insulin delivery. Proc Natl Acad Sci USA 115:7296–7301. https://doi.org/10.1073/pnas.172233 8115

Banerjee A, Ibsen K, Iwao Y, Zakrewsky M, Mitragotri S (2017) Transdermal protein delivery using choline and geranate (CAGE) deep eutectic solvent. Adv Healthc Mater 6:1–11. https://doi.org/ 10.1002/adhm.201601411

Berardi A, Bisharat L (2017) Nanotechnology systems for oral drug delivery: challenges and opportunities. Nanotechnol Drug Del Chapter 3:52–84.

Berton P, Bona KRD, Yancey D, Rizvi SAA, Gray M, Gurau G, Shamshina JL, Rasco JF, Rogers RD (2017) Transdermal bioavailability in rats of lidocaine in the forms of ionic liquids. Salts and Deep Eutectic 8:498–503. https://doi.org/10.1021/acsmedchemlett.6b00504

Bisht M, Mondal D, Pereira MM, Freire MG, Venkatesu P, Coutinho JAP (2017) Long-term protein packaging in cholinium-based ionic liquids: improved catalytic activity and enhanced stability of cytochrome c against multiple stresses. Green Chem 19:4900–4911. https://doi.org/10.1039/ c7gc02011b

Choonara BF, Choonara YE, Kumar P, Bijukumar D, Toit LC, Pillay V (2014) A review of advanced oral drug delivery technologies facilitating the protection and absorption of protein and peptide molecules. Biotechnol Adv 32:1269–1282. https://doi.org/10.1016/j.biotechadv.2014.07.006

Chowdhury MR, Moshikur RM, Wakabayashi R, Tahara Y, Kamiya N, Moniruzzaman M, Goto M (2019) Development of a novel ionic liquid-curcumin complex to enhance its solubility, stability, and activity. Chem Commun 55:7737–7740. https://doi.org/10.1039/c9cc02812a

Cone RA (2009) Barrier properties of mucus. Adv Drug Deliv Rev J 61:75–85. https://doi.org/10. 1016/j.addr.2008.09.008

Docherty KM, Kulpa CF (2005) Toxicity and antimicrobial activity of imidazolium and pyridinium ionic liquids. Green Chem 7:185–189. https://doi.org/10.1039/b419172b

Elgharbawy AA, Azizah F, Alam Z, Moniruzzaman M (2018) Ionic liquids as a potential solvent for lipase-catalysed reactions: a review. J Mol Liq 251:150–166. https://doi.org/10.1016/j.mol liq.2017.12.050

Freire MG, Cláudio AFM, Araújo JMM, Coutinho JAP, Marrucho IM, Canongia Lopes JN, Rebelo LPN (2012) Aqueous biphasic systems: a boost brought about by using ionic liquids. Chem Soc Rev 41:4966–4995. https://doi.org/10.1039/c2cs35151j

Hough WL, Rogers RD (2007) Ionic liquids then and now: from solvents to materials to active pharmaceutical ingredients. Bull Chem Soc Jpn 80:2262–2269. https://doi.org/10.1246/bcsj.80. 2262

Hye LJ (2002) Protein drug oral delivery the recent progress. Arch Pharm Res 25(5):572–584. https://doi.org/10.1007/BF02976925

Islam MR, Chowdhury MR, Wakabayashi R, Kamiya N, Moniruzzaman M, Goto M (2020a) Ionic liquid-in-oil microemulsions prepared with biocompatible choline carboxylic acids for improving the transdermal delivery of a sparingly soluble drug. Pharmaceutics 12(4):392. https://doi.org/ 10.3390/pharmaceutics12040392

Islam MR, Chowdhury MR, Wakabayashi R, Tahara Y, Kamiya N, Moniruzzaman M, Goto M
(2020b) Choline and amino acid based biocompatible ionic liquid mediated transdermal delivery
of the sparingly soluble drug acyclovir. Int J Pharm 582:119335. https://doi.org/10.1016/j.ijp
harm.2020.119335

Kragl U, Eckstein M, Kaftzik N (2002) Enzyme catalysis in ionic liquids. Curr Opin Biotechnol
13:565–571. https://doi.org/10.1016/S0958-1669(02)00353-1

Lennern H (1995) A theoretical basis for a biopharmaceutic drug classification: the correlation of
in vitro drug product dissolution and in vivo bioavailability. Pharma Res 12:413–420. https://
doi.org/10.1023/a:1016212804288

Li Z, Ye E, Lakshminarayanan R, Loh XJ (2016) Recent advances of using hybrid nanocarriers in
remotely controlled therapeutic delivery. Small 12:4782–4806. https://doi.org/10.1002/smll.201
601129

Mirhoseini F, Salabat A (2015) Ionic liquid based microemulsion method for the fabrication of poly
(methyl methacrylate)–TiO2 nanocomposite as a highly efficient visible light photocatalyst. RSC
Adv 5:12536–12545. https://doi.org/10.1039/C4RA14612C

Moniruzzaman M, Kamiya N, Goto M (2010a) Ionic liquid based microemulsion with pharmaceutically accepted components: formulation and potential applications. J Colloid Interface Sci
352:136–142. https://doi.org/10.1016/j.jcis.2010.08.035

Moniruzzaman M, Tamura M, Tahara Y, Kamiya N, Goto M (2010b) Ionic liquid-in-oil microemulsion as a potential carrier of sparingly soluble drug: characterization and cytotoxicity evaluation.
Int J Pharm 400:243–250. https://doi.org/10.1016/j.ijpharm.2010.08.034

Morgen M, Saxena A, Chen X, Miller W, Nkansah R, Goodwin A, Cape J, Haskell R, Su C,
Gudmundsson O, Hageman M, Kumar A, Singh G, Rao A, Holenarsipur VK (2017) European
journal of pharmaceutics and biopharmaceutics lipophilic salts of poorly soluble compounds to
enable high-dose lipidic SEDDS formulations in drug discovery. Eur J Pharm Biopharm 117:212–
223. https://doi.org/10.1016/j.ejpb.2017.04.021

Moshikur RM, Chowdhury MR, Moniruzzaman M, Goto M (2020) Biocompatible ionic liquids
and their applications in pharmaceutics. Green Chem. https://doi.org/10.1039/d0gc02387f

Peng K, Shi Y, Labarbiera A, Mitragotri S (2020) Mucoadhesive ionic liquid gel patches for oral
delivery. ACS Biomater Sci Eng. https://doi.org/10.1021/acsbiomaterials.0c01024

Pernak J, Niemczak M, Giszter R, Shamshina JL, Gurau G, Cojocaru OA, Praczyk T, Marcinkowska
K, Rogers RD (2014) Glyphosate-based herbicidal ionic liquids with increased efficacy. ACS
Sustain Chem Eng 2:2845–2851

Rehman M, Webster TJ (2018) Advancements in the oral delivery of Docetaxel: challenges, current
state-of-the-art and future trends. Int J Nanomedicine 13:3145–3161

Reslan M, Kayser V (2018) Ionic liquids as biocompatible stabilizers of proteins. Biophys Rev
10:781–793. https://doi.org/10.1007/s12551-018-0407-6

Rita C, Júlio A, Rosado C, Santos de Almeida T (2017) Applicability of ionic liquids in topical
drug delivery systems: a mini review. J Pharmacol Clin Res 4(5):555649. https://doi.org/10.
19080/JPCR.2018.04.555649

Rogers RD, Seddon KR (2003) Ionic liquids—solvents of the future? Science 80(302):792–793.
https://doi.org/10.1126/science.1090313

Sahbaz Y, Nguyen T, Ford L, Mcevoy CL, Williams HD, Scammells PJ, Porter CJH (2017) Ionic
liquid forms of weakly acidic drugs in oral lipid formulations: preparation, characterization,
in vitro digestion, and in vivo absorption studies. Mol Pharm 14:3669–3683. https://doi.org/10.
1021/acs.molpharmaceut.7b00442

Sahbaz Y, Williams HD, Nguyen T, Saunders J, Ford L, Charman SA, Scammells PJ, Porter CJH
(2015) Transformation of poorly water-soluble drugs into lipophilic ionic liquids enhances oral
drug exposure from lipid based formulations. Mol Pharm 12:1980–1991. https://doi.org/10.1021/
mp500790t

Santos de Almeida T, Júlio A, Saraiva N, Fernandes AS, Araújo MEM, Baby AR, Rosado C,
Mota JP (2017) Choline- versus imidazole-based ionic liquids as functional ingredients in topical

delivery systems: cytotoxicity, solubility, and skin permeation studies. Drug Dev Ind Pharm 43:1858–1865. https://doi.org/10.1080/03639045.2017.1349788

Shadid M, Gurau G, Shamshina JL, Chuang B-C, Hailu S, Guan E, Chowdhury SK, Wu J-T, Rizvi SAA, Griffin RJ, Rogers RD (2015) Sulfasalazine in ionic liquid form with improved solubility and exposure. Med Che 6:1837–1841. https://doi.org/10.1039/c5md00290g

Shamshina JL, Kelley SP, Gurau G, Rogers RD (2015) Chemistry: develop ionic liquid drugs. Nature. https://doi.org/10.1038/528188a

Shi Y, Zhao Z, Gao Y, Pan DC, Salinas AK, Tanner EEL, Guo J, Mitragotri S (2020) Oral delivery of sorafenib through spontaneous formation of ionic liquid nanocomplexes. J Control Release 322:602–609. https://doi.org/10.1016/j.jconrel.2020.03.018

Sivapragasam M, Moniruzzaman M, Goto M (2020) An overview on the toxicological properties of ionic liquids toward microorganisms. Biotechnol J 15:1–9. https://doi.org/10.1002/biot.201 900073

Sivapragasam M, Moniruzzaman M, Goto M (2016) Recent advances in exploiting ionic liquids for biomolecules: solubility, stability and applications. Biotechnol J 11:1000–1013. https://doi. org/10.1002/biot.201500603

Sun W, Wang Y, Cai M, Lin L, Chen X, Cao Z, Zhu K, Shuai X (2017) Codelivery of sorafenib and GPC3 siRNA with PEI-modified liposomes for hepatoma therap. Biomater Sci 5:2468–2479. https://doi.org/10.1039/c7bm00866j

Tahara Y, Morita K, Wakabayashi R, Kamiya N, Goto M (2020) Biocompatible ionic liquid enhances transdermal antigen peptide delivery and preventive vaccination effect. Mol Pharm 17:3845–3856. https://doi.org/10.1021/acs.molpharmaceut.0c00598

Tanner EEL, Ibsen KN, Mitragotri S (2018) Transdermal insulin delivery using choline-based ionic liquids (CAGE). J Control Release 286:137–144. https://doi.org/10.1016/j.jconrel.2018.07.029

Thanki K, Gangwal RP, Sangamwar AT, Jain S (2013) Oral delivery of anticancer drugs: challenges and opportunities. J Control Release. https://doi.org/10.1016/j.jconrel.2013.04.020

Uddin S, Chowdhury MR, Wakabayashi R, Kamiya N, Moniruzzaman M, Goto M (2020) Lipid based biocompatible ionic liquids: synthesis, characterization and biocompatibility evaluation. Chem Commun 56:13756–13759. https://doi.org/10.1039/d0cc04491a

Ventura SP, e Silva FA, Quental MV, Mondal D, Freire MG, Coutinho JA (2017) Ionic-liquid-mediated extraction and separation processes for bioactive compounds: past, present, and future trends. Chem Rev 117:6984–7052. https://doi.org/10.1021/acs.chemrev.6b00550

Williams HD, Sahbaz Y, Ford L, Nguyen TH, Scammells PJ, Porter CJH (2014) Ionic liquids provide unique opportunities for oral drug delivery: structure optimization and in vivo evidence of utility. Chem Commun 50:1688–1690. https://doi.org/10.1039/c3cc48650h

Yang Z, Pan W (2005) Ionic liquids: green solvents for nonaqueous biocatalysis. Enzyme Microb Technol 37:19–28. https://doi.org/10.1016/j.enzmictec.2005.02.014

York P (2013) No title design of dosage forms. In: Aulton ME, Taylor KMG (eds) Aulton's pharmaceutics: the design and manufacture of medicines, 4th edn. Churcill Livingstone, pp 7–19

Zakrewsky M, Banerjee A, Apte S, Kern TL, Jones MR, Sesto RED, Koppisch AT, Fox DT, Mitragotri S (2016) Choline and geranate deep eutectic solvent as a broad-spectrum antiseptic agent for preventive and therapeutic applications. Adv Healthc Mater 5:1282–1289. https://doi. org/10.1002/adhm.201600086

Chapter 7
Advances Brought by Ionic Liquids in the Development of Polymer-Based Drug Delivery Systems

Sónia N. Pedro, Carmen S. R. Freire, Armando J. D. Silvestre, and Mara G. Freire

Abstract The low solubility, permeation, and bioavailability of active pharmaceutical ingredients are the main factors affecting their delivery and therapeutic efficiency. Intensive research has pursuit new possibilities to overcome these drawbacks, in which polymers have played a major role in advances brought by drug delivery systems. Furthermore, the research of alternative solvents, such as ionic liquids (ILs), offered many advantages to improve polymer properties in drug delivery systems. ILs have been studied as solvent media in polymerization processes and applied as additives in the development of nanosystems. More recently, the tunable character of ILs has allowed the design of multi-responsive copolymers from IL monomers to enhance the delivery of different drugs. In this work, we review the latest applications of ILs toward the development of improved polymer-based drug delivery systems and highlight the fundamental knowledge required for the design of these systems.

Keywords Active pharmaceutical ingredients · Drug delivery · Ionic liquids · Polymers

Abbreviations

ILs

$[C_2C_1im][BF_4]$	1-Ethyl-3-methylimidazolium tetrafluoroborate
$[C_2C_1im][C_2H_5SO_4]$	1-Ethyl-3-methylimidazolium ethylsulfate
$[C_4C_1im][BF_4]$	1-Butyl-3-methylimidazolium tetrafluoroborate
$[C_4C_1im][CH_3CO_2]$	1-Butyl-3-methylimidazolium acetate
$[C_4C_1im]Cl$	1-Butyl-3-methylimidazolium chloride
$[C_4C_1im][HCOO]$	1-Butyl-3-methylimidazolium formate

S. N. Pedro · C. S. R. Freire · A. J. D. Silvestre · M. G. Freire (✉)
Department of Chemistry, CICECO-Aveiro Institute of Materials, University of Aveiro, 3810-193 Aveiro, Portugal
e-mail: maragfreire@ua.pt

© The Author(s), under exclusive license to Springer Nature Singapore Pte Ltd. 2021
M. Goto and M. Moniruzzaman (eds.), *Application of Ionic Liquids in Drug Delivery*,
https://doi.org/10.1007/978-981-16-4365-1_7

[C$_4$C$_1$im][N(CN)$_2$]	1-Butyl-3-methylimidazolium dicyanamide
[C$_4$C$_1$im][NTf$_2$]	1-Butyl-3-methylimidazolium bis(trifluoromethylsulfonyl)imide
[C$_4$C$_1$im][PF$_6$]	1-Butyl-3-methylimidazolium hexafluorophosphate
[C$_8$C$_1$im][NTf$_2$]	1-Octyl-3-methylimidazolium bis-(trifluoromethylsulfonyl)amide
[CH = C$_2$C$_1$im]Cl	1-Allyl-3-methylimidazolium chloride
[Ch][Glu]	Cholinium glutaminate
[Ch][Phe]	Cholinium phenylalaninate
[C$_4$C$_1$pyr][NTf$_2$]	1-Butyl-3-methylpyridinium bis-(trifluoromethylsulfonyl)imide
[C$_4$py][NTf$_2$]	1-Butyl-3-pyridinium bis-(trifluoromethylsulfonyl)imide
[C$_8$C$_1$pyr][NTf$_2$]	1-Octyl-3-methylpyridinium bis(trifluoromethylsulfonyl)imide
[N$_{1444}$][NTf$_2$]	Butyltrimethylammonium bis-(trifluoromethanesulfonyl)imide
[N$_{8881}$][NTf$_2$]	Tetraoctylammonium bis-(trifluoromethanesulfonyl)imide
[P$_{66614}$][NTf$_2$]	Trihexyltetradecylphosphonium bis(trifluoromethanesulfonyl)imi

7.1 Introduction

In the past decades, drug research and development of novel drug delivery strategies allowed to improve therapeutics and treatment outcomes (Bheemidi et al. 2011; Dugger et al. 2018). Before the 1950s, solid forms of drugs were normally formulated into pills or capsules that released the loaded active pharmaceutical ingredient (API) immediately upon contact with aqueous media, with no control of the release kinetics (Yun et al. 2015). These conventional drug delivery systems (first-generation) have been associated with undesirable side-effects, derived not only from the lack of drug-controlled release profiles but also due to the unspecified drug distribution (Liu and Gu 2016; Hrubý et al. 2015). These constraints, along with low bioavailability displayed by many drugs, required the use of higher doses and frequent applications.

In 1952, Smith Klein Beecham introduced, for the first time, the possibility to formulate a sustained release system for the delivery of dextroamphetamine (Dexedrine), a stimulant used to treat attention-deficit hyperactivity disorder (Lee and Li 2010). This new system was able to control the drug release over 12 h. After this pioneering system, several pharmaceutical companies intensified their research on the development of improved drug delivery systems, aiming at developing different administration routes for the existing APIs (Laffleur and Keckeis 2020). The research focused on drug delivery systems aimed at effectively reducing the dosage frequency while maintaining the drug concentration in targeted organs/tissues for longer periods of time (Tiwari et al. 2012). After years of advances in drug delivery technologies, the differences in the functions, release profiles, targets, and efficacy had a strong contribution from the advances faced directly in polymer science.

Since each drug has its own therapeutic range, above which it is toxic and below which it is ineffective, to achieve the desired results it is mandatory to select the most suitable administration route (Chenthamara et al. 2019). The administration route has an impact on the drug's bioavailability, especially if administrated via the systemic route due to incomplete absorption or degradation of many drugs. In this sense, the development of oral-controlled release formulations has been a particular challenge to pharmaceutical industries (Elgindy et al. 2011). The development of first-generation drug delivery systems led to an increase in the pursue of excipients that might allow regulating the release profiles. Among these excipients, polymers have played a key role in the progress of drug delivery technology (Tiwari et al. 2012).

Polymers are macromolecules formed by a high number of repeating units (monomers) covalently bonded, calling upon one of several polymerization mechanisms, existing in nearly limitless possibilities of combinations depending on the monomer's nature and combinations (yet, only a limited set is commercially available). The resulting polymers can be classified according to their origin (natural or synthetic), nature of the repeating unit, and (in the case of the synthetic ones) of the polymerization mechanism, structure, and supra-molecular/morphological arrangement, as summarized in Fig. 7.1. Natural-derived polymers are addressed in previous chapters of the current book.

Owing to the diversity and versatility of polymeric materials, their applications are nowadays widespread and essential across all sectors of human activity, spanning from extremely low value (yet essential) packaging materials to high-technology and added-value materials, in which biomedical and pharmaceutical applications are often included. Their use in the development of novel drug release devices, which include not only oral controlled release systems but also fast dispersing dosage forms, nano-systems, and site-specific delivery systems, has shown to be essential and with significant advantages (Liechty et al. 2010). These second-generation strategies, however, have experienced great challenges to reach and stay in the market, as often occurs with emerging technologies, due to the lack of regulatory regimes and uncertainty in the methodologies to apply (Yun et al. 2015; Te Kulve 2014). This type of system aims to overcome biological barriers to be more effective; however, the challenges addressed by the biological environment in pathological cases can be a drawback to these system's success. One example is Exubera, the first marketed inhaled insulin approved by the U.S. Food and Drug Administration (FDA) (Oleck et al. 2016). This delivery system was found to have similar pharmacokinetic and pharmacodynamic properties to regular insulin (Patton et al. 2004). Nevertheless, the unexpected side-effects, less preference by the patients and physicians, among other reasons, lead Pfizer to withdraw this product from the market in 2007. Unlike first-generation formulations, in which the pharmacokinetic profiles are dependent on the drug delivery systems, second-generation pharmacokinetics are determined by the body (Yun et al. 2014). Therefore, the efficacy of the system is not determined by the drug release kinetics of the formulation but by the body's biodistribution and

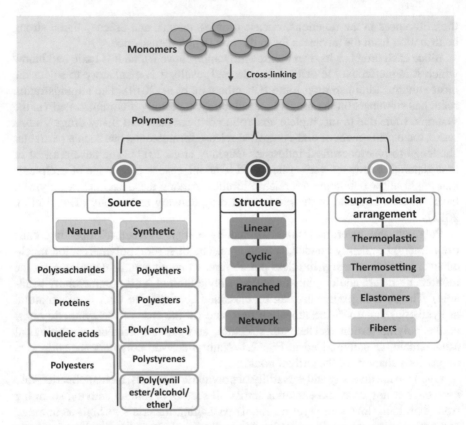

Fig. 7.1 Polymer's classification according to the respective structure, molecular forces, and source, with an emphasis on synthetic polymer classes

drug absorption at the target site (Park 2013). The drug delivery technologies to be developed in the present and next decade (third-generation) pursue to tackle the limitations of the previous systems, being designed to overcome both physicochemical and biological barriers. The specificity brought by these novel systems aims to reduce the drug's side-effect and improve the overall efficacy of the treatment, changing the drug delivery evolution. Figure 7.2 summarizes the historical development of drug delivery systems.

The use of polymers and polymer-based drug delivery systems may be advantageous to deliver both hydrophilic and lipophilic drugs (Amgoth et al. 2020). However, the use of organic solvents to solubilize sparingly soluble polymers and biopolymers, during polymers synthesis and/or formulation steps, usually contributes to address health and environmental concerns. Therefore, it is mandatory to explore novel solvents that can associate improved handling qualities, biocompatibility, and allow the development of adequate drug delivery systems. In this sense, ionic liquids (ILs) have received, in recent years, significant attention due to their intrinsic properties (Zhao 2006). ILs are molten salts, which among many other properties, and,

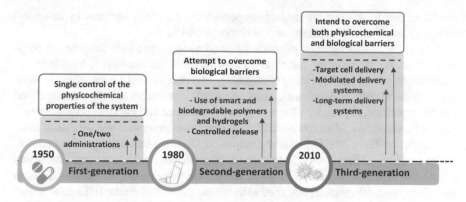

Fig. 7.2 Historical development of drug delivery systems

if properly designed, may possess low toxicity toward human cell lines, as well as high ionic conductivity, high electrochemical and thermal stability, nonvolatility, and nonflammability (Freire et al. 2012). ILs have been studied as solvents in polymer science not only for the solubilization of low water-soluble drugs and polymers but also as reaction media for polymerization or as polymerizable monomers for fabrication of drug delivery systems (Lu et al. 2009; Almeida et al. 2017). Thus, apart from the ILs use as solvents in materials preparation, polymeric ILs (PILs) are also extremely important in the present context, as they share many of ILs tunability features and often lead to stimuli-responsive materials (Chen et al. 2016).

Hitherto, ILs have been successfully used as solvent media and initiators for polymerizations, as polymers additives, as novel pharmaceutical solvents or functional excipients, and as monomers to obtain stimuli-responsive PILs aiming to offer efficient delivery options. The present chapter revises and focuses on the most relevant and recent advances found in the application of ILs in the development of polymer-based systems for drug delivery, while providing future perspectives and drawbacks of their use, toward the rational development of improved drug delivery systems. Accordingly, the current chapter is divided into two main subchapters, corresponding to the following topics: the use of ILs as media for polymer synthesis and applications of ILs in the development of polymer-based drug delivery systems. The latest subchapter presents a section dedicated to PILs and their application in the design of stimuli-responsive drug delivery systems.

7.2 Polymer Synthesis in Ionic Liquid Media

Polymers stand as promising candidates to solve numerous drawbacks associated with drug administration (Englert et al. 2018). The possibility to synthesize these materials in high amounts with no significant batch-to-batch variations and to perform modifications to achieve the desired properties can be easily attained by using the adequate polymerization technique (Shrivastava 2018).

One of the most important methods for producing polymers is free-radical polymerization, which accounts for *circa* 50% of all mass-produced polymers (mechanism illustrated in Fig. 7.3) (Hong et al. 2002). This process can advantageously work with a wide variety of monomers, being relatively insensitive to both monomer and media impurities, especially when compared with ionic and condensation polymerization mechanisms. Most free-radical polymerization processes use volatile organic solvents to facilitate mixing and heat dissipation, allowing better temperature control of the reaction (Odian 2004). Despite their compatibility with most monomers and easiness in separation, the use of organic solvents in these reactions comprises health and environmental concerns that need to be surpassed. In this context, ILs have drawn attention as alternative polymerization media, pursuing more environmentally benign processes (Hong et al. 2002; Benton and Brazel 2004; Strehmel et al. 2006).

The conventional free-radical polymerization of the methylmethacrylate monomer (MMA) into PMMA is usually performed at temperatures above 40°C in organic media using conventional initiators, such as azobisisobutyronitrile (AIBN) and benzoyl peroxide (BPO). The polymerization of MMA

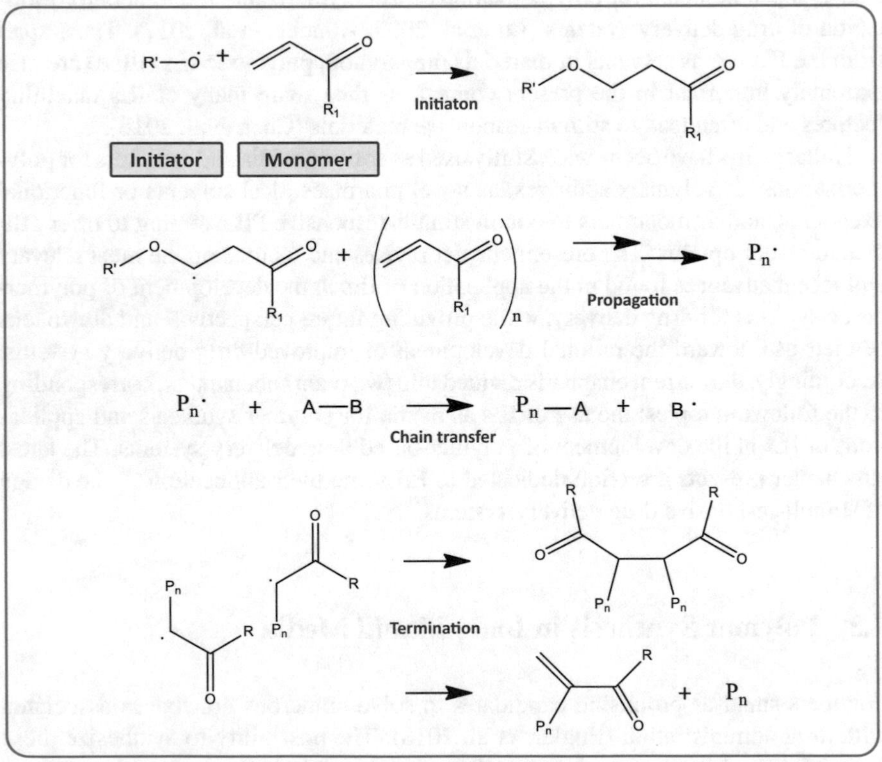

Fig. 7.3 Mechanism of free-radical polymerization using a peroxide initiator

and styrene (St) at room temperature in ILs, such as 1-butyl-3-methylimidazolium hexafluorophosphate ($[C_4C_1im][PF_6]$), has been studied by several authors (Hong et al. 2002; Benton and Brazel 2004). These studies have confirmed that reactions carried out in IL media present degrees of polymerization five times higher and takes place four times faster as compared to the reaction in organic solvents under similar conditions (Li et al. 2006).

ILs have been applied in the polymerization of several monomer types, and different polymerization techniques have been explored for the same polymer attempting to enhance the process performance. Major examples found in the literature correspond to the use of imidazolium, pyridinium, and ammonium-based ILs as media in the synthesis of poly(methacrylate) (PMA) and PMMA and some corresponding ester derivatives (Englert et al. 2018). PMA hydrogels have been applied in cell and drug delivery, while PMMA has been investigated in medical devices (Ajekwene 2020). The former polymer has been included in implants to deliver antibiotics, allowing to successfully decrease the risk of related prosthesis infection (Bistolfi et al. 2019). Although polystyrene (PS) does not present medical relevance, some derivates, such as poly(styrene sulfonate), have found drug delivery applications (Wibowo et al. 2020).

Although in conventional radical polymerizations the observed rates of polymerization are generally high and are thus interesting from a mechanistic and kinetic point of view, the study of reversible-deactivation radical polymerization (RDRP) processes, often referred to in the literature as controlled/living polymerizations, allows for superior control over the polymer chain characteristics and the end-group functionalization. Processes such as atom transfer radical polymerization (ATRP) (He et al. 2014), reversible addition-fragmentation chain transfer (RAFT) (Puttick et al. 2009), and polymerization and nitroxide-mediated polymerization (NMP) (Zhang et al. 2004) in IL media have been studied. However, RAFT reactions have found more advantages than the other processes regarding the use of ILs for MMA and BMA polymerizations (Puttick et al. 2009; Santha Kumar et al. 2018). Several RAFT agents have been reported to work efficiently in ILs; yet, their performance is dependent upon both the type of RAFT agent employed and the selected IL. Puttick et al. (2009) studied the effect of the IL anion on both the propagation rate and the control of molecular weight in MMA polymerization. The solubility of the monomer and of the polymer in the IL, the affinity of the RAFT agents [2-cyanoprop-2-yl dithiobenzoate (CTA1) and 2-dodecylsulfanyl thiocarbonyl sulfanyl-2-methyl propionic acid (CTA2)] for the solvent phase rather than the monomer swollen polymer, and the partitioning of the radical species in the ionic domains of the IL have an important role in the success of RAFT control. Regarding the IL ability to control the polymer molecular weight, slightly better control was noticed for CTA2 than CTA1 in 1-ethyl-3-methylimidazolium ethyl sulfate ($[C_2C_1im][C_2H_5SO_4]$), while with 1-ethyl-3-methylimidazolium tetrafluoroborate ($[C_2C_1im][BF_4]$) poor control was achieved for both RAFT agents. Additionally, by properly selecting the IL media and manipulating the RAFT:AIBN ratio, similar lower values of polydispersity to those achieved in toluene were observed (1.1–2.0). The study of ILs for these processes has evolved to the possibility to have RDRP-like results without the need for a rate mediating agent. Low et al. (2016) studied protic and aprotic ILs to improve the stability of carbon-based radicals. Kinetic data relating to propagating systems for several industrially common monomers, such as MMA, indicated that propagation rates can be increased or decreased (up to six orders of magnitude) depending on the monomer/IL combination. Specific interactions such as electrostatic, dispersion forces, or hydrogen bonding between IL components and the reactants (methyl radicals of the $CH_2 = CH(X)$ type) seem to play a crucial role in controlling reaction kinetics, being protic ILs preferable to achieve the same kinetic control in free-radical polymerizations (Low et al. 2016).

ILs have also been used as reaction media for cationic polymerization. The properties of ILs, such as viscosity and polarity, along with their tunability, can provide an adequate environment for both the solubilization of monomers and to stabilize the ensuing carbocations, which are essential in the cationic polymerization mechanism (Vijayaraghavan and MacFarlane 2004,2012; Han et al. 2016). The reactions carried out in ILs produce polymers with lower molecular weights and narrow polydispersity (>2.0) than those performed in organic media (Vijayaraghavan and MacFarlane 2004,2012). Despite being shortly investigated, cationic polymerizations carried out in imidazolium-based ILs, such as styrene, lead to similar yields and polymerization rates to those achieved with organic solvents like dichloromethane (DCM) (Han et al. 2016).

The polymer yields achieved in ILs using several polymerization techniques for monomers like MMA, BMA, BZMA, and St are summarized in Table 7.1, and compared with the results obtained in volatile organic solvents under the same reaction conditions. Generally, yields of polymerization are higher in IL media, particularly in conventional free-radical polymerizations. However, other yields achieved with other polymerization processes allow observing the need to adjust reaction conditions and reactants to obtain improved results.

Although polymerization reactions are well documented, few works deal with polycondensation reactions and enzymatic synthesis of polyesters in IL media. Dali et al. (2006) studied the polycondensation of glycolic acid using 1,3-dialkyl-imidazolium ILs as alternative solvents. Nevertheless, low yields in poly(glycolic acid) (PGA) were obtained (<38%) due to the monomer evaporation during the reaction, since high temperatures were applied (200–240°C). The efficiency of catalysts such as $Zn(CH_3CO_2)$ in these alternative media was low and attributed to the preferential interaction of Zn^{2+} with IL anions rather than with the polymer carboxylate end-groups. The synthesis of poly (L-lactic acid) (PLLA) and poly (L-lactic-co-glycolic acid) (PLGA) has also been studied in imidazolium-based ILs mediated by the enzyme lipase B from *Candida antarctica* at low-temperature conditions (65–90°C) (Chanfreau et al. 2010; Mena et al. 2010). PLGA with high crystallinity (up to 85%) and low polydispersity (<1.6) were obtained with a yield of 63% using such conditions (Chanfreau et al. 2010). Additionally, PLLA synthesized in 1-butyl-3-methylimidazolium-based ILs presented higher molecular weights than those synthesized in bulk or toluene as solvent. However, low yields were obtained (35%) presumably due to the solubility of PLLAs in these ILs (Yoshizawa-Fujita et al. 2008).

Table 7.1 Polymerization reactions for different monomers in ILs and volatile organic solvents

Monomer	Initiator	T_{react} (°C)	Time (min)	IL	Yield (%)	Organic solvent	Yield (%)	References
Conventional free-radical polymerizations								
MMA	AIBN	75	480	[C$_4$C$_1$im][PF$_6$]	98	Benzene	17	Hong et al. (2002)
St	BPO				99		9	
MMA/St	AIBN	60	30	[C$_2$C$_1$im][C$_2$H$_5$SO$_4$]	17	DMF	6	Bistolfi et al. (2019)
	BPO	80			72		17	
MMA/AN	AIBN	60			89		13	Schmidt-Naake et al. (2008)
	BPO	80			46		31	
BZMA	AIBN	70	1440	[C$_8$C$_1$pyr][NTf$_2$]	92	Toluene	–	Strehmel et al. (2014)
				[C$_4$C$_1$im][N(CN)$_2$]	89			
				[C$_8$C$_1$pyr][NTf$_2$]	57			
				[N$_{1444}$][NTf$_2$]	34			
				[N$_{8881}$][NTf$_2$]	39			
				[C$_4$C$_1$im][N(CN)$_2$]	43			
Reversible-deactivation radical polymerizations								
MMA	AIBN	RT	-	[C$_2$C$_1$im][C$_2$H$_5$SO$_4$]	92	DMF	98	Puttick et al. (2009)
				[C$_2$C$_1$im][BF$_4$]	99			
				[C$_4$C$_1$im][NTf$_2$]	76			
				[C$_4$C$_1$im][PF$_6$]	99			
St	TMPPA/TEMPO	115	68	[C$_4$C$_1$im][PF$_6$]	–	DCB	90	Zhang et al. (2004)
		125	240		41			
		–	–		42			

(continued)

Table 7.1 (continued)

Monomer	Initiator	T_{react} (°C)	Time (min)	IL	Yield (%)	Organic solvent	Yield (%)	References
MMA/St	SG1 nitroxide	90	450	[C$_4$C$_1$im][NTf$_2$]	86	–	–	Brusseau et al. (2011)
MA	AIBN	130	1200	[C$_4$C$_1$im][PF$_6$]	11	–	–	Ryan et al. (2004)
BMA		70	120	[C$_4$C$_1$pyr][NTf$_2$]	95	Toluene	71	Santha Kumar et al. (2018)
Cationic polymerizations								
St	AlCl$_3$[2]	0	120	[C$_2$C$_1$im][NTf$_2$]	88	DCM	95	Vijayaraghavan and MacFarlane (2004)
	HBOB	60	30	[P$_{66614}$][NTf$_2$]	96		15	
	HBOP				85	–	–	Vijayaraghavan and Macfarlane (2012)
					80			
	DMBCl[3]	– 15	–	[C$_4$C$_1$im][PF$_6$]	85			
	DMBCl[4]				99	DCM		Han et al. (2016)
	DMBCl[5]				89			

1 Achieved for 720 min of reaction. CPBDT was employed as a RAFT agent for both reaction media
2 Used as a reaction catalyst
3 Al$_2$(Et)$_3$Cl$_3$ was used as reaction co-initiator
4 BF$_3$OEt$_2$ was used as reaction co-initiator
5 TiCl$_4$ was used as reaction co-initiator

Despite being reported in most of the described studies, the possibility of easily isolating the polymerization products, as well as the recovery and reuse of the catalysts and the IL, has been seldom addressed. When appraised, the isolation of polymers from the IL is usually performed by the use of organic solvents, but the reuse of ILs is often not assessed Benton and Brazel (2004) and Ryan et al. (2004). Even so, the fundamental knowledge acquired hitherto on the use of ILs as solvent media in polymerization opens the possibility to apply ILs for the development of polymer-based drug delivery systems. In this field, efforts must be made to design more environmentally friendly processes and to properly select the most effective and advantageous reagents and techniques.

7.3 Applications of Ionic Liquids in the Development of Polymer-Based Drug Delivery Systems

Drug delivery systems offer the possibility to enhance pharmacokinetic profiles and therapeutic efficiency by improving drug permeation, offering controlled release, and allowing the drug to reach the target site. To this purpose, a wide variety of drug delivery systems has been explored, including solid lipid nanoparticles, nanostructured lipid carriers, micelles, and polymeric nanoparticles (Venkatraman et al. 2012), the last option being the one in the current chapter. These polymer-based nanoparticles (NPs) carrying the drug mainly comprehend particles with variable sizes from 1 to 100 nm in diameter (Venkatraman et al. 2012). The API molecules can be dissolved, adsorbed on the surface, or encapsulated in the NPs (Kumari et al. 2014). Several methods have been explored for the preparation of these systems, either based on assembly/self-assembly processes or by reducing large particles to the nanoscale. ILs have been explored in the development of nanoparticles for the delivery of poorly water-soluble APIs for different therapeutic purposes. In these studies, several applications of ILs in drug delivery can be highlighted, being possible to successfully apply ILs as solvents for the synthesis of novel polymer-based systems (Dong et al. 2008), as solvents for the solubilization of low water-soluble APIs (Júlio et al. 2020, 2019), and even as functional excipients compatible with combined therapies (Tang et al. 2016), as summarized in Fig. 7.4. The resultant properties of nanoparticles, such as particle size, size distribution, drug loading, and surface charge can be modulated by the reaction medium (Vauthier and Bouchemal 2009). In this sense, the study of polymerization reactions in IL media has been a hot topic of research.

Zhang and Zhu (2015) studied the co-polymerization of BMA, 2-hydroxyethyl methacrylate (HEMA), and St through chain extension from a trithiocarbonate-terminated PEG macro-RAFT agent, carried out in $[C_4C_1im][PF_6]$. The resulting block copolymers aggregated spontaneously in vesicular morphologies in IL media. Moreover, by regulating the formulation, nanoaggregates with multiple morphologies were obtained, showing a promising method for the development of drug delivery systems. In a different study, a water-soluble amphiphilic derivative of a

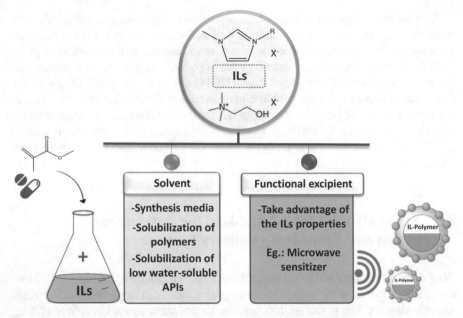

Fig. 7.4 IL applications in the development of polymer-based drug delivery systems

chitosan oligosaccharide modified with linoleic acid (LCOS) has been prepared in 1-butyl-3-methylimidazolium acetate ($[C_4C_1im][CH_3CO_2]$) with the goal of studying its self-assembly in an aqueous solution (Liu et al. 2015). The degree of substitution obtained in IL media was found to be higher than the one obtained in traditional solvents, such as DMSO. Moreover, the LCOS prepared in $[C_4C_1im][CH_3CO_2]$ revealed improved surface activity and could self-assemble into nanomicelles with better-defined spherical shape and a narrower particle size distribution (30–40 nm) in aqueous media. Although further studies on cytotoxicity are necessary to be conducted, the obtained micelles presented a high ability to incorporate poorly water-soluble drugs like ibuprofen (achieving 200 nm of size after the drug incorporation), being suitable for intravenous drug delivery (Liu et al. 2015).

Dong et al. (2008) developed a new copolymer comprising poly(L-lactic acid) (PLLA) and cellulose segments (cellulose-g-PLLA) for the delivery of prednisone acetate, an anti-inflammatory drug with low water solubility. The block copolymer was prepared using 1-butyl-3-methylimidazolium chloride ($[C_4C_1im]Cl$) and 1-allyl-3-methylimidazolium chloride ($[CH = C_2C_1im]Cl$), respectively, as polymer solvents. The resulting system was able to self-assemble into micelles in water, presenting the hydrophobic PLLA segments at the core of micelles and the hydrophilic cellulose segments in the outer shells. It was shown that these micelles exhibit nanospheric morphology within a size range of 30–80 nm. The delivery systems developed allowed the incorporation and controlled release of prednisone over 7 days, without the toxicity associated with 3T3 mouse fibroblast cells (Dong et al. 2008).

One of the most commonly applied co-polymers in the development of nanoparticles for drug delivery is poly(L-lactic-co-glycolic acid) (PLGA) (Danhier et al. 2012). PLGA is a biocompatible and biodegradable polymer, approved by the FDA and European Medicine Agency (EMA), that further presents tunable physicochemical properties appropriate for drug delivery (Bohrey et al. 2016; Zhang et al. 2006). Júlio et al. (2020, 2019) evaluated the potential of ILs mixtures with PLGA (in the ratio 50:50 and 75:25, lactic acid: glycoc acid) to develop IL-polymer nanoparticles for the delivery of poorly water-soluble APIs. The ILs cholinium L-phenylalaninate ([Ch][Phe]) and cholinium L-glutaminate ([Ch][Glu]) were used for the solubilization of rutin, a nutraceutical that displays antidiabetic, antihypertensive, and antilipidemic activities. The use of these ILs allowed not only incorporated higher amounts of the API (51–76%), without toxicity to HaCat cells but also allowed its sustained release. The nanoparticles obtained present a particle size between 250 and 300 nm, with good polydispersity and high colloidal stability, achieving 85% of rutin release after 72 h (Júlio et al. 2019).

In a different approach, polydopamine (PDA) particles loaded with doxorubicin and the IL [C_4C_1im][PF_6] were developed, aiming to provide alternative delivery systems for cancer treatment (Tang et al. 2016). The IL was employed as a microwave sensitizer, which allows combining chemotherapy by intravenous administration and microwave thermal therapy. The antitumor efficacy of doxorubicin-loaded IL-polydopamine nanoparticles was demonstrated in cancer cells in vitro and in vivo in mice, after intravenous injection via tail vein. Taking advantage of the IL's susceptibility to microwave radiation, the produced nanoparticles presented a high ability to induce the tumor's ablation without inducing significant toxicity to healthy tissues.

Overall, the size and shape of nanoparticles are dependent not only on the degree of substitution (Jiang et al. 2006; Kwon et al. 2003) but can also be adjusted by the manipulation of the IL chemical structure (Rajiv Gandhi et al. 2013; Kim et al. 2016). Despite the immense potential behind the combination of ILs and polymer-based drug delivery systems, different types of ILs and materials need to be deeply investigated toward the development of more efficient drug delivery systems. In this context, both the therapeutic efficacy in the target tissue and the biocompatibility of these systems are required to be assessed if the clinical application is envisaged.

7.3.1 Polymeric Ionic Liquids in the Design of Stimuli-Responsive Drug Delivery Systems

The complexity of biological systems and the environmental changes driven by pathological conditions challenge the development of stimuli-responsive systems (Alvarez-Lorenzo and Concheiro 2014). These systems require the selection of components that can act as "sensors" of the surrounding environment, presenting the ability to trigger the release of a certain drug. Ideally, such stimuli-responsive excipients, like "smart" polymers, allow the drug release to occur at the target tissues/organs, and proper rates at the action sites (Liu and Gu 2016). Smart polymers change physically or chemically in response to external stimuli, being separated into two classifications according to the nature of the stimuli (Karimi et al. 2016). Whereas polymers responsive to physical stimuli respond to factors such as temperature or light, those responsive to chemical stimuli respond by interactions (hydrophobic/hydrophilic balance) between the polymer and solvent or between polymer chains, resultant from changes of pH, ionic strength, among others (Torchilin 2014).

The growing interest in the development of alternative stimuli-responsive drug delivery systems has drawn attention to the study of polymeric forms of ILs (PILs) for this purpose. PILs can be prepared by the selection of polymerizable cations or anions, resulting in a polyelectrolyte (Ohno and Yoshizawa 2005). The functional materials obtained from the polymerization of IL monomers may present high stability, processability, and mechanical strength and offer a high flexibility of design (Appetecchi et al. 2010; Bansal et al. 2020). A variety of polymeric hydrogels and nanoparticles has been prepared from PILs, among which acryloyl- and vinyl-based moieties can be highlighted (structures provided in Fig. 7.5 with the schematic representation of the preparation of stimuli-responsive systems from IL monomers) (Green et al. 2009).

PILs can be copolymerized with other stimuli-sensitive monomers, such as MMA or N-isopropylacrylamide, allowing the development of new temperature and pH-responsive delivery systems with controlled drug release properties (Rasouli et al. 2014; Bielas et al. 2019; Mahkam and Pakravan 2013). Yu et al. (2013) polymerized

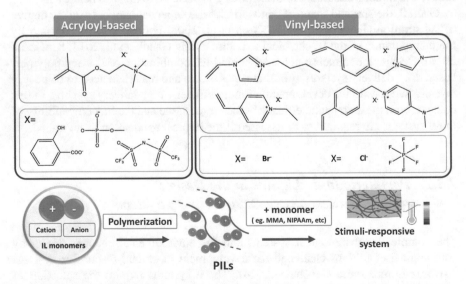

Fig. 7.5 Schematic illustration of the preparation of stimuli-responsive systems from acryloyl- and vinyl-based IL monomers and respective structures used in the preparation of copolymer-based delivery systems

a new IL monomer, 2-(methacryloyloxy)ethyl choline phosphate, by direct ATRP at room temperature with 100% functionalization. The obtained polymer presented high solubility in water and in vitro biocompatibility toward human red blood cells. The obtained polymer presents a promising application in tissue sealant formulations, and due to its ability to be rapidly taken into the cell's cytoplasm, appealing drug delivery options can be envisaged. More recently, copolymers based on the ILs [2-(methacryloyloxy)ethyl]trimethylammonium cation, with chloride, salicylate, or bis(trifluoromethanesulfonate)imide anions and MMA were synthesized by ATRP, and their characteristics were evaluated (Bielas et al. 2016). The IL monomer anion seems to present a variable influence on the thermal properties of the resulting polymers. Likewise, the polymer morphology can be adjusted by the introduction of a proper counterion. Such studies opened the possibility to correctly design PILs containing ammonium moieties to include pharmaceutical anions in their composition, allowing the development of novel polymer-based drug delivery systems such as a tissue sealant biomembrane.

Oral drug delivery is the most widely used and most readily accepted route of drug administration (Homayun et al. 2019). However, these formulations present some drawbacks to be tackled, such as the poor drug stability in the gastric environment, the low solubility and/or bioavailability of the API itself, and the drug absorption across biological membranes. To overcome such limitations, the study of nanoparticle formulations based on PILs and copolymers has been recently expanded. Promising systems in this field are summarized in Table 7.2. The synthesis of PILs from cholinium-based ILs containing polymerizable moieties through ATRP processes has been explored and applied to drug delivery, with promising ability to control their release profiles (Bielas et al. 2016).

The oral delivery of low-water soluble drugs such as naproxen, using pH-sensitive PIL-based nanocarriers, has been attempted (Hosseinzadeh et al. 2012; Mahkam et al. 2015). Two IL monomers, namely 1-(4-vinylbenzyl)-3-methylimidazolium hexafluorophosphate and 1-(4-vinylbenzyl)-4-(dimethylamino)pyridinium hexafluorophosphate, have been synthesized and copolymerized with methyl styrene (Hosseinzadeh et al. 2012). These copolymers were synthesized by free-radical polymerizations at 70 °C using AIBN as initiator and then loaded with naproxen. The amount of drug loading increased with increasing positive charge densities, resulting from the increasing ratio of the IL monomer. A controllable release of naproxen was obtained according to the pH of the medium, being able to target the intestine for drug release. The release rate decreased at pH 2–6 but increased at pH > 7 at the body's temperature. Thereby, the release of naproxen is dependent on the hydrolysis rate. By increasing ionic strength at pH > 6, the diffusion of the hydrolyzing agents on the polymer is increased and so the release of the API. In a different approach, cholinium-derivatives have been applied in the development of pH-responsive nanogels for the delivery of 5-fluorouracil, an anticancer drug (Mukesh et al. 2014). To this purpose, cholinium acrylate was polymerized N,N-methylene-bis-acrylamide as cross-linking agent. The obtained nanogels allowed the drug release at the stomach pH (pH = 1.2), over 10 days at human body temperature (37 °C). Despite the promising

Table 7.2 Stimuli-responsive drug delivery systems prepared from IL monomers and their characteristics and applications

IL monomers	Co-monomer	API	Stimuli	Drug delivery application	References
1-(4-vinylbenzyl)-3-methylimidazolium hexafluorophosphate	Methyl styrene	Naproxen	pH (>7)	Anti-inflammatory action; colon-specific delivery	Hosseinzadeh et al. (2012)
1-(4-vinylbenzyl)-4-(dimethylamino)-pyridinium hexafluorophosphate					
1-(4-vinylbenzyl)-4-(dimethylamino)-pyridinium chloride	Methyl methacrylate	Naproxen	pH (7.4)	Anti-inflammatory action; Colon-specific delivery	Mahkam et al. (2015)
1-(4-vinylbenzyl)-3-methylimidazolium chloride					
Cholinium acrylate	N,N-methylene-bis-acrylamide	5-fluorouracil	pH (1.2)	Anticancer activity; Stomach-specific delivery	Mukesh et al. (2014)
1-butyl-3-vinylimidazolium bromide	N-isopropylacrylamide	Model drug	Thermal (38–42°C)	Drug-releasing at a target cell	Seo et al. (2014)
1-ethylvinylpyridinium bromide	N-isopropylacrylamide	Doxorubicin	pH (5.2) Thermal (>37°C)	Anticancer activity; controlled delivery to pathological cells	Cui et al. (2012)

systems developed through these methods, further studies on the cytotoxicity, pharmacokinetics, and therapeutic efficacy are still necessary to envision their application in a clinical context. Such examples state the need for a correct design of these systems, relating the APIs' therapeutic effect with its local absorption.

Due to the changes in the environment of biological systems under pathological conditions, temperature- and/or pH stimuli-responsive polymers have been widely investigated (Cayre et al. 2011; Zhu et al. 2010). Poly(N-isopropylacrylamide) is an example of a well-known material with biomedical application owing to the reversible thermoresponsive phase transition (lower critical solution temperature, LCST, of 32°C) (Heskins and Guillet 1968). However, the low LCST and its low drug-carrying ability limit the extensive application of this polymer. To improve these characteristics and offer a more efficient drug delivery to a target site, N-isopropylacrylamide-based copolymers containing IL monomers have been prepared (Seo et al. 2014; Cui et al. 2012). For this purpose, Seo et al. (2014) reported the preparation of a copolymer with 1-butyl-3-vinylimidazolium bromide and N-isopropylacrylamide. Temperature-dependent micelles were obtained, with highly increased zeta potential values and optimal LCST (>38°C) in comparison to those obtained without the ionic moiety. The new delivery system allows the release of charged compounds at 38–42°C, without significant toxicity to human embryonic kidney cells (BSA protein was used as a model compound). Cui et al. (2012) reported the possibility to produce a polymeric system that combines both temperature- and pH-stimuli, for a more effective delivery of doxorubicin, an anticancer drug. Ionically assembled nanoparticles were prepared through the copolymerization of 1-ethylvinylpyridinium bromide and N-isopropylacrylamide. The obtained nanoparticles were able to release the API at lower pH values (pH = 5.2) and high temperatures (above 37 °C), accordingly with pathological conditions, as shown in Fig. 7.6. Moreover, drug-loaded particles exhibit biocompatibility and inhibitory effect on cell growth.

The combination of the advantages of ILs and responsive polymers represents a relevant progress in drug delivery. The works reported hitherto with these delivery systems present improved and more tunable performance than common polymeric systems. Nonetheless, to correctly design these materials, the interactions between IL monomers and copolymerization monomers must be better clarified. The number of APIs incorporated in PILs-based systems is still low, and to understand the possibility of modulating the respective release profiles distinct pharmacological classes need to be studied. Different IL monomers must also be explored and the final cytotoxicity of these systems must be carefully evaluated to produce competitive biocompatible delivery systems.

7.4   Conclusions and Future Perspectives

Given the works critically reviewed here, ILs seem to hold promise as polymers reaction media and as additives in the development of drug delivery systems, as well as as (co)monomers in applications of PILs as stimuli-responsive systems. Several

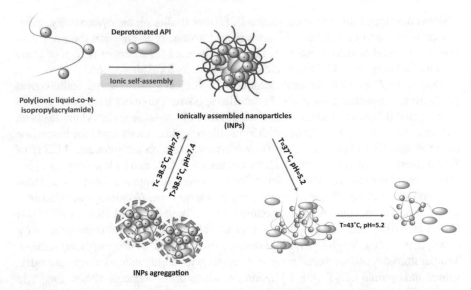

Fig. 7.6 Schematic illustration of the pH and thermal responsive behavior of the IL-based nanoparticles and respective release profile of doxorubicin (adapted from Cui et al. (2012))

works showed that ILs are promising alternatives to organic solvents for polymerization processes to occur. In this field, conventional and reversible-deactivation radical polymerizations processes have been applied for the production of several polymers. Polymerizations in ILs media showed enhanced yields and kinetic profiles. Efforts have been devoted to understand the mechanisms behind such differences, being radical protection and the ILs viscosity mainly highlighted. Nonetheless, the knowledge regarding the polymerization mechanisms in IL media is far from being complete. Furthermore, the influence of the IL structure and the selection of the correct polymerization agents and initiators should be further investigated. In this field, more sustainable processes need to be designed as well, encompassing the recovery of the polymer, as well as of the catalysts and IL media for further reuse. Rather than providing new examples of processes that are possible to be performed in IL media, research of the application of these systems to a specific purpose or attempting to unveil new insights in polymerization processes that enable their correct application seems to be a priority.

The drug delivery systems resulting from IL-polymer combinations have been expanded. ILs have enabled the design of nanosystems for the delivery of several APIs and aiming for different administration routes. The application of these solvents is compatible with different strategies in the development of (bio)polymer-based systems and has enabled to obtain controlled release. New IL-polymer combinations and understanding how to control the size and morphology of the nanoparticles will allow improving the design of these systems. Also, the physical stability of these systems and the stability of APIs in these nanoparticles need to be explored. However, problems of cytotoxicity exhibited by many nanoparticles hinder their application in the biomedical field. Therefore, the synthesis route and the design of IL-based nanoparticles loaded with drugs should contribute to minimizing the side-effects observed with conventional polymeric systems. In this vein, the in vivo study of these delivery systems and their therapeutic efficacy must be addressed if their clinical application is intended.

The recent application of PILs in the fabrication of multi-responsive systems highlights the versatility of the use of ILs in drug delivery and allows the possibility to obtain site-specific and target-oriented delivery of APIs. In this field, it is mandatory to gather systematic knowledge to be able to fully design the performance and stimuli-responsiveness of these delivery systems towards target medical applications. Much is yet to be understood in the interactions between ILs and polymers and IL monomers and multi-responsive systems.

Although there is a long path ahead, the possibilities investigated so far open a window of opportunity for novel, efficient, and designed drug delivery systems, where ILs seem to be a promising tool.

Acknowledgments This work was developed within the scope of the project CICECO-Aveiro Institute of Materials, UIDB/50011/2020 & UIDP/50011/2020, financed by national funds through the FCT/MEC and when appropriate co-financed by FEDER under the PT2020 Partnership Agreement. This work was prepared under the scope of the project POCI-01-0145-FEDER-031106 (IonCyt-Device) funded by FEDER, through COMPETE2020—Programa Operacional Competitividade e Internacionalização (POCI), and by national funds (OE), through FCT/MCTES. S. N. Pedro acknowledges the PhD grant SFRH/BD/132584/2017.

References

Ajekwene KK (2020) Properties and applications of acrylates. In: Serrano-Aroca Á, Deb S (eds) Acrylate polymers for advanced applications, Chapter 3. pp 35–56 https://doi.org/10.5772/intech open.89867

Almeida TS, Júlio A, Mota JP, Rijo P, Reis CP (2017) An emerging integration between ionic liquids and nanotechnology: general uses and future prospects in drug delivery. Ther Deliv 8:461–473

Alvarez-Lorenzo C, Concheiro A (2014) Smart drug delivery systems: from fundamentals to the clinic. Chem Commun 50:7743–7765

Amgoth C, Phan C, Banavoth M, Rompivalasa S, Tang G (2020) Polymer properties: functionalization and surface modified nanoparticles. In: Tyagi KR, Garg N, Shukla R, Singh Bisen P (eds) Role of novel drug delivery vehicles in nanobiomedicine. https://doi.org/10.5772/intech open.84424

Appetecchi GB, Kim G-T, Montanino M, Carewska M, Marcilla R, Mecerreyes D et al (2010) Ternary polymer electrolytes containing pyrrolidinium-based polymeric ionic liquids for lithium batteries. J Power Sources 195:3668–3675

Bansal M, Dravid A, Aqrawe Z, Montgomery J, Wu Z, Svirskis D (2020) Conducting polymer hydrogels for electrically responsive drug delivery. J Controlled Release 328:192–209

Benton MG, Brazel CS (2004) An investigation into the degree and rate of polymerization of poly(methyl methacrylate) in the ionic liquid 1-butyl-3-methylimidazolium hexafluorophosphate. Polym Int 53:1113–1117

Bheemidi VS, Tiruckovela M, Varanas P (2011) An imperative note on novel drug delivery systems. J Nanomedicine Nanotechnol 2(7):1000125. https://doi.org/10.4172/2157-7439.1000125

Bielas R, Mielańczyk A, Siewniak A, Neugebauer D (2016) Trimethylammonium-Based Poly-methacrylate Ionic Liquids with Tunable Hydrophilicity and Charge Distribution as Carriers of Salicylate Anions. ACS Sustain Chem Eng 4:4181–4191

Bielas R, Mielańczyk A, Skonieczna M, Mielańczyk Ł, Neugebauer D (2019) Choline supported poly(ionic liquid) graft copolymers as novel delivery systems of anionic pharmaceuticals for anti-inflammatory and anti-coagulant therapy. Sci Rep 9:14410

Bistolfi A, Ferracini R, Albanese C, Vernè E, Miola M (2019) PMMA-based bone cements and the problem of joint arthroplasty infections: status and new perspectives. Materials 12:4002

Bohrey S, Chourasiya V, Pandey A (2016) Polymeric nanoparticles containing diazepam: prepara-tion, optimization, characterization, in-vitro drug release and release kinetic study. Nano Converg 3:3

Brusseau S, Boyron O, Schikaneder C, Santini CC, Charleux B (2011) Nitroxide-Mediated Controlled/Living Radical Copolymerization of Methyl Methacrylate with a Low Amount of Styrene in Ionic Liquid. Macromolecules 44:215–220

Cayre OJ, Chagneux N, Biggs S (2011) Stimulus responsive core-shell nanoparticles: synthesis and applications of polymer based aqueous systems. Soft Matter 7:2211–2234

Chanfreau S, Mena M, Porras-Domínguez JR, Ramírez-Gilly M, Gimeno M, Roquero P et al (2010) Enzymatic synthesis of poly-l-lactide and poly-l-lactide-co-glycolide in an ionic liquid. Bioprocess Biosyst Eng 33:629–638

Chen F, Guo J, Xu D, Yan F (2016) Thermo- and pH-responsive poly(ionic liquid) membranes. Polym Chem 7:1330–1336

Chenthamara D, Subramaniam S, Ramakrishnan SG, Krishnaswamy S, Essa MM, Lin F-H et al (2019) Therapeutic efficacy of nanoparticles and routes of administration. Biomater Res 23:20

Cui W, Lu X, Cui K, Niu L, Wei Y, Lu Q (2012) Dual-Responsive Controlled Drug Delivery Based on Ionically Assembled Nanoparticles. Langmuir 28:9413–9420

Dali S, Lefebvre H, Gharbi RE, Fradet A (2006) Synthesis of poly(glycolic acid) in ionic liquids. J Polym Sci Part Polym Chem 44:3025–3035

Danhier F, Ansorena E, Silva JM, Coco R, Le Breton A, Préat V (2012) PLGA-based nanoparticles: An overview of biomedical applications. J Controlled Release 161:505–522

Dong H, Xu Q, Li Y, Mo S, Cai S, Liu L (2008) The synthesis of biodegradable graft copolymer cellulose-graft-poly(l-lactide) and the study of its controlled drug release. Colloids Surf B Biointerfaces 66:26–33

Dugger SA, Platt A, Goldstein DB (2018) Drug development in the era of precision medicine. Nat Rev Drug Discov 17:183–196

Elgindy N, Elkhodairy K, Molokhia A, ElZoghby A (2011) Biopolymeric nanoparticles for oral protein delivery: design and in vitro evaluation. J Nanomedicine Nanotechnol 2:3

Englert C, Brendel JC, Majdanski TC, Yildirim T, Schubert S, Gottschaldt M et al (2018) Pharmapolymers in the 21st century: synthetic polymers in drug delivery applications. Prog Polym Sci 87:107–164

Freire MG, Cláudio AFM, Araújo JMM, Coutinho JAP, Marrucho IM, Lopes JNC et al (2012) Aqueous biphasic systems: a boost brought about by using ionic liquids. Chem Soc Rev 41:4966–4995

Green O, Grubjesic S, Lee S, Firestone MA (2009) The Design of Polymeric Ionic Liquids for the Preparation of Functional Materials. Polym Rev 49:339–360

Han L, Wu Y, Dan Y, Wang H, Zhang X, Wei X et al (2016) Characteristics and mechanism of styrene cationic polymerization in 1-butyl-3-methylimidazolium hexafluorophosphate ionic liquid. RSC Adv 6:105322–105330

He H, Zhong M, Luebke D, Nulwala H, Matyjaszewski K (2014) Atom transfer radical polymer-ization of ionic liquid monomer: The influence of salt/counterion on polymerization. J Polym Sci Part Polym Chem 52:2175–2184

Heskins M, Guillet JE (1968) Solution Properties of Poly(N-isopropylacrylamide). J Macromol Sci Part - Chem 2:1441–1455

Homayun B, Lin X, Choi H-J (2019) Challenges and Recent Progress in Oral Drug Delivery Systems for Biopharmaceuticals. Pharmaceutics 11:129

Hong K, Zhang H, Mays JW, Visser AE, Brazel CS, Holbrey JD, et al (2002) Conventional free radical polymerization in room temperature ionic liquids: a green approach to commodity polymers with practical advantages. Chem Commun 1368–1369

Hosseinzadeh F, Mahkam M, Galehassadi M (2012) Synthesis and characterization of ionic liquid functionalized polymers for drug delivery of an anti-inflammatory drug. Des Monomers Polym 15:379–388

Hrubý M, Filippov SK, Štěpánek P (2015) Smart polymers in drug delivery systems on crossroads: which way deserves following? Eur Polym J 65:82–97

Jiang G-B, Quan D, Liao K, Wang H (2006) Novel polymer micelles prepared from chitosan grafted hydrophobic palmitoyl groups for drug delivery. Mol Pharm 3:152–160

Júlio A, Caparica R, Lima SAC, Fernandes AS, Rosado C, Prazeres DMF et al (2019) Ionic liquid-polymer nanoparticle hybrid systems as new tools to deliver poorly soluble drugs. Nanomaterials 9:1148

Júlio A, Lima SAC, Reis S, Santos de Almeida T, Fonte P (2020) Development of ionic liquid-polymer nanoparticle hybrid systems for delivery of poorly soluble drugs. J Drug Deliv Sci Technol 56:100915

Karimi M, Ghasemi A, Sahandi Zangabad P, Rahighi R, Moosavi Basri SM, Mirshekari H et al (2016) Smart micro/nanoparticles in stimulus-responsive drug/gene delivery systems. Chem Soc Rev 45:1457–1501

Kim SY, Hwang J-Y, Shin US (2016) Preparation of nano/macroporous polycaprolactone microspheres for an injectable cell delivery system using room temperature ionic liquid and camphene. J Colloid Interface Sci 465:18–25

Kumari A, Singla R, Guliani A, Yadav SK (2014) Nanoencapsulation for drug delivery. EXCLI J 13:265–286

Kwon S, Park JH, Chung H, Kwon IC, Jeong SY, Kim I-S (2003) Physicochemical characteristics of self-assembled nanoparticles based on glycol chitosan bearing 5β-cholanic acid. Langmuir 19:10188–10193

Laffleur F, Keckeis V (2020) Advances in drug delivery systems: work in progress still needed? Int J Pharm X 2:100050

Lee PI, Li J-X (2010) Evolution of oral controlled release dosage forms. In: Wen H, Park K (eds) Oral controlled release formulation design and drug delivery. John Wiley & Sons, Inc., pp 21–31

Li D, Zhang Y, Wang H, Tang J, Wang B (2006) Effect of the medium on the stereostructure of poly(methyl methacrylate) synthesized in ionic liquids. J Appl Polym Sci 102:2199–2202

Liechty WB, Kryscio DR, Slaughter BV, Peppas NA (2010) Polymers for drug delivery systems. Annu Rev Chem Biomol Eng 1:149–173

Liu D, Gu N (2016) The smart drug delivery system and its clinical potential. Theranostics 6:1306–1323

Liu Y, Huang Y, Boamah P-O, Cao L, Zhang Q, Lu Z et al (2015) Homogeneous synthesis of linoleic acid-grafted chitosan oligosaccharide in ionic liquid and its self-assembly performance in aqueous solution. J Appl Polym Sci 132:41727

Low K, Wylie L, Scarborough DLA, Izgorodina EI (2016) Is it possible to control kinetic rates of radical polymerisation in ionic liquids? Chem Commun 54:11226–11243

Lu J, Yan F, Texter J (2009) Advanced applications of ionic liquids in polymer science. Prog Polym Sci 34:431–448

Mahkam M, Latifpour A, Rafi AA, Gheshlaghi LM, Takfallah A (2015) Preparation of montmorillonite-pH-sensitive positive charges nanocomposites as a drug delivery system. Int J Polym Mater Polym Biomater 64:32–37

Mahkam M, Pakravan A (2013) Synthesis and characterization of pH-sensitive positive-charge silica nanoparticles for oral anionic drug delivery. J Chin Chem Soc 60:293–296

Mena M, Chanfreau S, Gimeno M, Bárzana E (2010) Enzymatic synthesis of poly-l-lactide-co-glycolide in the ionic liquid 1-butyl-3-methylimidazolium hexafluorophosphate. Bioprocess Biosyst Eng 33:1095–1101

Mukesh C, Bhatt J, Prasad K (2014) A polymerizable bioionic liquid based nanogel: a new nanocarrier for an anticancer drug. Macromol Chem Phys 215:1498–1504

Odian GG (2004) Principles of polymerization, 4th edn. Wiley-Interscience, Hoboken, N.J

Ohno H, Yoshizawa M (2005) Preparation and properties of polymerized ionic liquids as film electrolytes. In: Rogers RD, Seddon KR (eds) Ionic liquids IIIB fundamentals progress challenges and opportunities, vol 902. ACS Symposium Series, pp 159–170. https://doi.org/10.1021/bk-2005-0902.ch013

Oleck J, Kassam S, Goldman JD (2016) Commentary: why was inhaled insulin a failure in the market? Diabetes Spectr 29:180–184

Park K (2013) Facing the truth about nanotechnology in drug delivery. ACS Nano 7:7442–7443

Patton JS, Bukar JG, Eldon MA (2004) Clinical pharmacokinetics and pharmacodynamics of inhaled insulin. Clin Pharmacokinet 43:781–801

Puttick S, Irvine DJ, Licence P, Thurecht KJ (2009) RAFT-functional ionic liquids: towards understanding controlled free radical polymerisation in ionic liquids. J Mater Chem 19:2679

Rajiv Gandhi R, Gowri S, Suresh J, Sundrarajan M (2013) Ionic liquids assisted synthesis of ZnO nanostructures: controlled size, morphology and antibacterial properties. J Mater Sci Technol 29:533–538

Rasouli S, Davaran S, Rasouli F, Mahkam M, Salehi R (2014) Positively charged functionalized silica nanoparticles as nontoxic carriers for triggered anticancer drug release. Des Monomers Polym 17:227–237

Ryan J, Aldabbagh F, Zetterlund PB, Yamada B (2004) First nitroxide-mediated controlled/living free radical polymerization in an ionic liquid. Macromol Rapid Commun 25:930–934

Santha Kumar ARS, Roy M, Singha NK (2018) Effect of ionic liquids on the RAFT polymerization of butyl methacrylate. Eur Polym J 107:294–302

Schmidt-Naaake G, Schmalfuß A, Woecht I (2008) Free radical polymerization in ionic liquids— influence of the IL-concentration and temperature. Chem Eng Res Des 86:765–774

Seo J-W, Hwang J-Y, Shin US (2014) Ionic liquid-doped and p-NIPAAm-based copolymer (p-NIBIm): extraordinary drug-entrapping and -releasing behaviors at 38–42 °C. RSC Adv 4:26738–26747

Shrivastava A (2018) Polymerization. In: Introduction to plastics engineering. Elsevier, pp 17–48

Strehmel V, Laschewsky A, Wetzel H, Görnitz E (2006) Free radical polymerization of n -butyl methacrylate in ionic liquids. Macromolecules 39:923–930

Strehmel V, Wetzel H, Laschewsky A (2014) Ionic liquids as advantageous reaction media for free radical polymerization. Macromol Symp 342:78–85

Tang W, Liu B, Wang S, Liu T, Fu C, Ren X et al (2016) Doxorubicin-loaded ionic liquid–polydopamine nanoparticles for combined chemotherapy and microwave thermal therapy of cancer. RSC Adv 6:32434–32440

Te Kulve H (2014) Anticipating market introduction of nanotechnology-enabled drug delivery systems. In: Sezer AD (ed) Application of nanotechnology in drug delivery. InTech. https://doi.org/10.5772/57180

Tiwari G, Tiwari R, Bannerjee S, Bhati L, Pandey S, Pandey P et al (2012) Drug delivery systems: an updated review. Int J Pharm Investig 2:2–11

Torchilin VP (2014) Multifunctional, stimuli-sensitive nanoparticulate systems for drug delivery. Nat Rev Drug Discov 13:813–827

Vauthier C, Bouchemal K (2009) Methods for the preparation and manufacture of polymeric nanoparticles. Pharm Res 26:1025–1058

Venkatraman G, Ramya, Shruthilaya, Akila, Ganga, Suresh Kumar, et al (2012) Nanomedicine: towards development of patient-friendly drug-delivery systems for oncological applications. Int J Nanomedicine 1043–1060

Vijayaraghavan R, Macfarlane DR (2012) Novel acid initiators for the rapid cationic polymerization of styrene in room temperature ionic liquids. Sci China Chem 55:1671–1676

Vijayaraghavan R, MacFarlane DR (2004) Living cationic polymerisation of styrene in an ionic liquid. Chem Commun 700–701

Wibowo FR, Saputra OA, Lestari WW, Koketsu M, Mukti RR, Martien R (2020) pH-triggered drug release controlled by poly(styrene sulfonate) growth hollow mesoporous silica nanoparticles. ACS Omega 5:4261–4269

Yoshizawa-Fujita M, Saito C, Takeoka Y, Rikukawa M (2008) Lipase-catalyzed polymerization of L-lactide in ionic liquids. Polym Adv Technol 19:1396–1400

Yu X, Yang X, Horte S, Kizhakkedathu JN, Brooks DE (2013) ATRP synthesis of poly(2-(methacryloyloxy)ethyl choline phosphate): a multivalent universal biomembrane adhesive. Chem Commun 49:6831–6833

Yun Y, Lee BK, Park K (2014) Controlled drug delivery systems: the next 30 years. Front Chem Sci Eng 8:276–279

Yun YH, Lee BK, Park K (2015) Controlled drug delivery: historical perspective for the next generation. J Control Release 219:2–7

Zhang H, Cui W, Bei J, Wang S (2006) Preparation of poly(lactide-co-glycolide-co-caprolactone) nanoparticles and their degradation behaviour in aqueous solution. Polym Degrad Stab 91:1929–1936

Zhang H, Hong K, Mays J (2004) First report of nitroxide mediated polymerization in an ionic liquid. Polym Bull 52:9–16

Zhang Q, Zhu S (2015) Ionic Liquids: Ionic liquids: versatile media for preparation of vesicles from polymerization-induced self-assembly. ACS Macro Lett 4:755–758

Zhao H (2006) Innovative applications of ionic liquids as "green" engineering liquids. Chem Eng Commun 193:1660–1677

Zhu X, Liu Y, Huang J, Li G (2010) A pH-Responsive Gate Fabricated with Nanochannels and Nanoparticles. Chem Eur J 16:1441–1444

Chapter 8
Design and Selection of Ionic Liquids Via COSMO for Pharmaceuticals and Medicine

Huma Warsi Khan, Amal A. M. Elgharbawy, Azmi Bustam, and Muhammad Moniruzzaman

Abstract Ionic liquids (ILs) have been used significantly in pharmaceutics and medicine because of their exceptional "green" properties and tailor-made physico-chemical and biological properties. ILs can dissolve many important drug molecules at ambient conditions, which are sparingly soluble in water and other conventional organic solvents. However, the selection of potential ILs from thousands of anion–cation combinations is quite challenging. To address this limitation, various computation methods have been used to screen appropriate ILs for solubility and applications of a specific drug molecule. Among those methods, the conductor-like screening model (COSMO)—a continuum solvation model based on quantum chemistry—is found remarkably effective in screening suitable ILs for pharmaceutical applications due to its excellent capability in predicting physical and chemical properties. The advantage of these optimizing tools is that only the molecular structure is enough for prior predictions. Solubility, activity coefficient, selectivity, free energies, partition coefficient, and octanol–water coefficient are few properties that can be easily predicted. This chapter presents a detailed discussion on the COSMO screening of cations and anions of ILs for various applications and highlighting the advantages of this predictive tool.

Keywords Drug molecules · Ionic liquids · COSMO-RS · COSMO-SAC

H. W. Khan · A. Bustam · M. Moniruzzaman (✉)
Department of Chemical Engineering, Universiti Teknologi PETRONAS, 32610 Perak, Malaysia
e-mail: m.moniruzzaman@utp.edu.my

A. A. M. Elgharbawy
International Institute for Halal Research and Training (INHART), International Islamic University, Kuala Lampur, Malaysia

A. Bustam · M. Moniruzzaman
Centre of Research in Ionic Liquids, Universiti Teknologi PETRONAS, Seri Iskandar, Perak, Malaysia

© The Author(s), under exclusive license to Springer Nature Singapore Pte Ltd. 2021
M. Goto and M. Moniruzzaman (eds.), *Application of Ionic Liquids in Drug Delivery*,
https://doi.org/10.1007/978-981-16-4365-1_8

137

Abbreviations

AAD	Average absolute deviation
AC^{id}	Activity coefficient at infinite dilution
ACV	Acyclovir
AMT	Amitriptyline
API	Active pharmaceutical ingredients
BP	Becke and Perdew
COSMO-RS	Conductor-like screening model for real solvents
COSMO-SAC	Conductor-like screening model-segment activity coefficient
CYANEX	Trialkylphosphine
CYPHOS 105	Tetradecyltrihexylphosphonium dicynamide
CYPHOS 109	Tetradecyltrihexylphosphonium bistriflamide
DCP	Dichlorophenol
DDT	Dichlorodiphenyl trichloraethane
DES	Deep eutectic solvents
DFT	Density functional theory
DHA	Docosanohexaenoic acid
DLLME	Dispersive liquid–liquid microextraction
DMSO	Dimethyl sulfoxide
DTZ	Diltiazem
EPA	Eicosapentaenoic acid
ETO	Etodolac
GMV	Gamavuton-0
GSEq	General solubility equations (GSEq)
HBA	Hydrogen bonding ability
HBC	Hydrogen bonding capacity
HSP	Hansen solubility parameter
ILs	Ionic liquids
LA	Lactic acid
LLE	Liquid–liquid equilibrium
MSD	Mean square deviation
NRTL	Non-random two-liquid model
RMSD	Root mean square deviation
RMSE	Root mean square error
SLE	Solid–liquid equilibrium
TCP	Trichlorophenol
UNIFAC	Universal Quasi-Chemical Functional Group Model
UNIQUAC	Universal Quasi-Chemical Model
VLE	Vapor–liquid equilibrium

Cations

[BDMAAm]	Benzyldimethylalkylammonium
[BGu]	Benzyl-1-butylguanidinium
[BMIm]	1-Butyl-3-methyl imidazolium
[BMPip]	1-Butyl 3-methylpiperidinium
[BMPy]	1-Butyl-3-methylpyridinium
[DMIm]	1-Decyl-3-methylimidazolium
[DMPh]	Dimethylphosphate
[EMIm]	1-Ethyl-3-methylimidazolium
[EMPy]	1-Ethyl-3-methylpyridinium
[HMIm]	1-Hexyl-3 methyl imidazolium
[HTDPh]	Hexyltridecylphosphonium
[MAm]	Methylammonium
[OMIm]	1-Octyl-3-methylimidazolium
[TBPh]	Tetrabutylphosphonium
[TOAm]	Trioctylammonium

Anions

[AlC$_{l4}$]	Aluminium tetrachloride
[Br]	Bromide
[Cin]	Cinnamate
[Cl]	Chloride
[ClO$_4$]	Perchlorate
[Ffm]	Flufenamicate
[Lac]	Lactate
[Ntf$_2$]	Bis(trifluoromethylsulfonyl)imide
[Oct]	Octanoate
[Otf]	Trifluoromethanesulfonate
[PF$_6$]	Hexafluorophosphate
[Sa]	Salicylate
[TBP]	Tributylphosphate
[TOA]	Trioctylamine
[Tos]	Tosylate

8.1 Introduction

Pharmaceutical companies experience many challenges in the development of drug formulation and delivery of many newly invented compounds due to their poor solubility, bioavailability, stability, and polymorphic conversion (Adawiyah et al. 2020). These drawbacks are further exacerbated when drug molecules are insoluble or sparingly soluble in water and most pharmaceutically accepted organic solvents. It is reported that many newly invented drugs never reach the formulation process due to their solubility in conventional solvents (Stefanuto et al. 2020). Generally, conventional solvents/agents are used as solubility enhancers, including acetone, pyridine, and dimethyl sulfoxide. However, the recovery and reuse of such organic solvents are tough and cumbersome, leading to environmental hazards. In addition, the use of a huge amount of such organic solvents to formulate drugs may result in detrimental effects on human health and ecosystems.

Ionic liquids (ILs)—referred to as "architect solvents"—have been emerged as a solvent or material, or both, in many applications, including pharmaceutics and medicine due to their exceptional properties consisting of the combination of "green" properties with tunable physicochemical and biological properties. In fact, low vapor pressure, high melting point, thermal stability, and inflammability are the excellent characteristics that make them ideal for various applications in almost every sector. They are also known as environmentally friendly or "green solvents" because of their lower environmental impact and low toxicity (Claus et al. 2018). Since there is a pool of million ILs, the experimental detection of ILs with excellent dissolving properties is extremely challenging. The quantity of the drug is also limited in the early formulation phases, so it is not recommended to experimentally assess the solubility. Solubility prediction for pharmaceuticals is necessary for designing, separation, and purification (Wichmann et al. 2019). As a result, in the early formulation development, computational methods are the most effective way to find suitable ILs for any specific drug. The use of predictive simulation methods, such as the conductor-like screening model for real solvents (COSMO-RS) and COSMO-SAC, is an appropriate approach to screen the ILs. This simulation tool predicts solubility, activity coefficient at infinite dilution (AC^{id}), capacity, selectivity, solvation energies, and bonding energies of the associated drug molecules under study (Loschen et al. 2015).

This chapter aims to highlight the advances of using COSMO predictive tools to screen the suitable ILs for pharmaceuticals and medicine. The screened ILs can be used to enhance the solubility of drug molecules, in delivery, formulation, and recovery of drug molecules. COSMO makes these predictions assessing the chemical potential and calculating thermodynamic property's, reaction energies, enthalpies, etc. In addition, this chapter addresses COSMO-RS, COSMO-SAC, COSMO-EOS studies used to screen ILs, DES, and API-ILS to improve the solubility of different drug molecules.

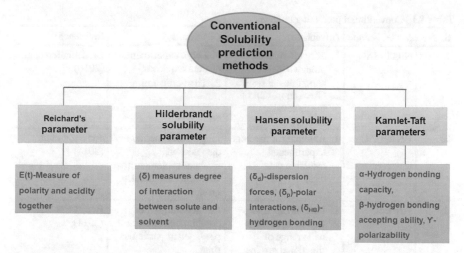

Fig. 8.1 Conventional solubility prediction methods

8.2 Solubility-Predicting Parameters and Tools

To screen the suitable solvents, the predictive methods used were based upon prag-
matic parameters. These parameters include the ionizing power of solvent, solvent
polarity, solvation behavior, and free enthalpy of reactions (Reichardt 1965). These
techniques are restricted to non-polar and slightly polar solvents, whereas they are
not efficient for polar solvents. The Hansen solubility parameter was designed to
address these disadvantages (Taft et al. 1976). Figure 8.1 shows the conventional
solubility prediction methods along with the parameters of the study. In addition,
new predictive models such as the non-random two-liquid model (NRTL), universal
quasi-chemical model (UNIQUAC), and universal quasi-chemical functional group
model (UNIFAC) were used for the selection of various conventional solvents (Klamt
2008). Table 8.1 shows the models used for prior predictions and their drawbacks.
These models were efficient for the prediction of experimental data but were ineffi-
cient in computing the solute–solvent interactions (Klamt 1995). Hence, there was
a need for advanced computational methods with better predicting ability. One such
predictive tool is conductor-like modeling (COSMO).

8.3 COSMO: An Efficient Predicting Tool

Conductor-like modeling (COSMO) is completely based on quantum chemical calcu-
lations. The various models include conductors like modeling for real solvents
(COSMO-RS) and conductor-like modeling segment activity coefficient (COSMO-
SAC). Additionally, the equation of state (EOS) was incorporated into COSMO-RS

Table 8.1 Conventional predicting models

S. No	Model Name	Advantages	Disadvantage	Reference
1.	NRTL-SAC	1. Very efficient to predict the experimental results 2. Gives an insight of the molecular interactions	1. Large experimental data required 2. Time-consuming	Diedenhofen et al. (2010)
2.	UNIQUAC	1. Efficient for the large experimental available data	Large experimental data needed Evaluates interaction based on functional groups	Alevizou et al. (2014)
3.	UNIFAC	1. Fast and accurate 2. Can be used for a wide range of functional groups	Limited number of combinations can be predicted at particular time. Unable to predict the hydrogen bonding ability For lesser values of solubilities (<0.5), it is inefficient	Bouillot et al. (2011)

and COSMO-SAC. Figure 8.2 shows the COSMO and its applications in various sectors employed for the selection of solvents. Box 8.1 shows the advantages of COSMO.

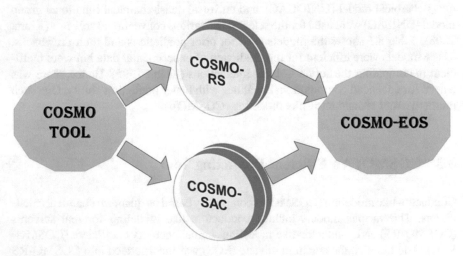

Fig. 8.2 COSMO tools

Box 8.1 Advantages of COSMO over conventional predicting tools

The use of COSMO is beneficial as it has following advantages as compared to conventional methods:

1. The only key in information needed is the molecular structure for the drug molecule and the solvents under the study.
2. Minimum adjustable parameters required for computation.
3. Gives the real insight of the interaction between solute and solvent.
4. Able to provide the intermolecular interaction associated with the bonding nature particularly hydrogen bonding.
5. Very helpful in the synthesis of new ILs for specific application.
6. Can evaluate various parameter for thousands of ionic combinations in a day.
7. Enormous cation-anion combinations can be estimated using only the structural ion files in COSMO

8.3.1 COSMO-RS

COSMO-RS is an efficient method widely used for finding the solubility and other properties of molecules. It can also be termed as "property explorer" as it is advantageous in predicting the properties of innumerous solvents and solutes. The application extends to solvent screening, liquid–liquid equilibrium (LLE), vapor–liquid equilibrium (VLE), solid–liquid equilibrium (SLE) studies, thermodynamic property calculation for drugs, organic solvents, ionic liquids, hydrocarbons, and deep eutectic solvents (DES) (Khan et al. 2020) (Hilmy et al. 2020). Additionally, COSMO provides significant results for evaluating the enthalpies, bonding energies, heat of reaction, and heat of fusion, which help in designing a hazard preventive process design (Wichmann et al. 2019). It has been used in chemical, petrochemical, agrochemical, pharmaceutical, and biochemical industries. Figure 8.3 shows the various properties that can be evaluated using COSMO-RS. COSMO-RS has been practiced in the pharmaceutical industries to predict the solubility of drug molecules, to screen solvents for drug development and processing, prediction of free energies for the reaction occurring in solutions, as well to screen cocrystals. In the design and production of drugs, solubility prediction is important because compounds in the early design process are often virtually taken into account by computational design methods (Wichmann et al. 2019).

In COSMO-RS calculation, the solute (drug molecule) is placed in the cavity formed in molecular form and the surrounding solvent is considered as a continuum. COSMO uses scaled conductor boundary conditions instead of using the dielectric boundary conditions and takes place in the ideal conductor. In the conductor, which is the reference state for COSMO-RS calculation, the solute electron density and geometry are thus converged to its energetically ideal state. The resulting surface screening charge densities "σ", energy is stocked up in the COSMO files (Diedenhofen et al. 2010). Sigma surface is represented in the form of colors red, blue, and green. These colors signify the negative, positive, and neutral sites on the drug

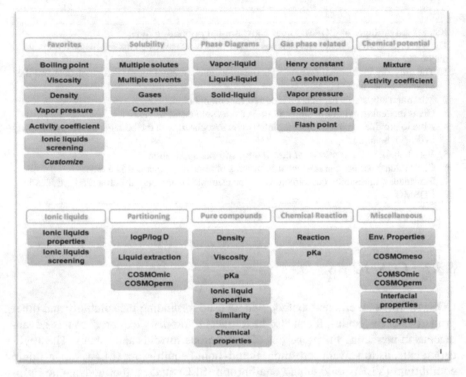

Fig. 8.3 COSMO-RS property prediction desktop view version COSMOtherm 18.0

molecule (Khan et al. 2020). The solvent is viewed as a mixture of interactive surface segments COSMO. In terms of screening charge densities σ and σ' in the respective surface segments, the interaction energies of surface pairs are defined.

These results are used for further COSMO predictions σ (σ $_{acceptor}$) and σ' (σ$_{donor}$), if the sections belong to H-bond donor or acceptor atom. The σ profile and potentials depict the H-bonding nature of the molecule under study (Diedenhofen et al. 2010). In COSMO-RS, the IL can be defined as the mixture of cation and anion, and calculations are made distinctly.

A study was conducted to evaluate the solubility of 31 pharmaceuticals considering sigma profiles. The sigma profiles gave an idea about the nature of bonding and this helped select appropriate solvents for the molecules under consideration. The results were further used in the determination of QSPR (Niederquell et al. 2018).

Hydrogen bond energy E_{HB}, misfit E_{MF}, and van der Waal energy E_{vdW} are the interaction energies that tell us about the bonding nature of molecules (Diedenhofen et al. 2010). These can be represented as:

$$E_{MF=\alpha_{eff}} \frac{\alpha'}{2}(\sigma + \sigma')^2 \tag{8.1}$$

$$E_{HB=\alpha_{eff}C_{HB}}\min(0;\ \min(0;\ \sigma_{donor} + \sigma_{HB)max}(0;\ \sigma_{acceptor} - \sigma_{HB}) \qquad (8.2)$$

$$E_{vdW} = \alpha_{eff}(\tau_{vdW} + \tau'_{vdW}) \qquad (8.3)$$

α' = an interaction parameter.

α_{eff} = the effective contact area.

C_{HB} = strength of hydrogen bond.

σ_{HB} = hydrogen bonding cutoff.

τ_{vdW} = interaction parameter for a specific element.

Using these interaction parameters, the chemical potential is calculated which is further used in the estimation of AC^{id}. The chemical potential and AC^{id} is evaluated using the following equation:

$$\mu_S^X = \mu_{C,S}^X + \int P^X(\sigma)\mu_{S(\sigma)}d\sigma \qquad (8.4)$$

Thermodynamic properties can be derived using Eq. 8.5. The activity coefficient of component "i" is strongly linked with solute's affinity and can be obtained by definition from the values of chemical potential in the mixture and pure state (Paduszyński 2017):

$$\gamma = exp\left(\frac{\mu - \mu^0}{RT}\right) \qquad (8.5)$$

μ = chemical potential of the solvent.

μ^0 = chemical potential in a compound.

The capacity is defined as the amount of IL needed for the removal of solute from solution during extraction (Rashid et al. 2018).

$$C_{12}^\infty = \left(\frac{1}{\gamma_1^\infty}\right)^{IL\ phase} \qquad (8.6)$$

The selectivity is defined as the ratio of the amount of solute IL-rich phase to the amount in the aqueous phase (IL lean phase) (Paduszynski 2017)

$$S_{12}^\infty = \frac{\gamma_2^\infty}{\gamma_1^\infty} \qquad (8.7)$$

Subscripts 1 and 2 refer to water and the solute (target solute), respectively.

The higher the capacity, the lesser is the amount of IL required. Likewise, selectivity reflects the extraction ability of the IL understudy for the recovery of the drug molecule. For the IL selection capacity and selectivity, values must be high whereas

AC^{id} must be as low. Another important property that is sufficient to determine the potency of IL is the performance index (PI). It is the product of capacity and selectivity and can be represented by equation

$$PI = C_{12}^{\infty} * S_{12}^{\infty} = \frac{\gamma_2^{\infty}}{\gamma_1^{\infty 2}} \qquad (8.8)$$

AC^{id}, capacity, selectivity, and PI are enough to determine the potency of IL for its selection as a carrier (Khan et al. 2020).

Using COSMO-RS, the bonding energies and thus excess enthalpies can be predicted. In the evaluation of the energetic contributions of all possible unique interactions formed by each species and their contribution to the total excess enthalpy, excess enthalpy is important (Gonfa et al. 2018). Excess enthalpy is advantageous to estimate the strength of solvent–drug molecule interaction in the mixtures. Furthermore, it also gives an idea about the cocrystal formation (Loschen et al. 2016). In COSMO-RS it is represented as follows:

$$H_m^E = H_{m,MF}^E + H_{m,HB}^E + H_{m,vdW}^E \qquad (8.9)$$

where
H_m^E=mixture excess enthalpy.
$H_{m,MF}^E$=misfit energy.
$H_{m,HB}^E$=hydrogen bonding energy.
$H_{m,vdW}^E$=van der Walls energy.

Table 8.2 shows some of the studies that used COSMO-RS for selecting solvents for pharmaceuticals.

Quantum calculations are to be performed on the basis of density functional theory (DFT) using resolution identity approximation. The energy calculations have to be accomplished using Becke and Perdew (BP) functional and triple-zeta valence polarized basis. For each molecule solvent (ILs) or solute (drug molecule), the quantum calculations are to be performed separately (Diedenhofen et al. 2010). The quantum calculations initially can be performed using various packages such as TURBO-MOLE (TMOLE-X), DMOL3, GAMESS, etc. Once these calculations are made the results are stored as COSMO files. Further estimations are made using COSMOtherm software. Figure 8.4 shows the step-by-step calculation procedure using the COSMO tool.

8.3.1.1 COSMO-RS in ILs Screening

ILs are composed of cation and anion. Figure 8.5 denotes the IL representation on COSMO-RS. Initially, the IL is treated as a compound consisting of σ-profiles, area,

Table 8.2 Application of COSMO-RS in pharmaceuticals

S. No.	Drug name	Solvent/ILs	Finding	Reference
1.	Quercetin Curcumin Gallilic acid	DES (combination of cholinium chloride and xylitol)	Positive correlations (+5) between predicted COSMO results and Pearson correlation analysis were observed	Oliveira et al. (2020)
2.	Alverine	[Alv][Tos] [Alv][Sa] [Alv][cin] [Alv][Oct] [Alv][amp] [Alv][ffm]	The predicted and experimental results were quite close with an error of ± 0.27	Stefanuto et al. (2020)
3.	5-Hydroxymethylfurfural	Conventional solvents	RMSE was found to be 2.85% between COSMO and experimental values	Esteban et al. (2020)
4.	Gamavuton-0	Conventional solvents	Both experimental and COSMO results showed that GMV possesses two pKa values	Kurnia et al. (2019)
5.	Ibuprofen Cinnarizine Naproxen paracetamol	API-DES	RMSE of 1.042 was observed between experimental and COSMO-RS solubility	Palmelund et al. (2019)
6.	Cyclosporin Loratadine Simvasatin Zafirlukast	Conventional solvents	R2 value for COSMO and observed results was found to be 0.855	Niederquell et al. (2018)
7.	Acyclovir	Biodegradable ILs	The experimental and predicted results differ only by 7%	(Lotfi et al. 2017)
8.	Aspirin Paracetamol Ibuprofen	Conventional solvents	RMSE of 4.98% was observed for COSMO-RS results	(Hahnenkamp et al. 2017)

(continued)

Table 8.2 (continued)

S. No.	Drug name	Solvent/ILs	Finding	Reference
9.	Methotrexate	[TMAm][Ac] [TMAm][Cl] [TMAm][Br]	-	(Lotfi et al. 2016)
10.	Danazol	Conventional solvents	The predicted and experimental solubility results were in good agreement	(Pozarska et al. 2013)
11.	Phenol	Cyanex 923 TBP TOA [CYPHOS 105] [CYPHOS 109]	K_D^{sim} and K_D^{exp} were in good correlation with each other	(Burghoff et al. 2008)
12.	Diclofenac	DMSO	The predicted and experimental results were satisfactory	(Zilink et al. 2007)
13.	Aspirin Caffeine Mannitol etc.	Conventional solvents	The results were helpful in predicting the solubility of drugs and carry out the study further	(Ikeda et al. 2005)

and volume of the ions involved. Then the calculations are made in the form of cation–anion pair at the COSMO level. But in COSMO-RS calculation the ions are supposed to form a neutral mixture [e.g. [TMAm][Cl][(50:50)]. Taking into consideration lactic acid (LA) as model drug molecule, [TMAm][Cl] as IL in COSMO calculation. Initially, the COSMO files for IL and LA are generated using TMOLE-X. The further calculation was performed using the COSMOtherm software (version 18.0.2).

After inserting the files into COSMOtherm, COSMO predictions by σ-surface and σ-profiles are made. The σ-profiles help in predicting the chemical nature of the molecule. The σ-profiles in the form of histograms depict the surface charge densities and polarity of the solute (Diedenhofen et al. 2010). Also, it gives an insight into the bonding nature and can be used for the determination of thermodynamic properties (Rashid et al. 2018). Figure 8.6 shows the σ-surface of the LA molecule (model compound) along with σ-profile. It reveals that LA consists of both positive, negative, and neutral sites, i.e., it can behave both as HBD and HBA nature. σ-profile is divided into three segments $\sigma < 1.0$ e/nm^2, $\sigma > 1.0$e/nm^2, and $-1.0 < \sigma < 1.0$ e/nm^2 signifying HBD, HBA, and neutral region (Khan et al. 2020). The HBD sites will be attracted toward HBA ILs while HBA sites toward HBD ILs. This trend is visible in Fig. 8.7 depicting the σ-potential of the LA molecule. These results are of great help in selecting suitable cation–anion combinations for the formation of ILs.

Table 8.3 COSMO-SAC in pharmaceuticals

S. No.	Drug molecule	Solvents/EOS	Parameter evaluated	Reference
1.	Thioglycolic acid	[OMIm][Otf] [OMIm][ClO$_4$] [OMIm][PF$_6$] [OMIm][AlCl$_4$]	RMSD was found to be less than 1%	(Zhou et al. 2018)
2.	Paracetamol Testosterone Benzil etc.	Conventional solvents	RMSE was found to be 0.76–0.78	(Shu et al. 2011)
3.	Fragrances	Alcohol Ketones Ether	AAD using COSMO-SAC and experimental was found to be 0.26–10.08%	(Xavier et al. 2020)
4.	Ibuprofen Paracetamol Benzoic acid Salicylic acid 4-Aminobenzoic acid Anthracene	Polar solvents	MSD was found to be 29–64%	(Bouillot et al. 2013)
5.	Paracetamol	ILs	RMSD was found to be 0.65 between predicted and COSMO results	(Lee et al. 2017)
6.	Ibuprofen Acetaminophen Benzoic acid Salicylic acid 4-Aminobenzoic acid	Conventional solvents	RMSE of 1.9 was observed using COSMO-SAC results	(Bouillot et al. 2011)

After selecting cations–anions for the screening, property calculation can be carried forward.

Solubility prediction using COSMO-RS

To estimate the solubility S_S^X of a liquid compound in solvent S, the data required only the chemical potential of the compound in a solvent and its pure state. The chemical potential in a solvent and pure liquid is represented by μ_S^X and μ_X^X. The logarithmic of solubility is represented by Eq. 8.10 (Wichmann et al. 2019):

$$log S_S^X = log\left(\frac{Molwt.^X \rho_S}{Molwt._S}\right) - \frac{ln(10)}{kT}\Delta_S^X \qquad (8.10)$$

where
ρ=density of solvent.

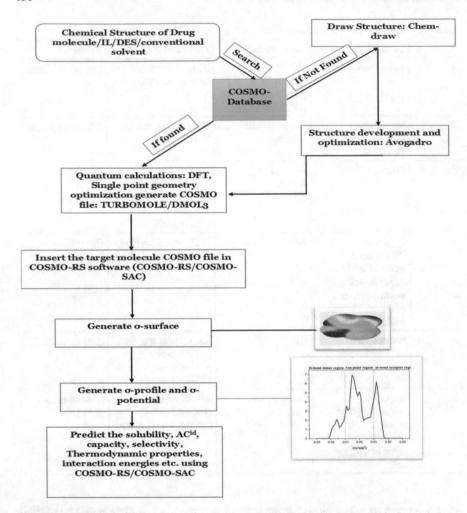

Fig. 8.4 Step-by-step COSMO procedure for property prediction in pharmaceuticals

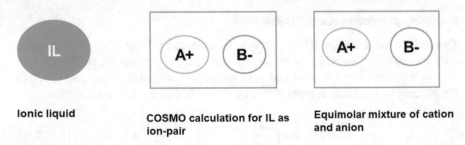

Ionic liquid   COSMO calculation for IL as   Equimolar mixture of cation
                                ion-pair   and anion

Fig. 8.5 COSMO-RS IL representation

Fig. 8.6 σ-surface and
σ-profile of the lactic acid
molecule

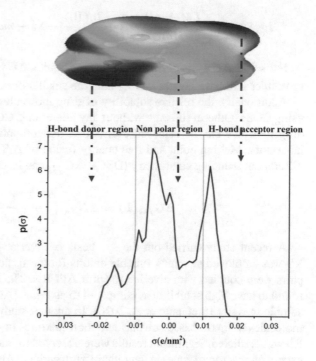

Fig. 8.7 σ-Potential of the
lactic acid molecule

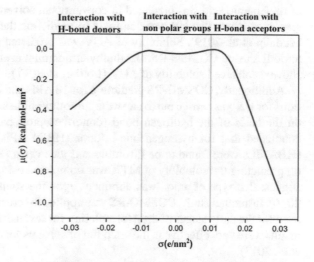

Δ_S^X=difference in the chemical potential.

Equation 8.10 is used to predict the solubility of the drug molecule.

Drugs are mostly solid at room temperature. The difference in phase transition is expressed by the Gibbs energy of fusion ΔG_{fus}. For solids solubility is given by the following equation 8.11:

$$logS_S^X = log\left(\frac{Molwt.^X \rho_S}{Molwt._S}\right) + \frac{ln(10)}{kT}[-\Delta_S^X - max(0, \Delta G_{fus}^X)] \qquad (8.11)$$

For solids ΔG_{fus}^X is positive whereas for liquids ΔG_{fus}^X is negative. Macro drug molecules will have greater ΔG_{fus}^X than the smaller ones (Wichmann et al. 2019).

Additionally, the relative solubility of drug molecules can be predicted directly using COSMOtherm software without any alterations. COSMO calculations for the drug molecule can be achieved even when the compound is not available in the laboratory. Melting point and free energy fusion of ΔH_{fus} can be obtained using differential scanning calorimetry (DSC). ΔG_{fus}^X can be calculated using Eq. 8.12 :

$$\Delta G_{fus}(T) = \Delta H_{fus}\left(1 - \frac{T}{T_{melt}}\right) \qquad (8.12)$$

A recent study carried out the synthesis of alverine-based API-ILs. COSMO-RS was employed to screen suitable anions for the IL combinations. The screened pairs were coupled with alverine to form API-ILs. The synthesized API-ILs were found to have high solubility as compared to alverine. [Av][Tos] proved to be highly soluble in water (Stefanuto et al. 2020). In another study, the solubility of 51 drug molecules in lipase-based solvents and their mixtures in glycerides using COSMO-RS was predicted. Significant results were observed for most of the drug components except for macromolecules in long-chain triglycerides (Alsenz et al. 2019). Solubility of mefenamic acid was predicted in conventional solvents using COSMO-RS. The results conclude that HB was dominant in predicting the solubility of MFA (Abdul Mudalip et al. 2019). Solubility of ACV was predicted in biodegradable and non-toxic ILs using COMSO-RS. ILs diethylammonium acetate and triethylammonium showed enhanced solubility of ACV (Lotfi et al. 2017).

Additionally, COSMO-RS prediction for 18 APIs and 240 cation–anion combinations for ILs was carried out to carry the solubility studies. The results were evaluated on the basis of the hydrogen bond (donor/acceptor) nature of APIs. The authors concluded that for hydrogen bond donor (HBD) APIs, hydrogen bond acceptor (HBA) ILs were found to be favorable and vice versa (Kruse et al. 2016). A study on predicting the solubility of MTX was completed using COSMO-RS. The results suggest that type of anion was dominant upon the solubility of MTX (Lofti et al. 2016). In another study, COSMO-RS was applied to screen solvents particularly to be used for the early stages of drug formulation for seven drug molecules. COSMO-RS results were very effective in the selection of solvents for drug formulation (Pozarska et al. 2013).

The traditional way to predict the solubility of drug molecules is the prediction of the octanol/water coefficient $K_{O,W}$. When pharmaceutical compounds dissolve in two immiscible liquid phases containing octanol and water, the components distribute between two phases. The distribution of component i between two phases, octanol (O) and water (W) at infinite dilution, measured by partition coefficient is as follows (Hsieh et al. 2011):

$$K_i^{O,W} = \frac{x_i^O}{x_i^W} = \frac{\gamma_i^W}{\gamma_i^O} \qquad (8.13)$$

Therefore, the octanol/water partition coefficient for component i was defined as follows (Hsieh et al. 2011):

$$\log K_{O,W,i} = \log\left(\frac{C_{O,W}\gamma_i^{W,\infty}}{C_{O,O}\gamma_i^{O,\infty}}\right) \qquad (8.14)$$

where

x_i^O = mole fraction on the octanol-rich phase.

x_i^W = mole fraction in the water-rich phase.

γ_i^W = AC^{id} in the water phase.

γ_i^O = AC^{id} in the octanol phase.

$C_{O,W}$ = concentration in the water-rich phase.

$C_{O,O}$ = concentration in the octanol-rich phase.

$K_{O,W}$ is a widely used parameter in the initial predictions of solubility. If the solvent (conventional/IL/API-IL/DES) possesses a value greater than 1, it indicates that the drug molecule would be soluble in the organic phase rather than water and vice versa (Stefanuto et al. 2020). Furthermore, to investigate the solubility of ILs derived from the highly insoluble drug ampicillin, $K_{O,W}$ was evaluated. The derived ILs with enhanced solubility were employed in the formulation of antibiotics possessing similar functions as ampicillin (Florindo et al. 2013). In another study $K_{O,W}$ was predicted for the drugs possessing acidic groups. The results conclude that even a minute change in the concentration of salt can affect the $K_{O,W}$ significantly (Ingram et al. 2011).

COSMO-RS can be used to predict the equilibrium results for LLE/VLE/SLE etc. It is used for predicting ILs as an extractant for drug molecules. In a recent study, COSMO-RS was used to screen 800 cation–anion combinations forming ILs. The results suggest that trioctylammonium [TOAm] with long-chain anions was a better extractant for phenol (Khan et al. 2021). Another study was carried out for the removal of lactic acid (LA) from aqueous streams. In this work, 140 cation–anion combinations were screened. The results suggest that quaternary ammonium cation [TMAm] along with [SO$_4$], [Ac] anions are promising extractants for LA extraction (Khan et al. 2020). COSMO-RS was used for the screening of docosahexaenoic acid (DHA) from microalgae. A total of 352 different cation–anion combinations were used to screen the potential IL combination. The results suggest that quaternary ammonium along with hydrophilic anions such as SO$_4$ and Cl are potential for the extraction of DHA (Motlagh et al. 2020). In a similar work, ILs were screened for the extraction of eicosapentaenoic acid (EPA) from microalgae. The results suggest that non-aromatic cations along with non-coordinating anions were favorable for the extraction of EPA (Motlagh et al. 2019).

Screening of ILs for the extraction of β-carotene was carried out using COSMO-RS. The authors conclude that [Ac]-based ammonium ILs are very promising for the extraction of β-carotene (Rajabi et al. 2017). COSMO-RS screening was carried out for the extraction of methylxanthines from plants. A total of 874 IL combinations comprising different cations and anions were screened to select the potential one with the best extraction efficiency (Jelinski et al. 2017). In another study, ILs for the extraction of DDT were screened. In total, 1015 IL combinations were screened using COSMO-RS. The results show that non-aromatic cations prove better extractants (Pilii et al. 2012). In a similar study, screening of ILs for the extraction of monoethylene glycol (MEG) was done. COSMO-RS was employed to predict the σ-profile, selectivity and distribution coefficient to screen the best IL for extraction (Chavez et al. 2012). In another work with a similar objective for the extraction of cresol, 360 IL combinations were screened. It was observed that ILs with long alkyl chains gave better selectivity and hence are favorable to be used as extractants (Kumar et al. 2011). COSMO-RS with molecular dynamics was used for the screening of liposomes and micelles. It was concluded that COSMO-RS is the potential for screening in complex systems (Ingram et al. 2013).

8.3.1.2 COSMO-RS-EOS

A study was carried out to predict the solubility of phenacetin, acetanilide, and paracetamol using non-random hydrogen bonding (NRHB) EOS. Also, the result of these models was compared with the COSMO-RS prediction. The EOS and COSMO model results were found to be in good agreement (Tsivintzelis et al. 2009).

8.3.1.3 COSMO-RS/cocrystal Formation

In accordance with the assumptions that the interactions in a crystal form are similar to a virtual supercooled liquid, the formation of cocrystals can be represented via liquid-phase thermodynamics. In comparison with pure reactants, the strength of cocrystal interactions can be calculated through the mixing of enthalpy ΔH_{mix}. It is the rough approximation to the free energy of cocrystal formation ΔG_{co} and can be represented by Eq. 8.15 (Loschen et al. 2015):

$$\Delta G_{co} = \Delta H_{mix} - T\Delta S_{mix} - \Delta\Delta G_{fus} \approx \Delta H_{mix} \qquad (8.15)$$

where

ΔS_{mix}=mixing entropy.

$\Delta\Delta G_{fus}$=difference between the free energy of fusion between cocrystal and reactants. It is assumed to be zero.

A study was conducted for the prediction of the solubility of several drug molecules and their cocrystal formers. The results for solvent and cocrystals screening were found to be accurate and efficient (Loschen et al. 2015). In another study, COSMO-RS was used to screen solvents for cocrystal formation in super-cooled liquids. COSMO-RS proved to be potential for the selection of solvents for cocrystal formation (Abramov et al, 2012).

8.3.1.4 COSMO-RS-DES

The application of COSMO-RS extends to a new class of solvents referred to as deep eutectic solvents (DES). A DES is a combination of HBD and HBA salts. The studies report DES as a favorable alternative to conventional solvents in pharmaceutical applications (Perna et al. 2020). In very recent research work, quaternary ammonium chlorides DES were coupled with selected APIs. COSMO-RS was employed to find the interactions between APIs and DES through SLE. COSMO-RS was helpful in predicting SLE points for DES-APIs (Martins et al. 2021). In another study, COSMO-RS was employed to screen HBD and HBA combinations to form DES. COSMO-RS and experimental results suggest that the solubility of selected drug molecules under study was higher in DES as compared to conventional solvents and water (Palmelund et al. 2019). In another study, choline chloride-based eutectic solvents were screened for the early formulation of several drug molecules. Using COSMO-RS the eutectic temperatures were calculated and the screening was carried out. The results were verified experimentally using SLE and were found to be promising (Abranches et al. 2019).

8.3.2 COSMO-SAC

Modification of COSMO-RS using the Staverman–Guggenheim equation was established by Lin and Stanley, resulting in an improved combinatorial contribution in the activity coefficient (Paese et al. 2020). They established a model focused on activity coefficient that solves the problems that arise in COSMO-RS. The advantage of this approach is it evaluates the representation of chemical potential for segments that meet the boundary conditions. The expression for the activity coefficient obtained is thermodynamically consistent. COSMO-SAC model based on surface segment contributions is useful for making a prior phase equilibrium prediction (Lin et al. 2002).

The main activity coefficient is given by Eq. 8.16 (Paese et al. 2020):

$$ln\gamma_i = ln\gamma_i^{res} + ln\gamma_i^{comb} \tag{8.16}$$

where
$ln\gamma_i^{res}$=residual activity coefficient.
$ln\gamma_i^{comb}$=combinatorial activity coefficient.
Using discrete segments, the residual contribution can be calculated as Eq. 8.17:

$$ln\gamma_i^{res} = \sum_{m\in i} \frac{Q_m^i}{a_{eff}} \cdot (ln\gamma_s^m - ln\gamma_i^m) \tag{8.17}$$

where

Q_m^i=segment area in molecule i.

a_{eff}=standard segment surface area.

$ln\gamma_s^m$=logarithmic of the activity coefficient for segment m.

$ln\gamma_s^i$=logarithmic of activity coefficient in pure liquid.

The probability of finding segment m in a pure liquid i (σ-profile of the pure compound) is:

$$p_m^i = \frac{Q_m^i}{Q^i} \tag{8.18}$$

where

$Q^i = \sum_{m\in i} Q_m^i$ is the total cavity surface area of molecule i.

The interaction energy in COSMO-SAC $\Delta W_{m,n}$ can be computed using different assumptions. Lin and Sandler computed it as a function of surface charge densities σ_m and σ_n, where m and n are the segments (Lin et al. 2002):

$$\Delta W_{m,n} = \frac{\alpha'}{2}(\sigma_m + \sigma_n)^2 + c_{hb}\max[0, \sigma_{acc} - \sigma_{hb}] * \min[0, \sigma_{don} - \sigma_{hb}] \tag{8.19}$$

where

α'=misfit energy constant.

c_{hb}=hydrogen bonding (HB) constant.

σ_{hb}=cut-off hydrogen bonding sigma value.

σ_{acc}=larger sigma value of segment m.

σ_{don}=smaller sigma value of segment n.

A study validated the sigma profile database for 2434 files consisting of 1356 drug molecule–solvents pair, 160 solvents, and 194 drug molecules and provided FORTRAN codes for sigma profiles and activity. This database was applied to predict SLE, VLE, and other thermodynamic properties (Mullins et al. 2008). In another study, the σ-profile database was created using GAMES software. The results helped predict AC^{id} and other properties (Ferrarini et al. 2018). In terms of solvation energy, the activity coefficient is assessed ΔG_x^{sol}, which is the change in free energy when a solute molecule x is transferred from a static position in an ideal gas to a solution S at constant temperature and pressure. Activity coefficient is determined by using the following equation (Lin et al. 2002):

$$ln\gamma_{x,S} = \frac{\Delta G_{x/S}^{sol} - \Delta G_{x/x}^{sol}}{RT} + ln\frac{C_s}{C_x} \tag{8.20}$$

where

C_x=molar concentration of fluid x (which may be pure fluid x).

C_S=molar concentration of solution S.

Calculation of ΔG_x^{sol} takes place in the stepwise form. Initially, the charges of the solute (drug molecule) under study are turned off, and the remaining core particle is placed into the solvent. After that, the charges are turned on, and the electronic configuration of the drug molecule is activated. The difference in the energy for these two steps is the cavity formation energy and ΔG^{cav} and the charging free energy ΔG^{chg}, respectively (Bouillot et al. 2011).

In a recent study, the authors investigated the efficacy of the COSMO-SAC model in SLE for different pharmaceuticals. The results predicted solubilities in parent and mixed solvents, $K_{O,W}$ and the formation of cocrystals. The results of the original COSMO-SAC 2002 and modified COSMO-SAC 2010 were compared and found to be compatible (Zarei et al. 2020). In another study, the solubility of paracetamols in 2624 cation–anion combinations and binary IL using the COSMO-SAC model was estimated. The results conclude that in pure ILs the solubility of paracetamol is higher as compared to IL mixtures (Lee et al. 2017). The effect of pH on the $K_{O,W}$ was predicted using COSMO-SAC for 41 ionizable drugs. The acid dissociation constant pK_a was calculated using the solvation model. The effect of pH was significant on $K_{O,W}$ (Chen et al. 2016).

To improve the accuracy of the COSMO-SAC model for the improvement of drug molecule solubility, some amendments have been done. In order to perform this analysis for 352 drug molecules, σ-profiles were modified using a probability function. In polar solvents, an increase in drug solubility was observed, which helped evaluate the drug's crystallization rate (Bouillot et al. 2013, part 2). In their previous work, the modification of COSMO-SAC was carried out for six drug molecules in 35 solvents and their mixtures. Mean quadratic error and temperature dependence were key parameters to analyze the efficacy of revised COSMO-SAC. The results proved relevant for SLE, but for HB, further modifications are required (Bouillot et al. 2013, part 1).

A study was conducted to estimate the solubility of six drug molecules using conventional solvents. The results show that for smaller molecules such as anthracene COSMO-SAC predictions were accurate but for larger molecules, due to the HB involvement, some deviation was observed (Bouillot et al. 2011). Another study aimed to combine solubility data obtained experimentally with COSMO-SAC. Along with 127 different solvent combinations, 33 drug molecules were selected.

It was found that the application of COSMO-SAC along with experimental data reduced the predictive error remarkably from 442 to 88% (Shu et al. 2011). In a similar work, the solubility of 51 drug molecules using 37 pure and mixed solvent combinations was predicted using COSMO-SAC. The results from COSMO-SAC proved advantageous and it was concluded that COSMO-SAC is an efficient tool for making prior predictions of the properties of drug which are not available initially (Hsieh et al. 2010).

Recently, a research work was conducted for the extraction of isopropyl alcohol (IPA) using [Im]-based ILs. The selection of ILs for the extraction of IPA was done through the analysis of σ-profile of IPA using COSMO-SAC. The authors concluded that the predicted and experimental results were in good agreement (Zhu et al. 2020).

In another study, COSMO-SAC was used to predict VLE, odor, and intensity for mixtures containing fragrances. Stevens' power law was used for calculating odor intensity. Promising results were obtained for VLE (Xavier et al. 2020). Another study was conducted to screen 285 cation–anion combinations for the separation of acetone from n-hexane. Using COSMO-SAC the σ-profiles for the cations and anions were evaluated. The results suggest that short-chain [Im] ILs are better extractants for acetone (Wang et al. 2020). A study was conducted for the screening of [Im]-based ILs using COSMO-SAC for the extraction of thioglycolic acid. AC^{id}, capacity, selectivity, and PI were predicted. It was found that ILs with a long alkyl chain were favorable for extraction (Zhou et al. 2018).

8.3.2.1 COSMO-SAC/EOS

Quantum chemistry models were added to COSMO-SAC and the estimation of different properties was carried out. In a recent study on ILs, the Pitzer-Huckel-Debye model was modified to predict the long-chain interactions of ILs. COSMO-SAC was used for predicting the initial values of parameters required. The results show an improvement in the thermodynamic properties of ILs using the extended PDH model (Chang et al. 2019). Another study applied Peng-Robinson (PR) EOS along with COSMO-SAC to predict the VLE for the binary mixtures (Lee et al. 2007). EOS was employed to predict the solubility of 46 drug molecules possessing ring structures employing supercritical CO_2 as a solvent. PR was applied along with COSMO-SAC for the predictions. The findings indicate that PR-EOS/COSMO-SAC is a beneficial tool for predicting the solubility of solid drug molecules prior to analysis (Wang et al. 2014).

8.3.2.2 COSMO-SAC/DES

In a study with the objective to remove isopropanol from an azeotropic mixture using DES, COSMO-SAC was used for screening. It was explored that [Ch][Cl] and triethylene glycol were a suitable combination to form DES. The results were verified using VLE for the isopropanol-azeotropic-DES mixture (Jiang et al. 2019). COSMO-SAC was used for the screening of hydrophobic DES to extract ethanol, propanol, butanol from azeotropic alcoholic mixtures. The findings suggest that predicted and experimental results were quite complimentary (Verma et al. 2018).

8.4 Limitations of COSMO-RS for Pharmaceuticals

The only constraint on using COSMO for solubility predictions and in newly developed drug molecules is the quantum chemistry calculations. This step requires time and skills which are not available in the early formulation stages. But after this process of creating a database profile is a success, it can be further used for predicting the solubilities of drug molecules. The chemical structure optimization particularly for complex drug molecules is difficult and hence requires extra effort. Once this constraint is solved COSMO-RS is successful for screening purposes.

8.5 Conclusions and Future Outlook

With the rapid growth and advancements in drug design and development in pharmaceutical industries, it is important to select task-specific ILs. The resulting ILs must possess good bonding properties, thereby increasing the solubility and avoiding cocrystal formation. To save time and energy, predictive methods such as COSMO-RS/COSMO-SAC are advantageous and reliable. The use of these methods results in the screening of potential cation–anion combinations from millions of ILs combinations. In this chapter, we discussed the application of COSMO in pharmaceutical and medicinal applications. This chapter summarizes COSMO-RS and COSMO-SAC applications carried out for the screening of ILs. An important advantage of COSMO is that it requires only the chemical structure of the molecule for making further predictions. Furthermore, the method can be considered reliable cause of its validation using experiments. The capability of COSMO in pharmaceuticals for various applications has been tested and proved practically. Hence it can be concluded that COSMO is an efficient and advantageous tool that screens suitable solvents, cocrystals with minimum information-theoretically. Figure 8.8 shows the summary of the use of COSMO for screening purposes in pharmaceuticals and medicine.

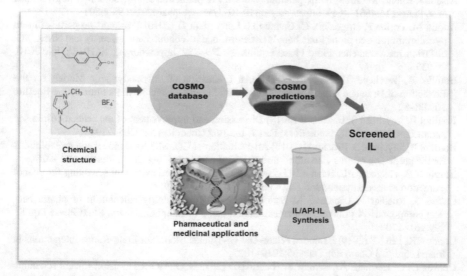

Fig. 8.8 COSMO application in pharmaceutics and medicine

However, there occurs some obstacles while making computations for complex drug molecules. These obstacles can be overcome easily by taking into account the essence of bonding. For drug molecules, especially for complex new formulations, the structure drawing and optimization must be done well so that further predictions are effective. The screening of σ-profile/σ-potential must be considered the key step for further predictions. Overall, the use of COSMO is favorable in pharmaceutical applications. This chapter highlighted the benefits and potential usage of COSMO for screening ILs as solubilizer, extractant, and also as preventing agent for any cocrystal formation.

References

Abdul Mudalip SK, Bakar MRA, Jamal P, Adam F (2018) Prediction of mefenamic acid solubility and molecular interaction energies in different classes of organic solvents and water. Ind Eng Chem Res 58:762–770. https://doi.org/10.1021/acs.iecr.8b02722

Adawiyah N, Moniruzzaman M, Hawatulaila S, Goto M (2016) Ionic liquids as a potential tool for drug delivery systems. Med. Chem. Commun. 7:1881–1897. https://doi.org/10.1039/C6MD00358C

Abramov YA, Loschen C, Klamt A (2012) Rational coformer or solvent selection for pharmaceutical cocrystallization or desolvation. J Pharm Sci 101:3687–3697. https://doi.org/10.1002/jps.23227

Abranches DO, Larriba M, Silva LP, Franco MM, Palomar JF, Pinho SP, Coutinho JAP (2019) Using COSMO-RS to design choline chloride pharmaceutical eutectic solvents. Fluid Phase Equilib 497:71–78. https://doi.org/10.1016/j.fluid.2019.06.005

Alevizou EI, Voutsas EC (2014) Evaluation of COSMO-RS model in binary and ternary mixtures of natural antioxidants, ionic liquids and organic solvents. Fluid Phase Equilib 369:55–67. https://doi.org/10.1016/j.fluid.2014.02.015

Alsenz J, Kuentz M (2019) From quantum chemistry to prediction of drug solubility in glycerides *Mol. Pharm* 16:4661–4669. https://doi.org/10.1021/acs.molpharmaceut.9b00801

Anouti M, Porion P, Brigouleix C, Galiano H, Lemordant D (2010) Transport properties in two pyrrolidinium-based protic ionic liquids as determined by conductivity, viscosity and NMR self-diffusion measurements. Fluid Phase Equilib 29:229–237. https://doi.org/10.1016/j.fluid.2010.09.035

Bouillot B, Teychené S, Biscans B (2011) An Evaluation of Thermodynamic Models for the Prediction of Drug and Drug-like Molecule Solubility in Organic Solvents Fluid Phase Equilib. 309:36–52

Bouillot B, Teychené S, Biscans B (2013) Discussion and improvement of the refined cosmo-sac parameters for solubility predictions: Part 2. Ind Eng Chem Res 52:9285–9294

Bouillot B, Teychené S, Biscans B (2013) An evaluation of COSMO-SAC model and its evolutions for the prediction of drug-like molecule solubility: part 1. Ind Eng Chem Res 52:9276–9284

Burghoff B, Goetheer EL, Haan ABD (2008) COSMO-RS-based extractant screening for phenol extraction as model system. Ind Eng Chem Res 47:4263–4269

Cassen J, Ruether F, Leonhard K, Sadowski G (2010) Solubility calculation of pharmaceutical compounds–A priori parameter estimation using quantum-chemistry. Fluid Phase Equilib 299:161–170

Chang CK, Lin ST (2019) Extended Pitzer–Debye–Hückel Model for Long-Range interactions in Ionic Liquids. J Chem Eng Data 65:1019–1027.

Chavez LYG, Hermans AJ, Schuur B, Haan ABD (2012) COSMO-RS assisted solvent screening for liquid–liquid extraction of mono ethylene glycol from aqueous streams. Sep Purif Technol 97:2–10

Chen CS, Lin ST (2016) Prediction of pH Effect on the Octanol-Water Partition Coefficient of Ionizable Pharmaceuticals. Ind Eng Chem Res 55:9284–9294. https://doi.org/10.1021/acs.iecr. 6b02040

Claus J, Sommer FO, Kragl U (2018) Ionic liquids in biotechnology and beyond. Solid State Ion 314:119–128. https://doi.org/10.1016/j.ssi.2017.11.012

Diedenhofen M (2010) Klamt A. COSMO-RS as a Tool for Property Prediction of IL Mixtures-a Review Fluid Phase Equilib. 294:31–38. https://doi.org/10.1016/j.fluid.2010.02.002

Esteban J, Vorholt AJ, Leitner W (2020) An overview of the biphasic dehydration of sugars to 5-hydroxymethylfurfural and furfural: a rational selection of solvents using COSMO-RS and selection guides. Green Chem 22:2097–2128. https://doi.org/10.1039/C9GC04208C

Ferrarini F, Flôres GB, Muniz AR, Soares RPD (2018) An open and extensible sigma-profile database for COSMO-based models. AIChE J 64:3443–3455. https://doi.org/10.1002/aic.16194

Florindo C, Araújo JMM, Fi A, Matos C, Ferraz R, Prudêncio C, Noronha JP, Petrovski Z, Branco L, Rebelo LPN, Marrucho IM (2013) Evaluation of solubility and partition properties of ampicillin-based ionic liquids. Int J Pharm 456:553–559. https://doi.org/10.1016/j.ijpharm.2013.08.010

Gonfa G, Muhammad N, Bustam MA (2018) Probing the interactions between DNA nucleotides and biocompatible liquids: COSMO-RS and molecular simulation study *Sep. and Purif*. Technol 196:237–243. https://doi.org/10.1016/j.seppur.2017.08.033

Hahnenkamp I, Graubner G, Gmehling J (2010) Measurement and prediction of solubilities of active pharmaceutical ingredients. Int J Pharm 388:73–81. https://doi.org/10.1016/j.ijpharm. 2009.12.036

Hilmy NIMF, Yahya WZN, Kurnia KA (2020) Eutectic ionic liquids as potential electrolytes in dye-sensitized solar cells: Physicochemical and conductivity studies. J Mol Liq 320:114381. https://doi.org/10.1016/j.molliq.2020.114381

Hsieh CM, Wang S, Lin ST, Sandler SI (2011) A predictive model for the solubility and octanol−water partition coefficient of pharmaceuticals. J Chem Eng Data 56:936–945. https://doi.org/10. 1021/je1008872

Ikeda H, Chiba K, Kanou A, Hirayama N (2005) Prediction of solubility of drugs by conductor-like screening model for real solvents. Chem Pharm Bull 53:253–255

Ingram T, Richter U, Mehling T, Smirnova I (2011) Modelling of pH dependent n-octanol/water partition coefficients of ionizable pharmaceuticals. Fluid Phase Equilib 305:197–203. https://doi. org/10.1016/j.fluid.2011.04.006

Ingram T, Storm S, Kloss L, Mehling T, Jakobtorweihen S, Smirnova I (2013) Prediction of micelle/water and liposome/water partition coefficients based on molecular dynamics simulations. COSMO-RS, and COSMOmic, Langmuir 29:3527–3537. https://doi.org/10.1021/la305035b

Jeliński T, Cysewski P (2017) Screening of ionic liquids for efficient extraction of methylxanthines using COSMO-RS methodology. Chem Eng Res Des 122:176–183. https://doi.org/10.1016/j. cherd.2017.04.015

Jiang H, Xu D, Zhang L, Ma Y, Gao J, Wang Y (2019) Vapor-Liquid Phase Equilibrium for Separation of Isopropanol from Its Aqueous Solution by Choline Chloride-Based Deep Eutectic Solvent Selected by COSMO-SAC Model. J Chem Eng Data 64:1338–1348. https://doi.org/10.1021/acs. jced.8b00895

Khan AS, Ibrahim TH, Rashid Z, Khamis MI, Nancarrow P, Jabbar, NA (2021) COSMO-RS based screening of ionic liquids for extraction of phenolic compounds from aqueous media. J Mol Liq 328:115387. https://doi.org/10.1016/j.molliq.2021.115387

Khan HW, Reddy AVB, Nasef MME, Bustam MA, Goto M, Moniruzzaman M (2020) Screening of ionic liquids for the extraction of biologically active compounds using emulsion liquid membrane: COSMO-RS prediction and experiments. J Mol Liq 309:11312. https://doi.org/10.1016/j.molliq. 2020.113122

Klamt A (2008) Comments on Performance of COSMO-RS with Sigma Profiles from Different Model Chemistries. Ind Eng Chem Res 47:987–988. https://doi.org/10.1021/ie0712535

Klamt A (1995) Conductor-like screening model for real solvents: a new approach to the quantitative calculation of solvation phenomena. J Phys Chem 99:2224–2235. https://doi.org/10.1021/j10000 7a062

Kruse AJ, Weber CC, Rogers RD, Myerson AS (2017) The a priori design and selection of ionic liquids as solvents for active pharmaceutical ingredients. Chem Eur J 23:5498–5508. https://doi.org/10.1002/chem.201605704

Kumar L, Banerjee T, Mohanty K (2011) Prediction of selective extraction of cresols from aqueous solutions by ionic liquids using theoretical approach. Sep Sci Technol 46(13):2075–2087. https://doi.org/10.1080/01496395.2011.589421

Kurnia KA, Harimurti S, Yung HK, Baraheng A, Alimin MAS, Dagang NSM, Fadhilah A, Rosyadi R, Yahya WZN, Bustam MA (2019) Understanding the effect of pH on the solubility of Gamavuton-0 in the aqueous solution: Experimental and COSMO-RS modelling. J Mol Liq 296:111845. https://doi.org/10.1016/j.molliq.2019.111845

Lee BS, Lin ST (2017) Prediction and screening of solubility of pharmaceuticals in single-and mixed-ionic liquids using COSMO-SAC model. AIChE J 63:3096–3104. https://doi.org/10.1002/aic.15595

Lee MT, Lin ST (2007) Prediction of mixture vapor–liquid equilibrium from the combined use of Peng-Robinson equation of state and COSMO-SAC activity coefficient model through the Wong-Sandler mixing rule. Fluid Phase Equilib 254:28–34. https://doi.org/10.1016/j.fluid.2007.02.012

Lin ST, Sandler SI (2002) A priori phase equilibrium prediction from a segment contribution solvation model. Ind Eng Chem Res 41:899–913. https://doi.org/10.1021/ie001047w

Loschen C, Klamt A (2015) Solubility prediction, solvate and cocrystal screening as tools for rational crystal engineering. J Pharm Pharmacol 67:803–811. https://doi.org/10.1111/jphp.12376

Loschen C, Klamt A (2016) Computational screening of drug solvates *Pharm*. Res 33:2794–2804. https://doi.org/10.1007/s11095-016-2005-2

Lotfi M, Moniruzzaman M, Mutalib MIA, Wilfred CD, Alitheen NB, Goto M (2016) Analysis of multiple solvation interactions of methotrexate and ammonium based ionic liquids using COSMO-RS. Procedia Eng 148:459–466. https://doi.org/10.1016/j.proeng.2016.06.464

Lotfi M, Moniruzzaman M, Sivapragasam M, Kandasamy S, Mutalib MIA, Alitheen NB, Goto M (2017) Solubility of acyclovir in nontoxic and biodegradable ionic liquids: COSMO RS prediction and experimental verification. J Mol Liq 243:124–131. https://doi.org/10.1016/j.molliq.2017.08.020

Martins MA. Silva LP, Jorge PS, Abranches DO, Pinho SP, Coutinho JA (2021) The role of ionic vs. non-ionic excipients in APIs-based eutectic systems. Eur J Pharm Sci 156:105583. https://doi.org/10.1016/j.ejps.2020.105583

Motlagh SR, Harun R, Biak DRA, Hussain SA, Elgharbawy AA, Khezri R, Wilfred CD (2020) Prediction of Potential Ionic Liquids (ILs) for the Solid-Liquid Extraction of Docosahexaenoic Acid (DHA) from Microalgae Using COSMO-RS Screening Model. Biomolecules 10:1149. https://doi.org/10.3390/biom10081149

Motlagh SR, Harun R, Biak DRA, Hussain SA, Wan AW, Ghani AK, Khezri R, Wilfred CD, Elgharbawy AA (2019) Screening of suitable ionic liquids as green solvents for extraction of eicosapentaenoic acid (EPA) from microalgae biomass using COSMO-RS model. Molecules 24:713. https://doi.org/10.3390/molecules24040713

Mullins E, Liu YA, Ghaderi A, Fast SD (2008) Sigma profile database for predicting solid solubility in pure and mixed solvent mixtures for organic pharmacological compounds with COSMO-based thermodynamic methods *Ind*. Eng Chem Res 47:1707–1725. https://doi.org/10.1021/ie0711022

Niederquell A, Wyttenbach N, Kuentz M (2018) New prediction methods for solubility parameters based on molecular sigma profiles using pharmaceutical materials *Int*. J Pharm 546:137–144. https://doi.org/10.1016/j.ijpharm.2018.05.033

Oliveira G, Wojeicchowski JP, Farias FO, Igarashi-Mafra L, de Pelegrini Soares R, Mafra MR (2020) Enhancement of biomolecules solubility in aqueous media using designer solvents as additives:

An experimental and COSMO-based models' approach. J Mol Liq 318:114266. https://doi.org/10.1016/j.molliq.2020.114266

Paduszyński K (2017) An overview of the performance of the COSMO-RS approach in predicting the activity coefficients of molecular solutes in ionic liquids and derived properties at infinite dilution. Phys Chem Chem Phys 19:11835–11850. https://doi.org/10.1039/C7CP00226B

Paese LT, Spengler RL, Soares RDP, Staudt PB (2020) Predicting phase equilibrium of aqueous sugar solutions and industrial juices using COSMO-SAC. J Food Eng 274:109836. https://doi.org/10.1016/j.jfoodeng.2019.109836

Palmelund H, Andersson MP, Asgreen CJ, Boyd BJ, Rantanen J, Löbmann K (2019) Tailor-made solvents for pharmaceutical use? Experimental and computational approach for determining solubility in deep eutectic solvents (DES). Int J Pharm 1:100034. https://doi.org/10.1016/j.ijpx.2019.100034

Perna FM, Vitale P, Capriati V (2020) Deep eutectic solvents and their applications as green solvents. Curr Opin Green and Sustain Chem 21:27–33. https://doi.org/10.1016/j.cogsc.2019.09.004

Pilli SR, Banerjee T, Mohanty K (2012) Extraction of pentachlorophenol and dichlorodiphenyl-trichloroethane from aqueous solutions using ionic liquids. Ind Eng Chem Res 18:1983-1996. https://doi.org/10.1016/j.jiec.2012.05.017

Pozarska A, Mathews CDC, Wong M, Pencheva K (2013) Application of COSMO-RS as an excipient ranking tool in early formulation development. Eur J Pharm Sci 4:505–511. https://doi.org/10.1016/j.ejps.2013.04.021

Rajabi MS, Moniruzzaman M, Mahmood H, Sivapragasam M, Bustam MA (2017) Extraction of β-carotene from organic phase using ammonium based ionic liquids aqueous solution.J. Mol Liq 227:15–20. https://doi.org/10.1016/j.molliq.2016.12.008

Rashid Z, Wilfred CD, Gnanasundaram N, Arunagiri A, Murugesan T (2018) Screening of ionic liquids as green oilfield solvents for the potential removal of asphaltene from simulated oil:COSMO-RS model approach. J Mol Liq 255:492–503. https://doi.org/10.1016/j.molliq.2018.01.023pal

Reichardt C (1965) Empirical parameters of the polarity of solvents. Angew Chem, Int Ed Engl 4:29–40. https://doi.org/10.1002/anie.196500291

Shu CC, Lin ST (2011) Prediction of drug solubility in mixed solvent systems using the COSMO-SAC activity coefficient model. Ind Eng Chem Res 50:42–147. https://doi.org/10.1021/ie100409y

Stefanuto VF, Esteiro P, Santiago R, Moreno D, Palomar J, Tojo E (2020) Design and synthesis of alverine-based ionic liquids to improve drug water solubility, New. J Chem 46:20428–20433. https://doi.org/10.1039/D0NJ05216G

Taft RW, Kamlet MJ (1976) The solvatochromic comparison method. 2. The. alpha.-scale of solvent hydrogen-bond donor (HBD) acidities. J Am Chem 98:2886–2894. https://doi.org/10.1021/ja00426a036

Tsivintzelis I, Economou IG, Kontogeorgis GM (2009) Modeling the solid–liquid equilibrium in pharmaceutical-solvent mixtures: Systems with complex hydrogen bonding behavior. AIChE J 55:756–770. https://doi.org/10.1002/aic.11716

Verma R, Banerjee T (2018) Liquid–liquid extraction of lower alcohols using menthol-based hydrophobic deep eutectic solvent: experiments and COSMO-SAC predictions. Ind Eng Chem Res 57:3371–3381. https://doi.org/10.1021/acs.iecr.7b05270

Wang LH, Lin ST (2014) A predictive method for the solubility of drug in carbon dioxide. J Supercrit Fluids 85:81–88. https://doi.org/10.1016/j.supflu.2013.10.019

Wang Y, Yang X, Bai W, Zhang J, Zhou X, Guo X, Peng J, Qi J, Zhu Z (2020) Screening of imidazole ionic liquids for separating the acetone–n-hexane azeotrope by COSMO-SAC simulations and experimental verification, ACS Sustain. Chem Eng 8:4440–4450. https://doi.org/10.1021/acssuschemeng.9b07358

Wichmann K, Loschen C, Klamt A (2019) Drug solubility, reaction thermodynamics, and co-crystal screening. Chem Eng Pharm Ind: Act Pharm Ing Chapter 20:67–491. https://doi.org/10.1002/9781119600800.ch20

Xavier VB, Staudt PB, Soares RDP (2020) Predicting VLE and odor intensity of mixtures containing fragrances with COSMO-SAC *Ind*. Eng Chem Res 59:2145–2154. https://doi.org/10.1021/acs. iecr.9b05474

Zarei MS, Gholamreza P (2020) Investigation of COSMO-SAC model for solubility and cocrystal formation of pharmaceutical compounds. Sci Rep 10:19879. https://doi.org/10.1038/s41598-020-76986-3

Zhou Y, Xu D, Zhang L, Ma Y, Ma X, Gao J, Wang Y (2018) Separation of thioglycolic acid from its aqueous solution by ionic liquids: Ionic liquids selection by the COSMO-SAC model and liquid-liquid phase equilibrium. J Chem Thermodyn 118:263–273

Zhu Z, Xu Y, Li H, Shen Y, Meng D, Cui P, Ma Y, Wang Y, Gao J (2020) Separation of isopropyl alcohol and isopropyl ether with ionic liquids as extractant based on quantum chemical calculation and liquid-liquid equilibrium experiment. Sep Purif Technol 247:116937. https://doi.org/10.1016/j.seppur.2020.116937

Zilnik LF, Jazbinšek A, Hvala A, Vrecer F, Klamt A (2007) Solubility of sodium diclofenac in different solvents. Fluid Phase Equilib 261:140–145. https://doi.org/10.1016/j.fluid.2007.07.020

Chapter 9
Surface-Active Ionic Liquids for Medical and Pharmaceutical Applications

Md Korban Ali⃝, Rahman Md Moshikur⃝, and Masahiro Goto⃝

Abstract Surfactant-based systems have been used in various fields such as food, cosmetics, detergents, and pharmaceuticals. However, many traditional surfactants are not eco-friendly or biodegradable. Recently, surface-active ionic liquids (SAILs) have attracted significant attention as alternatives to traditional surfactants owing to their non-volatility, negligible vapor pressure, high thermal stability, and superior surface activity. The physicochemical properties of SAILs can be tuned by controlling the nature of both the cations and anions. Four main categories of cations—imidazolium, ammonium, pyridinium or pyrrolidinium, and phosphonium—are often paired with various anions in the design of SAILs. These SAILs generally have better surface-active properties than traditional surfactants and have also been used to create molecular assemblies such as micelles, reverse micelles, and vesicles. In addition, SAILs can be used as surfactants/co-surfactants that help to form stable microemulsions (MEs) in aqueous or non-aqueous media. SAIL-based MEs are potential smart drug delivery vehicles for topical and transdermal drug delivery owing to their ability to dissolve water-insoluble drugs. SAILs have also been used as surfactants for the extraction and separation of various proteins and enzymes owing to the facile adaptability of their properties. Finally, this chapter presents SAILs as a potent biocompatible alternative to conventional surfactants and details why SAIL-based drug formulations represent favorable target drug delivery systems for pharmaceutical applications.

M. K. Ali · R. Md Moshikur · M. Goto (✉)
Department of Applied Chemistry, Graduate School of Engineering, Kyushu University, 744
Motooka, Nishi-ku, Fukuoka 819-0395, Japan
e-mail: m-goto@mail.cstm.kyushu-u.ac.jp

M. K. Ali
Department of Chemistry, Jashore University of Science and Technology, Jashore, Bangladesh

M. Goto
Advanced Transdermal Drug Delivery System Centre, Kyushu University, 744 Motooka,
Nishi-ku, Fukuoka 819-0395, Japan

Division of Biotechnology, Centre for Future Chemistry, Kyushu University, 744 Motooka,
Nishi-ku, Fukuoka 819-0395, Japan

Keywords Surface-active ionic liquid · Micellar system · Microemulsion · Vesicles · Permeation enhancer · Drug delivery

Abbreviations

[C$_4$MIM] [C$_{12}$SO$_4$]	1-Butyl-3-methylimidazolium dodecyl sulfate
[C$_4$MP][C$_{12}$SO$_4$]	N-butyl-N-methylpyrrolidinium dodecyl sulfate
SDS	Sodium dodecyl sulfate
[C$_{12}$MIM][Br]	1-Dodecyl-3-methylimidazolium bromide
[C$_{12}$MP][Br]	N-dodecyl-N-methylpyrrolidinium bromide
[C$_{12}$mim][Br]	1-Dodecyl-3-methylimidazolium bromide
[C$_{14}$MIM][Br]	1-Tetradecyl-3-methylimidazolium bromide
[C$_{16}$MIM][Br]	1-Hexadecyl-3-methylimidazolium bromide
[C$_4$MIM][(C$_{12}$H$_2$)$_5$SO$_3$]	Dodecyl sulfonate
SDDS	Sodium dodecyl sulfonate
[C$_4$MIM][(C$_{12}$H$_2$)$_5$SO$_4$]	1-Butyl-3-methylimidazolium dodecyl sulfonate
[Na][DBS]	Sodium dodecyl benzene sulfonate
[N1114][DBS]	N-butyltrimethylammonium dodecyl benzene sulfonate
[C$_4$MIM][DBS]	1-Butyl-3-methylimidazolium dodecyl benzene sulfonate
[C$_4$Py]DBS	N-Butylpyridinium dodecyl benzene sulfonate
[C$_4$MIM][BF$_4$]	1-Butyl-3-methylimidazolium tetrafluoroborate
DH	Dopamine hydrochloride
AC	Acetylcholine chloride
TTAB	Tetradecyltrimethylammonium bromide
ME	Microemulsion
[C$_1$MIM][(MeO)$_2$PO$_2$]	1-Methyl-3-methylimidazolium dimethyl phosphate
[C$_6$MIM][Cl]	1-Hexyl-3-methylimidazolium chloride
[C$_4$MIM][PF$_6$]	1-Butyl-3-methylimidazolium hexafluorophosphate
[C$_4$MIM][Br]	1-Butyl-3-methylimidazolium bromide
[C$_{10}$MIM][Br]	1-Decyl-3-methylimidazolium bromide
[C$_4$MIM][PF$_6$]	1-Butyl-3-methylimidazolium hexafluorophosphate
[C$_{14}$MIM][Br]	1-Tetradecyl-3-methylimidazolium bromide
[C$_8$MIM][C$_{12}$OSO$_3$]	1-Methyl-3-octylimidazolium dodecylsulfate
[C$_{10}$MIM][Cl]	1-Decyl-3-methylimidazolium chloride
[C$_{16}$MIM][Br]	1-Hexadecyl-3-methylimidazolium bromide

9.1 Introduction

Ionic liquids (ILs) are extraordinary chemical compounds that have novel applications in various fields of modern sciences. Because of their extraordinary physicochemical properties and highly tunable nature, ILs have been developed as important components in the areas of synthesis, catalysis, electrochemistry, extraction, analytics, and biotechnology (Earle and Seddon 2007). However, the biological activities of ILs and their application in pharmaceutics and medicine are newly emerging (Ghandi 2014; Moshikur et al. 2019). It has been well described that most pharmaceutical industries are facing unprecedented challenges for designing the smart drug delivery systems of poorly water-soluble drugs over the years due to their limited solubility, stability, bioavailability, and polymeric transformation (Dobler et al. 2013a; Kalepu and Nekkanti 2015; Ratti 2014; Zhang et al. 2014; Moshikur et al. 2020a). To solve this issue, ILs are used as solvents, co-solvents, surfactants, and/or for the formation of API-ILs (Moshikur et al. 2020b; Moshikur et al. 2018). ILs have subsequently been reported as tunable green candidates for biological applications such as enzymatic processes, whole-cell biocatalysis, and protein stabilization (W. Jin et al. 2015; C.-Z. Liu et al. 2012; Q.-P. Liu et al. 2012). Recently, surface-active ILs (SAILs) have attracted significant interest as possible alternatives to traditional surfactants owing to their controllable properties. In some cases, these ILs show better surface-active behavior than traditional surfactants and are used to form supramolecular nano to giant aggregates such as vesicles, multilayered vesicles, micelles, and wormlike micelles, and are also used in chemical, industrial, and pharmaceutical applications (Gehlot et al. 2017a; Ghosh and Dey 2015; Wang et al. 2013a; Ali et al. 2021). SAIL compounds are also finding extensive applications in DNA stabilization, micellar catalysis, drug encapsulation, and enzyme/protein stabilization in detergent development.

SAILs can be synthesized by modifying the structure of the anion or cation. This tunability has enhanced the development of numerous surfactants as it allows for significant structural variation leading to control of the general aggregation behavior. It has been shown that when the carbon number of the anion or cation exceeds eight ($n > 8$), SAILs show surfactant-like behavior because of their intrinsic amphiphilicity (Chen et al. 2010). The amphiphilicity of SAILs can be improved by linking the medium-to-long alkyl chains to cations and/or anions and using them as surfactants. These long chain-containing SAILs can form micelles by self-assembly in aqueous solutions. The aggregation into micelles is observed when SAILs with long alkyl chains as substituents of the cationic head group are dissolved in water (Fig. 9.1). In recent years, several research groups have established and examined

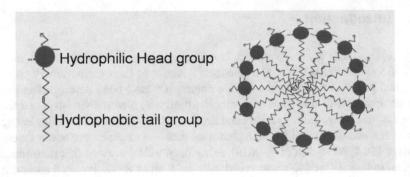

Fig. 9.1 Formation of a micelle in an aqueous medium

the self-assembly in water and IL media of various types of SAILs that are similar to conventional surfactants such as cationic, anionic, zwitterionic and bi-amphiphilic, or catanionic surfactants (Banjare et al. 2017; Y. Jin et al. 2015; Srinivasa Rao et al. 2014; Zhou and Zhao 2009). SAILs of imidazolium have been used to stabilize metal nanoparticles by covering the surface charge (Banjare et al. 2017; Jiao et al. 2012). The head group of imidazolium has been stabilizing capability in the chemical and photochemical synthesis of gold nanoparticles through oxidizing itself to a ketone (Banjare et al. 2017; Liu et al. 2013; Wang et al. 2013b). Bio-based amino acid-derived SAILs have shown numerous advantages; for example, in the synthesis of shape-controlled nano-compounds and the reduction of unsafe algal blooms. These types of SAILs have been used to improve the emulsifying capacity—particularly in the preparation of IL-containing microemulsions—in cases where traditional anionic surfactants are insoluble (Bordes et al. 2009; Bordes and Holmberg 2011).

The impetus to investigate SAILs stemmed from their potential for application in the medical and pharmaceutical fields. The development of such applications depends on the relationship between the molecular structure of the surfactant and the characteristics of its solution. SAILs have been effectively used as alternatives to traditional organic solvents in extraction procedures and as modifiers of the sorbent molecular structure, with excellent performance (Heidarizadeh and Majdi-nasab 2015). SAILs have also attracted attention as potential smart drug delivery vehicles for topical and transdermal drug delivery due to their exceptional ability to dissolve a wide range of compounds such as hydrophobic and hydrophilic molecules for skin delivery of water-soluble drugs (Goindi et al. 2015; Mahajan et al.2012; Zhou et al. 2014; Tourné-Péteilh et al. 2014). SAILs are also considered potential surfactants for the extraction (Mao et al. 2014; Sun et al. 2014; Wang et al. 2015) and separation (Hasan et al. 2015; Ventura et al. 2012) of various biomacromolecules such as proteins because of their structure, which can be easily altered to introduce tunable characteristics. Thus, SAILs are considered potential candidates for the development and refinement of pharmaceutics, for delivery of low solubility drugs, and as DDSs. This chapter presents the current progress in the medical and pharmaceutical application of SAILs in the previous decade. We also believe that our summary will provide a platform for new readers and researchers to obtain a grounding in the biomedical application of SAILs in the pharmaceutical industry.

9.2 Surface-Active Ionic Liquids

Surface-active compounds or surfactants are an interesting class of chemical compounds that are used in different areas of industry, including pharmaceuticals, food, petroleum, and cosmetics. The main properties of these substances are their ability to reduce interfacial and surface tension and to form and stabilize oil-in-water or water-in-oil or non-aqueous emulsions (Gehlot et al. 2017b; Pillai et al. 2017). Surfactants are generally amphiphilic molecules that contain both hydrophilic and hydrophobic groups; the long alkyl carbon chain-containing part is described as apolar, while the polar part is more variable and can be ionic (cationic or anionic) or non-ionic (Galgano and El Seoud 2011; Gehlot et al. 2017b; Pillai et al. 2017). The high market demand for surfactants is presently met by several synthetic, principally petroleum-based or chemical surfactants. They are generally known to be toxic to the environment and are not biodegradable. These compounds also form unstable micelles in aqueous media when using a large amount of surfactant (Shah et al. 2018a). To address these limitations, researchers have attempted to enhance their surface-active behavior by changing the physicochemical conditions. SAILs are environmentally friendly surfactants that have attracted considerable attention as promising substitutes for traditional surfactants owing to their favorable properties, for example, high thermal stability, negligible vapor pressure, biodegradability, non-volatility, and non-flammability (Gehlot et al. 2017b; Trivedi et al. 2011). It is well known that various ILs are dependent on mixtures of cations and anions that show surfactant-like behavior, which places them in the category of ionic surfactants, so they are usually termed SAILs (Y. Jin et al. 2015). These SAILs occasionally have better surface-active properties than conventional surfactants and have also been used for creating various kinds of molecular assemblies such as reverse micelles, normal micelles, and vesicles (Chauhan et al. 2017; Galgano and El Seoud 2011; Singh and Kumar 2007). In 2004, Sirieix-Plenet and Bowers et al. first investigated the amphiphilic nature of ILs in water (Bowers et al. 2004; Le Vot et al. 2014). Sirieix-Plenet et al. also described the formation of micelles of IL $C_{10}MIM$-Br in water. They found that the IL aggregated and formed micelles at low concentrations, but at higher concentrations, it assumed a multifaceted structure with interpenetrated domains of the water and electrolyte. SAILs have therefore been used as possible alternatives to conventional surfactants for different industries. SAILs also have numerous advantages over conventional surfactants of analogous alkyl chain length (Box 9.1).

Box 9.1 Advantages of SAILs in Comparison to Traditional Surfactants
- ILs are user-friendly because of their melting temperatures or glass transition temperatures below 100 °C
- SAILs are tunable substances owing to the potential variety of cation and anion combinations
- Some SAILs have a comparatively higher viscosity than surfactant solutions, which is beneficial to change in the ratio of mobility
- SAILs are more thermally and chemically stable over wide liquid ranges
- Sometimes co-surfactants are needed with surfactants to improve micelle stability. However, SAILs are able to form stable micelles without co-surfactants because of their strong cohesive forces
- Generally, SAILs have a wide range of miscibility and solubility
- SAILs are considered non-flammable
- The nature of the micellar solution can be easily changed by altering the molecular structure of the cation–anion pair.

9.3 Self-Assembly Properties of SAILs

Self-assembled and self-organized structures of molecules are considered an essential research area. SAILs are amphiphilic molecules with one or more tails comprising long alkyl hydrophobic chains and a hydrophilic charged head group like traditional surfactants. The self-assembly behavior of SAILs into different aggregated forms is controlled by their dual nature of polar and apolar groups (Bowers et al. 2004; Vanyur et al. 2007; Miskolczy et al. 2004; Jungnickel et al. 2008; Geng et al. 2010). Different SAILs with various cations such as morpholinium, imidazolium, pyridinium, and pyrrolidinium have been widely investigated to determine their micellar and interfacial behavior owing to their improved effectiveness in numerous fields such as food, cosmetics, detergents, and pharmaceuticals. The self-organizing characteristics of ILs have been studied using computational simulations (Smirnova and Safonova 2010; Jiang et al. 2007). Simulation studies indicate the formation of two kinds of domain—positively charged imidazolium rings and anions organized in a three-dimensional arrangement aided by strong electrostatic interactions and alkyl groups that form non-polar regions with short-ranged van der Waals interactions. A brief description of the numerous modes of self-assembly of SAIL systems, including micellar formation, vesicle/gel formation, and IL-microemulsion systems, will be given.

9.3.1 SAILs in Micellar Systems

The aggregation behavior of SAILs in water plays a vital role in micellar formation. The self-aggregation of surfactants in solutions produces micelles, which only form when the concentration of surfactant reaches the critical micelle concentration

(CMC). According to colloidal and surface chemistry, the concentration of surfactants above which micelles are formed and all added surfactants to the system will form micelles is defined as the critical micelle concentration (CMC). Characterization of the CMC allows exact features of SAILs that affect the properties of these ionic liquids to be identified. The CMCs of SAILs in water have been investigated using different experimental and computational techniques such as conductivity measurements, surface tension (ST), potentiometry, fluorescent probes, nuclear magnetic resonance spectroscopy (NMR), isothermal titration calorimetry, mass spectrometry, dynamic light scattering (DLS), molecular dynamics simulations (MD), and small-angle X-ray and neutron scattering (SAXS and SANS), among others. All of the techniques indicate that the CMC is extremely dependent on the structure of the ionic liquid, and a linear connection between the number of carbon atoms in the alkyl chain and the logarithm of the CMC has been established (Gehlot et al. 2017b; Pillai et al. 2017; Shah et al. 2018b; Singh et al. 2016; Singh and Kumar 2007). For example, 1-alkyl-3-methylimidazolium salts formed spherical or elongated micelles in water at low concentrations but the shape of the micelles altered as the concentrations increased (Galgano and El Seoud 2011; Poh et al. 2019). Jingjing et al. (2012) reported that the SAILs $[C_4MIM][C_{12}SO_4]$ and $[C_4MP][C_{12}SO_4]$ had superior surface activity at a given temperature compared with the conventional ionic surfactant SDS. The formation of micelles in both SAILs was assisted by the reduced hydration of the large imidazolium or pyrrolidinium cations, which minimized the electrostatic repulsion between the IL head groups. Rao et al. (2015) made the same observation for SAILs containing three different cationic moieties with dodecylbenzenesulfonate—$[N_{1114}]DBS$, $[C_4MIM]DBS$, and $[C_4Py]DBS$—when compared with [Na]DBS.

The CMC values of gemini dicationic SAILs are much lower than those of monocationic SAILs owing to the presence of two hydrophobic alkyl chains. Baltazar et al. (2007) reported that the CMC values of imidazolium gemini surfactants are twofold lower than those of the monocationic surfactants. Hence, the nature and structure of gemini SAILs have a significant effect on CMC values. The CMC values decrease with the increasing alkyl chain length of the spacer because the spacers penetrate into the hydrophobic core of the micelles as well as increasing the intramolecular hydrophobic interaction (Zana et al. 1996). In a recent study, Goto and co-workers synthesized several cholinium fatty acid-based SAIL surfactants and investigated their aggregation behavior and cytotoxicity (Ali et al. 2019). They reported that the CMC and surface tension values of SAILs are significantly lower than those of the traditional surfactants sodium dodecylbenzene sulfate (SDS) and sodium dodecylbenzene sulfate (SDBS), implying that the cholinium fatty acid-based SAILs are highly efficient surfactants. This result is relatively interesting for designing systems using limited amounts of surfactants. The formation of micelles in SAILs is summarized in Table 9.1.

Table 9.1 Critical micelle concentration (CMC) (mM) and surface tension γ_{cmc} (mN/m) measurement in aqueous solution at $T = 298$ K

No	SAILs	CMC (mM)	γ_{cmc} (mN/m)	References
1	[C$_{12}$MIM]Br	10.9	39.4	Bin et al. (2008)
2	[C$_{14}$MIM]Br	2.8	39.2	
3	[C$_{16}$MIM]Br	0.55	39.1	
4	[C$_4$MIM] [C$_{12}$SO$_4$]	1.8	31.9	Jingjing et al. (2012)
5	[C$_4$MP] [C$_{12}$SO$_4$]	2.7	34.3	
6	[C$_{12}$MP]Br	13.5	42.4	
7	[C$_4$MIM][C$_{12}$H$_{25}$SO$_3$]	4.4	36.9	Yanxin et al. (2015)
8	SDDS	7.4	37.6	
9	[C$_4$MIM][C$_{12}$H$_{25}$SO$_4$]	1.8	31.9	
10	[Na]DBS	2.75	32.49	Rao et al. (2015)
11	[N$_{1114}$]DBS	1.34	29.11	
12	[C4MIM]DBS	1.08	29.18	
13	[C$_4$Py]DBS	0.92	29.69	
14	SALSIL 1	2.42	39.6	Mustahil et al. (2019)
15	SALSIL 2	2.44	37.1	
16	SALSIL 3	2.62	36.31	
17	SALSIL 4	2.67	36.24	
18	SALSIL 5	2.79	31.72	
19	SALSIL 6	2.8	33.40	
20	SALSIL 7	2.46	27.3	
21	SALSIL 8	1.15	29.81	
22	SALSIL 9	1.02	30.63	
23	[Cho][Ole]	1.7	24.4	Ali et al. (2019)
24	[Cho][Lin]	2.0	26.6	
25	[Cho][Eru]	0.8	23.2	
26	[Cho][NLS]	3.5	37.1	Moshikur et al. (2020c)
27	[Cho][NLG]	2.4	36.5	
28	[Cho]$_2$[NLA]	1.5	34.5	
29	[C$_{12}$-2-C$_{12}$IM]Br$_2$	0.55	33.6	Ao et al. 2009
30	[C$_{12}$-4-C$_{12}$IM]Br$_2$	0.72	35.7	
31	[C$_{12}$-6-C$_{12}$IM]Br$_2$	0.78	39.5	
32	[C$_{10}$-4-C$_{10}$IM]Br$_2$	4.5	35.2	Ao et al. 2008
33	[C$_{14}$-4-C$_{14}$ IM]Br$_2$	0.1	37.2	
34	[12-(S-2-S)-12]IM	0.4	39.7	Bhadani et al. 2011
35	[12-(S-4-S)-12]IM	0.31	40.8	
36	C$_6$[C$_{10}$ArOCH$_2$IM]$_2$•2Br	0.00302	39.8	Sunitha et al. 2011

(continued)

Table 9.1 (continued)

No	SAILs	CMC (mM)	γ_{cmc} (mN/m)	References
37	$C_6[C_{12}ArOCH_2IM]_2 \bullet 2Br$	0.0002	42.8	

9.3.2 SAILs in Vesicle Systems

Bilayer surfactant systems are known as vesicles and are formed by dispersing surface-active molecules in an aqueous phase; various vesicle sizes can be achieved depending on the production method. Vesicular systems have been widely used in pharmaceutical research owing to their similarity to biological membranes and are mainly used in drug delivery systems and drug encapsulation (Huiyong et al. 2013; Rao et al. 2015; Qi and Shiping 2015). Classical vesicular systems were first introduced in the 1960s but they were not used for drug delivery purposes until the 1980s. Vesicles (i.e., liposomes) have been reported to exist as small or large unilamellar vesicles, oligolamellar and multilamellar vesicles, or multivesicular liposomes. Liposomes generally comprise one or more phospholipid bilayers that surround an aqueous core. The Kimizuka and Nakashima (2001) groups demonstrated an unprecedented vesicle that is formed owing to the tendency of ILs to spontaneously assemble. Because of the variety and a huge number of possible cation and anion combinations, these ILs are known as designer solvents and have the potential to encapsulate pharmaceutically active compounds or facilitate the encapsulation of drugs in the micellar phase before vesicle formation (Rao et al. 2015). Sarkar and co-workers (2017) synthesized cholesterol-based SAILs that were used in the preparation of vesicles in water and reported that these SAILs have potential as biomimetic models and drug carriers. Moshikur et al. (2020) synthesized amino acid-based SAILs—choline N-lauroyl glycinate ([Cho][NLG]) and dicholine N-lauroyl aspartate ([Cho]$_2$[NLA])—which exhibited efficient hydrophobic ion pairing between the SAILs and the macromolecular drug heparin by primarily forming as unilamellar vesicles in water. The formation of micelles and vesicles using SAIL-based surfactants is depicted in Fig. 9.2.

9.3.3 SAILs in Microemulsion Systems

Microemulsions (MEs) are of great interest because of their remarkable properties such as the ability to solubilize both polar and non-polar molecules, poor interfacial tension, a large interfacial region, spontaneous formation, and good microstructures, which render them promising candidates for various applications (Hejazifar et al. 2020; Kandasamy et al. 2018; Kuchlyan et al. 2016; Moniruzzaman et al. 2010a, b). MEs are microheterogeneous systems consisting of two immiscible fluids stabilized through an interfacial film of surfactants, often with co-surfactants (Hejazifar et al. 2020). The Winsor classification (Fig. 9.3) groups them into four different categories

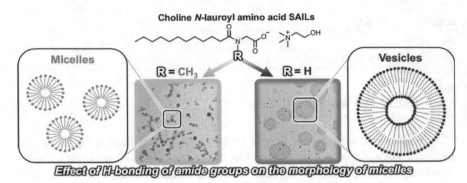

Fig. 9.2 Formation of micelles and vesicles in an aqueous phase using choline N-lauroyl amino acid-based SAILs. Figure reproduced with permission from (Moshikur et al. 2020c)

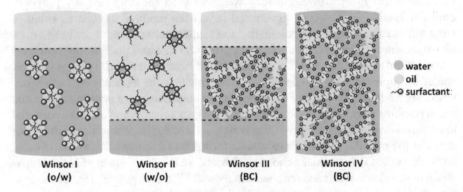

Fig. 9.3 The Winsor classification of microemulsions. Figure reproduced with permission from (Hejazifar et al. 2020)

depending on their microstructure—such as oil-in-water (Winsor I), water-in-oil (Winsor II), and bicontinuous (Winsor III)—which is influenced by their physicochemical properties and the proportions of their components (Hejazifar et al. 2020; Moniruzzaman et al. 2010a). In addition, Winsor IV MEs can be prepared where the components are not in excess. In recent years, an increasing number of studies have focused on MEs in which an ionic liquid is utilized as the oil phase, the aqueous phase, and the surfactant. Moreover, ILs are considered, as a rule, to be amphiphilic, which aids the development of MEs in aqueous or non-aqueous phases because of their special surface activity. Recently, SAIL-based MEs have been developed that are receiving increased interest compared with the other two kinds of MEs.

Water-in-oil MEs with the aerosol OT (AOT) surfactant have been comprehensively reported in the literature (Srinivasa Rao et al. 2014). Nevertheless, it is very unusual to form IL/oil MEs using the AOT surfactant. The formation of w/o microemulsions using the AOT surfactant is facile because of the favorable interaction between the water molecules and the inorganic cation of AOT (Na^+, NH_4^+, etc.). When water is replaced by an IL, this kind of interaction does not exist (Kuchlyan et al. 2016; Srinivasa Rao et al. 2014). However, when the inorganic cation containing AOT surfactant is substituted by an organic cation, the formation of an IL in oil (IL/o) ME is possible. Utilizing this idea, Sarkar et al. used an anion exchange reaction to synthesize [C$_4$MIM][AOT] by combining NaAOT and 1-butyl-3-methylimidazolium bromide [C$_4$MIM]Br and they succeeded in formulating an IL/o ME using [C$_4$MIM][AOT] as the surfactant, various types of IL as polar solvents, and benzene as the non-polar solvent (Rao et al. 2012a). DLS measurements, a phase diagram, and Coumarin-480 emission spectra were used to characterize the ternary formulate ME system. Rao et al. further reported that the uptake efficacy of double-chained SAIL was improved by extending the chain length of the SAIL (Rao et al. 2013b, 2012a). The loading capacity of the IL also increased as the anion/cation interaction strength of the added ILs was reduced and followed the order [C$_4$MIM][TF$_2$N] > [C$_4$MIM][PF$_6$] > [C$_4$MIM][BF$_4$]. Consequently, they established that the quantity of IL in the center of an ME can be easily controlled by altering the behavior of the IL (Rao et al. 2012b). One of the biggest advantages of IL/o MEs is their high thermal stability at a wide range of temperatures. Sarkar et al. prepared a high temperature stable ME using a SAIL of N,N-dimethylethanolammonium 1,4-bis(2-ethylhexyl) sulfosuccinate (DAAOT) where the IL was the polar medium and IPM was the non-polar solvent (Rao et al. 2013a). In addition, Banerjee and co-workers prepared the triple chain SAIL [BHD][AOT] using the anion exchange reaction between benzyl-nhexadecyldimethylammonium chloride (BHDC) and Na-AOT(Rao et al. 2012b). The two triple chain-containing SAILs synthesized effectively created an IL-based ME in IPM and two hydrophobic ILs, N-methyl-N-propylpyrrolidinium bis(trifluoromethanesulfonyl) imide ([P$_{13}$][TF$_2$N]) and N,N,N trimethyl-N-propylammonium bis-(trifluoromethanesulfonyl) imide ([N$_{3111}$][TF$_2$N]). Moreover, these SAILs also formed large unilamellar vesicles (LUVs) in aqueous media with non-polar solvents (Kuchlyan et al. 2016). AOT-free SAIL-based MEs have also been reported in the literature. Kunz et al. prepared MEs using a SAIL-based surfactant (1-hexadecyl-3-methylimidazolium chloride ([C$_{16}$MIM]Cl)), decanol co-surfactant, RTILs (ethyl-ammonium nitrate (EAN), and ([B$_{MIM}$]BF$_4$)) as the polar medium, and dodecane as the non-polar medium (Zech et al. 2009). The formation of spherical aggregates was demonstrated using SAXS measurements. The rigidity of the MEs altered when the IL was changed from EAN to [B$_{MIM}$]-BF$_4$. Like cationic SAILs, long chain-containing anion-based SAILs could also be used to formulate IL MEs. A recent study showed that IL/o MEs can be prepared using the SAIL 1-butyl-3-methylimidazolium dodecyl sulfate ([B$_{MIM}$][DodSO$_4$]), 1-ethyl-3-methylimidazolium ethylsulfate ([E$_{MIM}$][EtSO$_4$]) as the polar solvent, and toluene as a non-polar solvent (Rojas et al. 2013). The use of polar solvents such as [E$_{MIM}$][EtSO$_4$] has certain advantages over various ILs including low toxicity, a relatively wide window of fluidity, and low melting temperature. The swelling of MEs with the gradual addition of IL can be demonstrated using SAXS and DLS experiments (Kuchlyan et al. 2016). Goto and co-workers have developed non-aqueous MEs IL[C$_1$MIM][(CH$_3$O)$_2$PO$_2$]/(SAIL + Span-20)/IPM and IL/(SAIL[Cho][Ole] + Span-20)/IPM systems using SAILs. They reported that the SAILs are promising alternatives to traditional surfactants for biomedical applications (Ali et al. 2020). Hence, SAILs that have a certain surface activity due to long alkyl chain groups and that can self-aggregate have increased the use of ILs in MEs. In addition, MEs containing ILs show flexibility acting as polar or non-polar phases, and the amphiphilic nature of ILs allows their use as additives or surfactants that stabilize the microemulsion systems. Owing to their superior surface activity and tunable properties, IL and SAIL-based microemulsions have potential applications in a wide range of pharmaceutical and biomedical fields.

9.4 Application of SAILs in Medical and Pharmaceutical Systems

9.4.1 Pharmaceutical Applications of SAILs

The development of efficient drug delivery systems (DDSs) for managing diseases without compromising the safety and efficacy of drugs is a challenge. Researchers are continuing to develop new formulations and improve current formulations so that the therapeutic potential of drugs can be realized. Nearly 40% of the top 200 oral drugs marketed in the US and Europe; almost 90% of the developed pipeline drugs; 75% of compounds under development; and 33% of FDA drugs have been reported to be poorly soluble compounds (Gao et al. 2017; Rodriguez-Aller et al. 2015; Shamshina et al. 2013). In addition to poor solubility, some marketed drugs have shown low permeability, rapid metabolism and elimination from the body, as well as poor safety and tolerability (Hodgson 2001; Homayun et al. 2019; Kalepu and Nekkanti 2015). Thus, it is essential to formulate drugs with organized polymorphic forms as well as precise crystal size and product solubility. ILs are increasingly being considered environmentally safe and green solvents and can comprise biodegradable, biocompatible, and natural ions, such as amino acids and choline, or ions with recognized biological activity (Klein et al. 2013; Li et al. 2013; Trivedi et al. 2011). Therefore, ILs have shown promise for addressing the potential challenges of solid pharmaceuticals for effective drug delivery. Because of their surprising solvating power and superior surface activity, ILs can improve the solubility of various sparingly soluble drugs and significantly improve the drug penetration over biological obstacles to improve the therapeutic efficacy.

Numerous reports have discussed the interaction between drugs and SAIL surfactants (Enache and Volanschi 2011; Sharma and Mahajan 2012). The solubility of sparingly water-soluble drugs can be improved using SAILs because of their ability to reduce the interfacial tension between the solution and the drug. In addition, SAILs have been used as permeation enhancers in the skin delivery of poor lipid solubility drugs, acting as an essential component of emulsion-based formulations that increased drug permeation and absorption. For example, Patra and Barakat (2011) reported the association of drug-surfactants where the SAIL [C$_4$MIM][BF$_4$] acts as a modulator. The association between anionic surfactants such as sodium dodecyl sulfate (SDS) and curcumin was greater than that for SDS in the absence of SAILs. Without SAILs, there was repulsion between the negative charge of the deprotonated form of curcumin and the negative charge of the head group (sulfate ion) of SDS. In addition, the association between curcumin and neutral Triton X-100 (TX100) was improved by the addition of SAIL. This may be because of the presence of hydrogen bond and dipole–dipole interactions in the positive charge group of the SAIL with TX100, which increase the interactions or intense associations between the curcumin and neutral surfactant solution. Mahajan et al. (2012) demonstrated the drug binding ability between the SAIL [C_{14}MIM][Br] and acetylcholine chloride (AC) and dopamine hydrochloride (DH)—drugs that are widely used to treat various heart diseases. The results were compared with those for a cationic surfactant, tetradecyltrimethylammonium bromide (TTAB), which is structurally similar, and found that SAIL [C_{14}MIM][Br] was a better carrier for delivery of AC and DH. DH binds more strongly with [C_{14}MIM][Br] than AC because of the π-cation interactions between the aromatic part of DH and the positive charge of the surfactant components. In addition, the binding ability of SAIL [C_{14}MIM][Br] with DH was higher than that of TTAB owing to the presence of π–π interactions between the π system of the imidazolium ring of [C_{14}MIM][Br] and DH. However, the CMC value of SAIL [C_{14}MIM][Br] was lower than that of TTAB because of the variation in their head groups. Moreover, the positive charge of TTAB behaves as a point charge, whereas the charge is delocalized in the imidazolium ring of SAIL [C_{14}MIM][Br], which creates a steric hindrance to the formation of aggregates.

MEs have recently attracted significant attention in pharmacological research because of their ability to act as drug delivery vehicles, as well as their biocompatibility, long-term stability, nanometer-sized aggregations, and their straightforward preparation. Moniruzzaman et al. (2010) reported IL-based ME systems (IL-in-oil microemulsions) in which the drug-loaded IL was dispersed in oil in the presence of a surfactant and assessed their ability to solubilize drugs such as acyclovir, methotrexate, and dantrolene sodium, which are insoluble and/or poorly soluble in aqueous media. The solubility study demonstrated that the solubility of acyclovir increased 500 times in dimethylimidazolium dimethylphosphate IL and increased the transdermal delivery six times compared with those of commercial acyclovir creams. Dobler et al. (2013) subsequently investigated the impact of ILs—[C_4MIM][PF_6] and [C_6MIM][Cl]—on traditional water-in-oil (w/o) MEs and oil-in-water (o/w) MEs and showed that both ILs were successfully absorbed into the core of MEs and formed stable formulations. In addition, [C_6MIM][Cl] caused a significant reduction of the surface tension of the ME, which depended on its concentration in water, and the addition of [C_4MIM][PF_6] enhanced the excellent penetration of the drug into the skin as a result of its hydrophobicity.

IL-based MEs have shown greater efficacy in skin cancer treatment than commercially available DDSs. An effective drug, 5-fluorouracil (5-FU), which is currently available in an ointment form, exhibits low skin permeation efficacy and disagreeable dermal side effects that lead to the non-compliance of patients in treatment (Hejazifar et al. 2020). Unwanted skin reactions, like erythema and pain, were avoided using IL-based MEs because the drugs were encapsulated in the internal core of the ME. Goindi et al. (2014) established an IL/o ME system, which was used in ex vivo permeation studies of a poorly water-soluble drug, 5-fluorouracil, for transdermal delivery using 1-butyl-3-methylimidazolium bromide IL [C4MIM][Br]. They demonstrated 2.3- and 1.6-fold greater transport of the drugs than by oily solution and o/w ME, respectively. Furthermore, Goindi et al. (2015) reported an IL/w ME formulation that was used in ex vivo permeation studies of a sparingly water-soluble drug, etodolac (ETO), for topical delivery using [C4MIM][PF6]. They showed that the IL/w ME effectively increased the solubility and permeability of ETO. In vivo anti-inflammatory and anti-arthritic studies of these IL/w formulations showed that ETO containing IL/w ME was more effective for controlling inflammation than o/w ME, oily solutions, and a marketed formulation of ETO. Moreover, histopathological studies showed that the system did not cause any anatomical or pathological abnormalities in the skin. In a recent study, the Goto group proposed a SAIL-based IL/o ME system for the transdermal delivery of poorly water-soluble drugs such as celecoxib, acyclovir, methotrexate, and dantrolene sodium (Ali et al. 2020). The IL/o MEs consisted of the SAIL cholinium oleate, SAIL [Cho][Ole], as a surfactant with Span-20 as a co-surfactant (Fig. 9.4a). The IL/o ME created a twofold greater ME area than a conventional Tween-80-based IL/o ME, as well as 4.7- and 5-fold higher encapsulation of CLX and ACV, respectively. The higher encapsulation efficiency for SAIL-based MEs is due to the presence of hydrogen-bonding interaction between the electron-donating groups of the drugs and the positively charged group of cholinium ([Cho +]) (Bhat et al. 2019; Singla et al. 2018). H-bonding, as well as π-type interactions, plays essential roles in encapsulating groups in the core of MEs owing to the interaction between the electron-rich conjugated π-system of the drugs and the electron-deficient π-system of the [Cho +] (Matthews et al. 2014; Schulz et al. 2014; Singla et al. 2018; Zheng et al. 2007).

Skin is considered the most important barrier for protecting the human body; however, it also offers a suitable route for pharmaceutical delivery as a result of sustained and controlled delivery, good patient compliance, greater local concentration, and avoidance of first-pass metabolism (Aboofazeli et al. 2002; Monti et al. 2017). Nevertheless, the quantity of drug that infiltrates through the skin is extremely low because of multilayers of dermal obstacles, particularly the stratum corneum (SC) (Monti et al. 2017; Wang et al. 2018). The SC—known to be a formidable barrier to the delivery of drug molecules—is made up of structurally well-organized corneocytes with lipid layers. Therefore, many hydrophilic or macromolecule drugs (> 500 Da) cannot penetrate through the skin without reducing the barrier properties by disturbing the SC and varying the structure of lipids through the use of physical (device-based) or chemical (formulation-based) permeation improvement techniques. Recently, ILs and IL microemulsions have been found to be able to improve the penetration of drugs across the skin barrier; therefore, many studies have been carried out to determine their fundamental mechanisms of action (Kubota et al. 2016; Moshikur et al. 2019; Wang et al. 2018; Zakrewsky et al. 2014). The physicochemical properties of ILs are the key factor related to permeation enhancement (Zhang et al. 2017). Many of the proposed mechanisms depend primarily on the chemical make-up

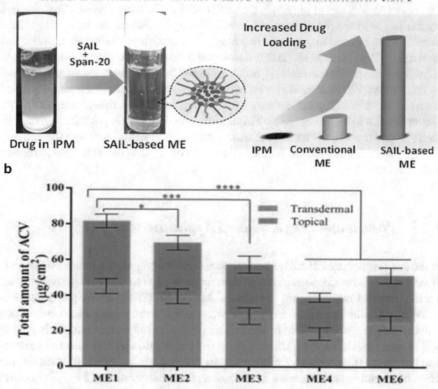

Fig. 9.4 **a** Formation of poorly soluble drug-loaded microemulsions with SAIL-based surfactant that improved the solubility of drugs. **b** Transdermal delivery of acyclovir (ACV) using various IL/o MEs where ME1 and ME6 are SAIL and conventional surfactant-based MEs, respectively. Figure reproduced with permission from (Ali et al. 2020; Islam et al. 2020)

of the ILs. Monti et al. (2017) reported that the degree of penetration depends on the structure of the IL. Conversely, the mechanism of permeation is tremendously wide-ranging and does not represent all IL permeation improvement profiles. All permeation enhancer ILs are classified as either hydrophilic or hydrophobic. Hydrophobic ILs can promote separation into the epithelial membrane by providing channels, advancing transcellular transport in the lipid sections, whereas hydrophilic ILs act by opening tight connections inside the SC, improving paracellular transport by increasing fluidization principally within protein and lipid sections (Agatemor et al. 2018). Among the best reported is 1-octyl-3-methylimidazolium-based ILs, which act by disrupting the structural integrity as a result of inserting into the membrane (Lim et al. 2015). It has also been shown that ILs enhance the fluidization of cell membranes. This was particularly observed for hydrophilic imidazolium-based ILs and lipid removal in the SC. Recently, choline and geranic acid-based ILs have shown high penetrability, antibacterial activity, and assisted transdermal delivery of proteins, and an RNAi-like cholinium-malic acid-based IL acts as a skin penetration enhancer (Aboofazeli et al. 2002; Dobler et al. 2013b; Singh 2019). Toxicity is a crucial factor in the therapeutic or biomedical application of any technology. Bio-based ILs can be used to overcome this hurdle owing to their biodegradation and low toxicity profiles. Islam et al. (2020) developed biocompatible cholinium propionate [Cho][Pro] and cholinium oleate [Cho][Ole]-based IL/o MEs to study the transdermal delivery of ACV. These SAIL-based MEs showed less skin irritation and increased the transdermal and topical delivery by 9- and 8-fold, respectively, compared with conventional w/o MEs (Fig. 9.4b). This finding will doubtlessly pave the way for ME formulations that are not traditional surfactant-based systems, although the delivery of large molecules and the mechanism of delivery of the drug through the skin require further investigation.

9.4.2   Protein and Enzyme-Based Applications of SAILs

In the last few decades, ILs have been shown to act as potential solvents/co-solvents or surfactants/co-surfactants to dissolve substances such as proteins, salts, fats, and a wide range of other organic substances. Mao et al. (2014) developed a dual-IL-based microemulsion where B_{MIM} PF6 was used instead of organic solvents and 1-decyl-3-methylimidazolium bromide (D_{MIM}Br) acted as a surfactant. This type of ME system was shown to be efficient for protein extraction with an excellent extraction efficiency (EF) for hemoglobin compared with pure B_{MIM}PF6. In addition, the EF of hemoglobin was increased by increasing the concentration of D_{MIM}Br because the aggregation number of D_{MIM}Br increased, which was effective for separating the hemoglobin from human blood. A copper-incorporating enzyme such as laccase showed catalytic activity when it was dissolved in water/ionic liquid (w/IL) ME. Sun et al. (2014) used a mixed surfactant blend of Triton X-100 and 1-tetradecyl-3-methylimidazolium bromide ([C_{14}MIM]Br) for the preparation of w/IL MEs. The formation of w/IL ME was detected by measuring electrical conductivity. In this ME system, the electrical conductivity increased linearly with increasing water concentration. Subsequently, Adak et al. (2015) investigated the thermal stability and activity of lipase from *Rhizopus oryzae* using imidazolium-based SAILs such as 1-hexadecyl-3-methylimidazolium bromide (C_{16}MIM)Br. The deactivation temperature of the *Rhizopus oryzae* lipase (rhL) increased as a result of the introduction of (C_{16}MIM)Br, and the thermal stability of the enzyme increased. At low (C_{16}MIM)Br concentration (100 μM), the enzyme activity was enhanced up to 80%.

Gene therapy by replacement of imperfect genes at the transcriptional level has become more attractive as an effective therapeutic approach as a result of the development of potential safe delivery systems (Qasba et al. 1971; Robbins et al. 1998). The efficiency of gene transfection can be improved by changing the structure of amphiphiles of self-assembled delivery systems. Recently, a novel IL-based gemini surfactant has received particular attention owing to its interaction with DNA molecules. Pyridinium-based gemini surfactants contain two positive charges and show high charge/mass ratios and high molecular adaptability. The Bhadani and Singh (2009) groups reported that pyridinium-based gemini surfactants with an ethane-1,2-dithiol spacer had low cytotoxicity by MTT assay and chain length-dependent DNA binding properties by agarose gel electrophoresis. The imidazolium-based gemini surfactants showed more propensity for self-aggregation and strong $\pi-\pi$ interaction between the aromatic rings and surfactants. In addition, the insertion of histidine molecules into the DNA condensing groups was shown to assist gene transfection by improving the proton-sponge effect. For example, with the addition of SAIL surfactants like $[C_n\text{-}4\text{-}C_nIM]Br_2$ ($n = 10, 12, 14$), the DNA molecules experience the action from densification to multimolecular condensation by changing the conformation, which is then compared with surfactants containing various chain lengths to describe the significant contribution of the interaction of hydrophobicity (Zhou et al. 2012). In addition, the $[C_{12}\text{-}4\text{-}C_{12}IM]Br_2$ can condense with DNA molecules in the exact charge ratio to give spherical particles with a diameter of 100 nm. We have also observed that $[C_{12}\text{-}4\text{-}C_{12}IM]Br_2$ shows less cytotoxicity at the concentration of transfection point, which indicates that $[C_{12}\text{-}4\text{-}C_{12}IM]Br_2$ acts as an efficient gene vector (Zhou et al. 2013). Recently, S. Mitragotri and co-workers showed effective dermal delivery of therapeutic RNAi robed with IL-moieties for treating skin diseases (Zakrewsky and Mitragotri 2016).

9.5 Conclusion

In recent years, ILs have increasingly been considered to be among the most promising emerging materials and technologies. In addition to providing substitutes in established procedures, ILs are also regarded as important candidates for solving critical problems such as achieving clean and efficient energy. Researchers are facing unparalleled challenges in the development of effective drug delivery systems for many drugs owing to their polymorphism, and limited solubility, permeability, and bioavailability. IL-based formulations provide a promising approach for addressing these limitations through the design of smart delivery systems. In particular, ILs have been found to play a unique role in the pharmaceutical field owing to their clean and multifunctional properties including negligible vapor pressure at comparatively ambient conditions, high thermal and chemical stability, high surface activity and electrochemical stability, and broadly tunable properties in terms of polarity, hydrophobicity, and solvent miscibility. One of the major problems in the pharmaceutical industry is the increasing number of active pharmaceutical ingredients (APIs) with low water solubility, class II drugs in the biopharmaceuticals classification system. In recent years, ILs have been considered solvents and/or materials with the potential to enhance therapeutic performance by increasing solubility and stability and improving drug delivery. In the past decade, IL-based microemulsions have been investigated as promising nanocarriers for TDD. ILs are considered tunable designer solvents and play a significant role in all phases of microemulsion systems by providing alternatives to the water, oil, and surfactant components.

Recently, researchers have prepared surface-active ILs (SAILs) to act as surfactants by combining cations and anions containing long alkyl chains. These SAILs allow the formation of various organized assemblies and improve the physicothermal stabilities of MEs compared with traditional aqueous systems. In this chapter, we primarily focused on surface-active ionic liquids (SAILs) and evaluated their feasibility for medical and pharmaceutical applications.

Acknowledgments The authors thank Sarah Dodds, PhD, from Edanz (https://jp.edanz.com/ac) for editing a draft of this manuscript.

References

Aboofazeli R, Zia H, Needham TE (2002) Transdermal Delivery of Nicardipine: An Approach to In Vitro Permeation Enhancement. Drug Deliv 9:239–247. https://doi.org/10.1080/10717540260397855

Agatemor C, Ibsen KN, Tanner EEL, Mitragotri S (2018) Ionic liquids for addressing unmet needs in healthcare. Bioeng. Transl. Med. 3:7–25. https://doi.org/10.1002/btm2.10083

Ali MK, Moshikur RM, Wakabayashi R, Moniruzzaman M, Goto M (2021) Biocompatible Ionic Liquid-Mediated Micelles for Enhanced Transdermal Delivery of Paclitaxel. ACS Appl Mater Interfaces 13:19745–19755. https://doi.org/10.1021/acsami.1c03111

Ali MK, Moshikur RM, Wakabayashi R, Moniruzzaman M, Kamiya N, Goto M (2020) Biocompatible Ionic Liquid Surfactant-Based Microemulsion as a Potential Carrier for Sparingly Soluble Drugs. ACS Sustain. Chem. Eng. 8:6263–6272. https://doi.org/10.1021/acssuschemeng.9b07773

Ali MK, Moshikur RM, Wakabayashi R, Tahara Y, Moniruzzaman M, Kamiya N, Goto M (2019) Synthesis and characterization of choline–fatty-acid-based ionic liquids: A new biocompatible surfactant. J Colloid Interface Sci 551:72–80. https://doi.org/10.1016/j.jcis.2019.04.095

Ao M, Xu G, Pang J, Zhao T (2009) Comparison of aggregation behaviors between ionic liquid-type imidazolium gemini surfactant [C12-4-C12im]Br 2 and its monomer [C12mim]Br on silicon wafer. Langmuir 25:9721–9727. https://doi.org/10.1021/la901005v

Ao M, Xu G, Zhu Y, Bai Y (2008) Synthesis and properties of ionic liquid-type Gemini imidazolium surfactants. J Colloid Interface Sci 326:490–495. https://doi.org/10.1016/j.jcis.2008.06.048

Bhat AR, Wani FA, Alzahrani KA, Alshehri AA, Malik MA, Patel R (2019) Effect of rifampicin on the interfacial properties of imidazolium ionic liquids and its solubility therein. J Mol Liq 292:111347. https://doi.org/10.1016/j.molliq.2019.111347

Banjare MK, Kurrey R, Yadav T, Sinha S, Satnami ML, Ghosh KK (2017) A comparative study on the effect of imidazolium-based ionic liquid on self-aggregation of cationic, anionic and nonionic surfactants studied by surface tension, conductivity, fluorescence and FTIR spectroscopy. J Mol Liq 241:622–632. https://doi.org/10.1016/j.molliq.2017.06.009

Bhadani A, Singh S (2011) Synthesis and Properties of Thioether Spacer Containing Gemini Imidazolium Surfactant. Langmuir 27:14033–14044. https://doi.org/10.1021/la202201r

Bin D, Xueyan Z, Liqiang Z, Jin Z, Na L, Tohru I (2008) Aggregation behavior of long-chain imidazolium ionic liquids in aqueous solution: Micellization and characterization of micelle microenvironment. Colloids Surf A Physicochem Eng Asp 317:666–672. https://doi.org/10.1016/j.colsurfa.2007.12.001

Bordes R, Holmberg K (2011) Physical chemical characteristics of dicarboxylic amino acid-based surfactants. Colloids Surf A Physicochem Eng Asp 391:32–41. https://doi.org/10.1016/j.colsurfa.2011.03.023

Bordes R, Tropsch J, Holmberg K (2009) Counterion specificity of surfactants based on dicarboxylic amino acids. J Colloid Interface Sci 338:529–536. https://doi.org/10.1016/j.jcis.2009.06.032

Bowers J, Butts CP, Martin PJ, Vergara-Gutierrez MC, Heenan RK (2004) Aggregation Behavior of Aqueous Solutions of Ionic Liquids. Langmuir 20:2191–2198. https://doi.org/10.1021/la0 35940m

Chauhan S, Kaur M, Singh K, Chauhan MS, Kohli P (2017) Micellar and antimicrobial activities of ionic surfactants in aqueous solutions of synthesized tetraalkylammonium based ionic liquids. Colloids Surfaces A Physicochem. Eng. Asp. 535:232–241. https://doi.org/10.1016/j.colsurfa. 2017.09.042

Dobler D, Schmidts T, Klingenhöfer I, Runkel F (2013) Ionic liquids as ingredients in topical drug delivery systems. Int J Pharm 441:620–627. https://doi.org/10.1016/j.ijpharm.2012.10.035

Dobler D, Schmidts T, Klingenhöfer I, Runkel F (2013b) Ionic liquids as ingredients in topical drug delivery systems. Int J Pharm 441:620–627. https://doi.org/10.1016/j.ijpharm.2012.10.035

Earle MJ, Seddon KR (2007) Ionic liquids. Green solvents for the future. Pure Appl Chem 72:1391–1398. https://doi.org/10.1351/pac200072071391

Galgano PD, El Seoud OA (2011) Surface active ionic liquids: Study of the micellar properties of 1-(1-alkyl)-3-methylimidazolium chlorides and comparison with structurally related surfactants. J Colloid Interface Sci 361:186–194. https://doi.org/10.1016/j.jcis.2011.04.108

Gao Z, Rohani S, Gong J, Wang J (2017) Recent Developments in the Crystallization Process: Toward the Pharmaceutical Industry. Engineering 3:343–353. https://doi.org/10.1016/J.ENG. 2017.03.022

Gehlot PS, Kulshrestha A, Bharmoria P, Damarla K, Chokshi K, Kumar A (2017) Surface-active ionic liquid cholinium dodecylbenzenesulfonate: Self-assembling behavior and interaction with cellulase. ACS Omega 2:7451–7460. https://doi.org/10.1021/acsomega.7b01291

Ghandi K (2014) A Review of Ionic Liquids, Their Limits and Applications. Green Sustain. Chem. 04:44–53. https://doi.org/10.4236/gsc.2014.41008

Ghosh R, Dey J (2015) Aggregation behavior of poly(ethylene glycol) chain-containing anionic amphiphiles: Thermodynamic, spectroscopic and microscopic studies. J Colloid Interface Sci 451:53–62. https://doi.org/10.1016/j.jcis.2015.03.054

Hejazifar M, Lanaridi O, Bica-Schröder K (2020) Ionic liquid based microemulsions: A review. J Mol Liq 303:112264. https://doi.org/10.1016/j.molliq.2019.112264

Hodgson J (2001) ADMET - Turning chemicals into drugs. Rapidly resolving the pharmacokinetic and toxicological properties of drug candidates remains a key challenge for drug developers. Nat Biotechnol 19:722–726. https://doi.org/10.1038/90761

Homayun B, Lin X, Choi HJ (2019) Challenges and recent progress in oral drug delivery systems for biopharmaceuticals. Pharmaceutics 11(3):129. https://doi.org/10.3390/pharmaceutics11030129

Jiao J, Dong B, Zhang H, Zhao Y, Wang X, Wang R, Yu L (2012) Aggregation behaviors of dodecyl sulfate-based anionic surface active ionic liquids in water. J Phys Chem B 116:958–965. https://doi.org/10.1021/jp209276c

Jin W, Yang Q, Zhang Z, Bao Z, Ren Q, Yang Y, Xing H (2015) Self-assembly induced solubilization of drug-like molecules in nanostructured ionic liquids. Chem Commun 51:13170–13173. https://doi.org/10.1039/c5cc03463a

Jin Y, Wang L, Wang T, Chen P, Bi Y, Yu L (2015) Aggregation behavior of dodecylsulfonate-based surface active ionic liquids in water. J Mol Liq 212:23–29. https://doi.org/10.1016/j.molliq.2015. 08.049

Jingjing J, Bin D, Huina Z, Yingyuan Z, Xiaoqing W, Rui W, Li Y (2012) Aggregation behaviors of dodecyl sulfate-based anionic surface active ionic liquids in water. J Phys Chem B 16:958–965. https://doi.org/10.1021/jp209276c

Kalepu S, Nekkanti V (2015) Insoluble drug delivery strategies: Review of recent advances and business prospects. Acta Pharm. Sin. B 5:442–453. https://doi.org/10.1016/j.apsb.2015.07.003

Kandasamy S, Moniruzzaman M, Sivapragasam M, Shamsuddin MR, Mutalib MIA (2018) Formulation and characterization of acetate based ionic liquid in oil microemulsion as a carrier for acyclovir and methotrexate. Sep Purif Technol 196:149–156. https://doi.org/10.1016/j.seppur. 2017.08.044

Klein R, Müller E, Kraus B, Brunner G, Estrine B, Touraud D, Heilmann J, Kellermeier M, Kunz W (2013) Biodegradability and cytotoxicity of choline soaps on human cell lines: Effects of chain length and the cation. RSC Adv 3:23347–23354. https://doi.org/10.1039/c3ra42812e

Kubota K, Shibata A, Yamaguchi T (2016) The molecular assembly of the ionic liquid/aliphatic carboxylic acid/aliphatic amine as effective and safety transdermal permeation enhancers. Eur J Pharm Sci 86:75–83. https://doi.org/10.1016/j.ejps.2016.03.002

Kuchlyan J, Kundu N, Sarkar N (2016) Ionic liquids in microemulsions: Formulation and characterization. Curr Opin Colloid Interface Sci 25:27–38. https://doi.org/10.1016/j.cocis.2016.05.011

Le Vot S, Dambournet D, Groult H, Ngo A, Petit C, Rizzi C, Salzemann C, Sirieix-Plenet J, Borkiewicz OJ, Raymundo-Piñero E, Gaillon L (2014) Synthesis of tin nanocrystals in room temperature ionic liquids. Dalt. Trans. 43:18025–18034. https://doi.org/10.1039/C4DT02289K

Li Y, Holmberg K, Bordes R (2013) Micellization of true amphoteric surfactants. J Colloid Interface Sci 411:47–52. https://doi.org/10.1016/j.jcis.2013.08.048

Lim GS, Jaenicke S, Klähn M (2015) How the spontaneous insertion of amphiphilic imidazolium-based cations changes biological membranes: a molecular simulation study. Phys Chem Chem Phys 17:29171–29183. https://doi.org/10.1039/C5CP04806K

Liu CZ, Wang F, Stiles AR, Guo C (2012) Ionic liquids for biofuel production: Opportunities and challenges. Appl Energy 92:406–414. https://doi.org/10.1016/j.apenergy.2011.11.031

Liu QP, Hou XD, Li N, Zong MH (2012) Ionic liquids from renewable biomaterials: synthesis, characterization and application in the pretreatment of biomass. Green Chem 14:304–307. https://doi.org/10.1039/C2GC16128A

Liu X, Hu J, Huang Y, Fang Y (2013) Aggregation behavior of surface active dialkylimidazolium ionic liquids [C12Cnim]Br (n = 1–4) in aqueous solutions. J Surfactants Deterg 16:539–546. https://doi.org/10.1007/s11743-012-1409-1

Matthews RP, Welton T, Hunt PA (2014) Competitive pi interactions and hydrogen bonding within imidazolium ionic liquids. Phys Chem Chem Phys 16:3238–3253. https://doi.org/10.1039/c3cp54672a

Moniruzzaman M, Kamiya N, Goto M (2010) Ionic liquid based microemulsion with pharmaceutically accepted components: Formulation and potential applications. J Colloid Interface Sci 352:136–142. https://doi.org/10.1016/j.jcis.2010.08.035

Moniruzzaman M, Tamura M, Tahara Y, Kamiya N, Goto M (2010) Ionic liquid-in-oil microemulsion as a potential carrier of sparingly soluble drug: Characterization and cytotoxicity evaluation. Int J Pharm 400:243–250. https://doi.org/10.1016/j.ijpharm.2010.08.034

Monti D, Egiziano E, Burgalassi S, Chetoni P, Chiappe C, Sanzone A, Tampucci S (2017) Ionic liquids as potential enhancers for transdermal drug delivery. Int J Pharm 516:45–51. https://doi.org/10.1016/j.ijpharm.2016.11.020

Moshikur RM, Chowdhury MR, Fujisawa H, Wakabayashi R, Moniruzzaman M, Goto M (2020) Design and Characterization of Fatty Acid-Based Amino Acid Ester as a New "green" Hydrophobic Ionic Liquid for Drug Delivery. ACS Sustain. Chem. Eng. 8:13660–13671. https://doi.org/10.1021/acssuschemeng.0c03419

Moshikur RM, Chowdhury MR, Moniruzzaman M, Goto M (2020) Biocompatible ionic liquids and their applications in pharmaceutics. Green Chem 22:8116–8139. https://doi.org/10.1039/d0gc02387f

Moshikur RM, Ali MK, Wakabayashi R, Moniruzzaman M, Goto M (2020c) Formation and potential application of micelles composed of biocompatible N-lauroyl-amino acid ionic liquids surfactant. J Mol Liq 320:114424. https://doi.org/10.1016/j.molliq.2020.114424

Moshikur RM, Chowdhury MR, Wakabayashi R, Tahara Y, Moniruzzaman M, Goto M (2019) Ionic liquids with methotrexate moieties as a potential anticancer prodrug: Synthesis, characterization and solubility evaluation. J Mol Liq 278:226–233. https://doi.org/10.1016/j.molliq.2019.01.063

Moshikur RM, Chowdhury MR, Wakabayashi R, Tahara Y, Moniruzzaman M, Goto M (2018) Characterization and cytotoxicity evaluation of biocompatible amino acid esters used to convert salicylic acid into ionic liquids. Int J Pharm 546:31–38. https://doi.org/10.1016/j.ijpharm.2018. 05.021

Mustahil NA, Baharuddin SH, Abdullah AA, Reddy AVB, Mutalib MIA, Moniruzzaman M (2019) Synthesis, characterization, ecotoxicity and biodegradability evaluations of novel biocompatible surface active lauroyl sarcosinate ionic liquids. Chemosphere 229:349–357. https://doi.org/10. 1016/j.chemosphere.2019.05.026

Pillai P, Pal N, Mandal A (2017) Synthesis, Characterization, Surface Properties and Micellization Behaviour of Imidazolium-based Ionic Liquids. J Surfactants Deterg 20:1321–1335. https://doi. org/10.1007/s11743-017-2021-1

Poh Y, Ng S, Ho K (2019) Formulation and characterisation of 1-ethyl-3-methylimidazolium acetate-in-oil microemulsions as the potential vehicle for drug delivery across the skin barrier. J Mol Liq 273:339–345. https://doi.org/10.1016/j.molliq.2018.10.034

Rao VG, Banerjee C, Ghosh S, Mandal S, Kuchlyan J, Sarkar N (2013) A step toward the development of high-temperature stable ionic liquid-in-oil microemulsions containing double-chain anionic surface active ionic liquid. J Phys Chem B 117:7472–7480. https://doi.org/10.1021/jp4 03265p

Rao VG, Ghosh S, Ghatak C, Mandal S, Brahmachari U, Sarkar N (2012) Designing a new strategy for the formation of IL-in-oil microemulsions. J Phys Chem B 116:2850–2855. https://doi.org/ 10.1021/jp2110488

Rao VG, Mandal S, Ghos S, Banerjee C, Sarkar N (2013) Phase boundaries, structural characteristics, and NMR spectra of ionic liquid-in-oil microemulsions containing double chain surface active ionic liquid: A comparative study. J Phys Chem B 117:1480–1493. https://doi.org/10.1021/ jp310616p

Rao VG, Mandal S, Ghosh S, Banerjee C, Sarkar N (2012) Ionic liquid-in-oil microemulsions composed of double chain surface active ionic liquid as a surfactant: Temperature dependent solvent and rotational relaxation dynamics of coumarin-153 in [Py][TF2N]/[C 4mim][AOT]/benzene microemulsions. J Phys Chem B 116:8210–8221. https://doi.org/10.1021/ jp304668f

Rao KS, Gehlot PS, Gupta H, Markus D, Arvind K (2015) Sodium bromide induced micelle to vesicle transitions of newly synthesized anionic surface-active ionic liquids based on dodecylbenzenesulfonate. J Phys Chem B 119:4263–4274. https://doi.org/10.1021/jp512805e

Ratti R (2014) Ionic Liquids: Synthesis and Applications in Catalysis. Adv Chem 2014:1–16. https:// doi.org/10.1155/2014/729842

Rodriguez-Aller M, Guillarme D, Veuthey JL, Gurny R (2015) Strategies for formulating and delivering poorly water-soluble drugs. J. Drug Deliv. Sci. Technol. 30:342–351. https://doi.org/ 10.1016/j.jddst.2015.05.009

Rojas O, Tiersch B, Rabe C, Stehle R, Hoell A, Arlt B, Koetz J, Golm P (2013) Nonaqueous Microemulsions Based on N,N′-Alkylimidazolium alkylsulfate ionic liquids. Langmuir 29(23):6833−6839. https://doi.org/10.1021/la401080q

Schulz A, Jaksch S, Schubel R, Wegener E, Di Z, Han Y, Meister A, Kressler J, Kabanov AV, Luxenhofer R, Papadakis CM, Jordan R (2014) Drug-induced morphology switch in drug delivery systems based on poly(2-oxazoline)s. ACS Nano 8:2686–2696. https://doi.org/10.1021/ nn406388t

Shah MUH, Sivapragasam M, Moniruzzaman M, Talukder MMR, Yusup SB, Goto M (2018) Aggregation behavior and antimicrobial activity of a micellar system of binary ionic liquids. J Mol Liq 266:568–576. https://doi.org/10.1016/j.molliq.2018.06.101

Shamshina JL, Barber PS, Rogers RD (2013) Ionic liquids in drug delivery. Expert Opin Drug Deliv 10:1367–1381. https://doi.org/10.1517/17425247.2013.808185

Singh G, Singh G, Kang TS (2016) Micellization Behavior of Surface Active Ionic Liquids Having Aromatic Counterions in Aqueous Media. J Phys Chem B 120:1092–1105. https://doi.org/10. 1021/acs.jpcb.5b09688

Singh SK (2019) Solubility of lignin and chitin in ionic liquids and their biomedical applications. Int J Biol Macromol 132:265–277. https://doi.org/10.1016/j.ijbiomac.2019.03.182

Singh T, Kumar A (2007) Aggregation behavior of ionic liquids in aqueous solutions: Effect of alkyl chain length, cations, and anions. J Phys Chem B 111:7843–7851. https://doi.org/10.1021/jp0726889

Singla P, Singh O, Chabba S, Mahajan RK (2018) Pluronic-SAILs (surface active ionic liquids) mixed micelles as efficient hydrophobic quercetin drug carriers. J Mol Liq 249:294–303. https://doi.org/10.1016/j.molliq.2017.11.044

Srinivasa Rao K, Gehlot PS, Trivedi TJ, Kumar A (2014) Self-assembly of new surface active ionic liquids based on Aerosol-OT in aqueous media. J Colloid Interface Sci 428:267–275. https://doi.org/10.1016/j.jcis.2014.04.062

Sunitha, Reddy PS, Kanjilal PRBN, , S (2011) Synthesis and evaluation of new imidazolium-based aromatic ether functionalized cationic mono and gemini surfactants. Eur J Lipid Sci Technol 113:756–762. https://doi.org/10.1002/ejlt.201000437

Trivedi TJ, Rao KS, Singh T, Mandal SK, Sutradhar N, Panda AB, Kumar A (2011) Task-specific, biodegradable amino acid ionic liquid surfactants. Chemsuschem 4:604–608. https://doi.org/10.1002/cssc.201100065

Wang C, Zhu J, Zhang D, Yang Y, Zheng L, Qu Y, Yang X, Cui X (2018) Ionic liquid – microemulsions assisting in the transdermal delivery of Dencichine: Preparation, in-vitro and in-vivo evaluations, and investigation of the permeation mechanism. Int J Pharm 535:120–131. https://doi.org/10.1016/j.ijpharm.2017.10.024

Wang X, Liu J, Yu L, Jiao J, Wang R, Sun L (2013) Surface adsorption and micelle formation of imidazolium-based zwitterionic surface active ionic liquids in aqueous solution. J Colloid Interface Sci 391:103–110. https://doi.org/10.1016/j.jcis.2012.09.073

Yanxin J, Lushan W, Tao W, Peng C, Yanhui B, Li Y (2015) Aggregation behavior of dodecylsulfonate-based surface active ionic liquids in water. J Mol Liq 212:23–29. https://doi.org/10.1016/j.molliq.2015.08.049

Zakrewsky M, Lovejoy KS, Kern TL, Miller TE, Le V, Nagy A, Goumas AM, Iyer RS, DelSesto RE, Koppisch AT, Fox DT, Mitragotri S (2014) Ionic liquids as a class of materials for transdermal delivery and pathogen neutralization. Proc Natl Acad Sci U S A 111:13313–13318. https://doi.org/10.1073/pnas.1403995111

Zakrewsky M, Mitragotri S (2016) Therapeutic RNAi robed with ionic liquid moieties as a simple, scalable prodrug platform for treating skin disease. J Control Release 242:80–88. https://doi.org/10.1016/j.jconrel.2016.09.003

Zech O, Thomaier S, Bauduin P, Rück T, Touraud D, Kunz W (2009) Microemulsions with an ionic liquid surfactant and room temperature ionic liquids as polar pseudo-phase. J Phys Chem B 113:465–473. https://doi.org/10.1021/jp8061042

Zhang D, Wang HJ, Cui XM, Wang CX (2017) Evaluations of imidazolium ionic liquids as novel skin permeation enhancers for drug transdermal delivery. Pharm Dev Technol 22:511–520. https://doi.org/10.3109/10837450.2015.1131718

Zhang S, Sun J, Zhang X, Xin J, Miao Q, Wang J (2014) Ionic liquid-based green processes for energy production. Chem Soc Rev 43:7838–7869. https://doi.org/10.1039/C3CS60409H

Zheng L, Guo C, Wang J, Liang X, Chen S, Junhe M, Yang B, Jiang Y, Liu H (2007) Effect of ionic liquids on the aggregation behavior of PEO-PPO-PEO block copolymers in aqueous solution. J Phys Chem B 111:1327–1333. https://doi.org/10.1021/jp066727c

Zhou T, Zhao J (2009) Synthesis and thermotropic liquid crystalline properties of zwitterionic gemini surfactants containing a quaternary ammonium and a sulfate group. J Colloid Interface Sci 338:156–162. https://doi.org/10.1016/j.jcis.2009.06.009

Chapter 10
Ionic Liquids for Transdermal Vaccination

Yoshiro Tahara

Abstract Ionic liquids (ILs) are attractive vehicles for the effective delivery of biopharmaceuticals. This chapter focuses on transdermal vaccination using ILs. Transdermal vaccination is the administration of vaccines through the skin. Because it is needleless and avoids the requirement for professional medical support, transdermal vaccination has become an important research field. However, in this form of vaccination, the antigens need to be captured by antigen presenting cells in the skin upon delivery. It is difficult to deliver antigens through the skin because antigens consist of proteins and peptides, which are hydrophilic and high molecular weight species. Recently, ILs have been reported as solvents for sparingly soluble drugs or skin penetration enhancers, and thus, transdermal vaccination using ILs has attracted much attention.

Keywords Ionic liquid · Transdermal delivery · Vaccination · Antigen peptide

10.1 Introduction

An ionic liquid (IL) is a salt that exists in a liquid state at room temperature and atmospheric pressure (Wasserscheid and Keim 2000). Research into ILs is an attractive field that started in the 1990s (Wilkes and Zaworotko 1992). High vapor pressure and designability are well-known advantages of ILs. Pharmaceutical applications of ILs have attracted much attention since the 2000s (Hough et al. 2007). One of the most important properties of ILs for pharmaceutical applications is their solubilization of sparingly soluble drugs. Moniruzzaman et al. reported that acyclovir, which is difficult to dissolve in water and various organic solvents, can be dissolved in ILs such as dimethylimidazolium dimethylphosphate. Additionally, acyclovir was successfully delivered through the skin using a microemulsion system (Moniruzzaman et al. 2010a; b). After these pioneering studies, drug delivery systems using ILs were reported and quickly regarded as one of the most attractive IL research fields

Y. Tahara (✉)
Doshisha University, Kyotanabe, Kyoto 610-0394, Japan
e-mail: ytahara@mail.doshisha.ac.jp

© The Author(s), under exclusive license to Springer Nature Singapore Pte Ltd. 2021 187
M. Goto and M. Moniruzzaman (eds.), *Application of Ionic Liquids in Drug Delivery*,
https://doi.org/10.1007/978-981-16-4365-1_10

(Adawiyah et al. 2016; Md Moshikur et al. 2020). Recently, ILs have been used for drug delivery systems involving not only small molecular drugs but also biopharmaceuticals such as siRNA (Zakrewsky and Mitragotri 2016), peptides (Tahara et al. 2020), and proteins (Araki et al. 2015; Banerjee et al. 2017, 2018; Tanner et al. 2018). These biopharmaceuticals are usually delivered as an aqueous solution and thus, deactivation by the IL is a major concern. However, an IL-robed siRNA targeted against an esterase down regulated the target enzyme in artificial skin tissues (Zakrewsky and Mitragotri 2016), and an insulin-loaded IL decreased the blood glucose level after oral administration (Banerjee et al. 2018; Tanner et al. 2018). Therefore, these studies suggested that ILs do not always deactivate biopharmaceuticals. Another use of ILs in the biopharmaceuticals is for transdermal vaccination as summarized at the end of this chapter.

10.2 Transdermal Drug Delivery and Vaccination

Transdermal delivery is drug delivery into the skin (Zaffaroni 1981). Compared with the injection route, transdermal delivery is a non-invasive administration method that avoids the accident risk related with needles. In the case of oral delivery, which is another painless method, after drug administration through the mouth, the degradation and loss of drugs in the gastrointestinal tract and liver need to be considered. First-pass metabolism by the liver is the major disadvantage of oral drug delivery. By comparison, during transdermal administration, drugs are directly delivered to the blood or lymph vessels under the skin making it is a simple and ideal route for drug administration.

Preventive vaccines containing viral or bacterial antigens are administered to healthy people to induce immunity against the virus or bacteria causing the infection. The COVID-19 pandemic (2020–2021) has globally reconfirmed the importance of vaccines. Transdermal vaccination is the delivery method of vaccines through the skin. Its advantages include avoiding the use of needles and the involvement of the healthcare professionals and is the one of the best methods for widespread vaccine administration. In the 1970s, Langerhans cells in the epidermis of the skin (Fig. 10.1) were reported to be stimulated by adjuvants (Silberberg et al. 1976) and research into the immunology of the skin was developed. Subsequent studies indicated that the Langerhans cells in the epidermis and dermal dendritic cells show robust antigen presentation effects, and thus, transdermal vaccinations have attracted much attention (Bal et al. 2010).

Transdermal drug delivery, including skin vaccination, is known as an excellent drug delivery system. Nevertheless, it is difficult to deliver drugs through the skin. The stratum corneum (SC) is the outer most layer and functions as the main barrier of the skin (Fig. 10.1). The SC consists of the lipids derived from keratinocytes and as a hydrophobic barrier, and it protects the skin from exogenous materials (Madison 2003). Therefore, transdermal delivery of biopharmaceuticals including vaccine antigens (peptides, proteins and RNAs) are difficult because most of them

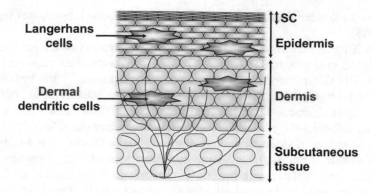

Fig. 10.1 Skin structure

are hydrophilic macromolecules. In 1998, Glenn et al. reported that it was possible to produce antigen-specific antibodies by transdermal vaccination (Glenn et al. 1998). At this stage, antigen delivery into the skin was accomplished through the removal of the SC by tape stripping. Over the following two decades, various approaches including ultrasound (Tezel et al. 2005), jet immunizations (Roberts et al. 2005), microneedles (Sullivan et al. 2010), liposomes (Paul et al. 1995), and hydrogel patches (Ishii et al. 2008) were reported as transdermal vaccine delivery systems. In the following sections, transdermal drug delivery and transdermal vaccination using ILs are summarized.

10.3 Ionic Liquids as Skin Penetration Enhancers

In the research field of ILs for pharmaceutical applications, transdermal delivery is one of the most studied areas. In 2010, Moniruzzaman et al. reported that acyclovir, a sparingly soluble drug, could be dissolved in several ILs (Moniruzzaman et al. 2010a). Hydrophilic ILs are more suitable for increasing the solubility of acyclovir than hydrophobic ones. Although acyclovir is a drug used against a herpes virus that causes skin shingles, it is usually administered orally. The low solubility of acyclovir in not only water but also organic solvents is one reason why acyclovir is not used as a skin external medicine. ILs are novel solvents for acyclovir because of their high polarity. However, the skin permeability of acyclovir by delivery using hydrophilic ILs is low. Thus, acyclovir in an IL was emulsified with an oil phase and surfactants, which are well-known as skin penetration enhancers, and transdermal delivery of acyclovir was achieved. In the companion report, the skin irritation of the IL and the IL-in-oil emulsion were investigated using an artificial skin model, and cytotoxicity of the hydrophilic IL alone to skin cells was relatively high while that of the IL-in-oil emulsion containing 4% IL was low (Moniruzzaman et al. 2010b). Similar results were obtained using methotrexate where it was not delivered across the skin by IL alone or oil enhancer alone, but it was delivered by an IL-in-oil emulsion (Yoshiura et al. 2013).

In the 2010s, the effects of ILs as skin penetration enhancers were investigated. In 2013, Doblar et al. prepared an emulsion loaded with model small molecular drugs and ILs. The skin permeability of the drugs were improved by the hydrophobic IL (1-butyl-3-methylimidazolium hexafluorophosphate) (Dobler et al. 2013). In 2018, Wang et al. reported that dencitine was delivered through the skin by emulsions containing ILs (1-hydroxyethyl-3-methylimidazolium chloride and 1-butyl-3-methylimidazolium dodecanesulfate) (Wang et al. 2018). One of the most attractive and well-studied ILs in 2010 was choline-geranate reported by the Mitragotri group. It was first recognized as a deep eutectic solvent because it consisted of a 1:2 molar ratio of choline and geranic acid, but is now recognized as an IL. The skin penetration enhancement effect provided by choline-geranate was very high. It can be used to deliver not only small molecules such as mannitol and cefadroxil (Zakrewsky et al. 2014) but also proteins including insulin and albumin (Banerjee et al. 2017; Tanner et al. 2018).

10.4 Skin Vaccination by Ionic Liquids Using Antigen Protein and a Solid-in-Oil Nanodispersion

10.4.1 The Solid-in-Oil Nanodispersion

In the 1990s, the improvement of enzyme activity in non-aqueous solutions was studied. For example, the catalytic effect of enzymes for a typical esterification to form an ester bond from an alcohol and an acid is very low in aqueous solution because water shifts the esterification equilibrium to hydrolysis. Therefore, approaches to dissolve enzymes in organic media is important technology. Goto et al. reported a method for modification of enzymes with hydrophobic surfactants (Tahara et al. 2012). An oil-based solution containing solid complexes consisting of enzymes and surfactants was called a solid-in-oil (S/O) nanodispersion, and this technology was used in transdermal delivery systems in the 2000s (Piao et al. 2008; Tahara et al. 2008). Transdermal vaccination using a S/O nanodispersion was first reported in 2010 and shown to be successful for delivery of antigens of cancer immunotherapy (Wakabayashi et al. 2018) and treatment of Japanese cedar pollinosis (Kitaoka et al. 2015).

10.4.2 Collaboration of Ionic Liquids with a Solid-in-Oil Nanodispersion

In 2015, Araki et al. reported that the transdermal vaccination of a S/O nanodispersion was improved by an IL (1-dodecyl-3-methylimidazolium bis(trifluoromethylsulfonyl) amide) (Araki et al. 2015). Transcutaneous delivery of a model antigen (ovalbumin, molecular weight: 45 kDa) into the skin was enhanced using ILs. The in vitro release ratio of antigens from the S/O nanodispersion to the water phase is the most important factor in transdermal protein delivery into the skin using S/O nanodispersions (Kitaoka et al. 2014). However, the antigen release ratios were not significantly increased with ILs. This result suggested that ILs acted as skin penetration enhancers. Finally, a transdermal vaccination study in vivo was performed using an antigen-loaded S/O nanodispersion with or without ILs. The antigen-specific antibody production was enhanced by the ILs. This is early evidence that the biological response by ILs was reported in vivo. The potential of using ILs as a transdermal vaccination enhancer was confirmed.

10.5 Skin Vaccination of Antigen Peptides Enhanced by Ionic Liquids

In 2020, Tahara et al. reported transdermal vaccination using an antigen peptide and biocompatible ILs (Tahara et al. 2020). The molecular weight of an antigen peptide is usually over 500 Da, and thus, transdermal antigen peptide delivery is difficult. Imidazolium-based ILs were known as skin penetration enhancers (Dobler et al. 2013; Araki et al. 2015; Wang et al. 2018), but their cytotoxicity is concerning (Moniruzzaman et al. 2010b). Tahara et al. prepared ILs consisting of choline-fatty acids (C8, C10, C14, C16, C18, and C18:1), in which both the cation and anion are derived from natural products. The investigation of skin irritation and penetration enhancement effect suggested that choline-oleate (Fig. 10.2a) is suitable for transdermal peptide delivery. The model antigen peptide SIINFEKL (an ovalbumin (OVA) epitope peptide) is soluble in choline-oleate, an IL, and is miscible with ethanol (EtOH) and isopropyl myristate (IPM). The in vitro skin penetration of SIINFEKL was dramatically enhanced, and the flux generated by the IL/EtOH/IPM system was 28-fold higher than with EtOH/PBS alone. The preventive cancer vaccination effect was evaluated in vivo using E.G7-OVA cells (OVA-expressing cancer cells). One week after two vaccinations, E.G7-OVA cells were inoculated into mice. The tumor growth was prevented by the SIINFEKL antigen peptide-loaded IL/EtOH/IPM system, while it was not by injection of the antigen peptide alone (Fig. 10.2b). These results suggested that the biocompatible IL choline-oleate is a promising enhancer for transdermal peptide delivery and vaccination.

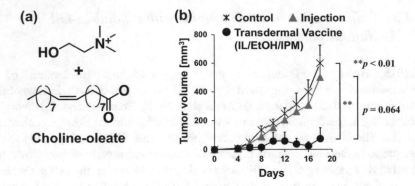

Fig. 10.2 Skin vaccination of antigen peptides enhanced by ionic liquids. **a** Choline-oleate is a suitable IL for transdermal drug delivery. **b** Tumor growth was prevented by transdermal vaccination with the SIINFEKL antigen peptide-loaded IL/EtOH/IPM system (●) but not with the antigen peptide alone by injection (▲) compared with the control (✗). Reproduced with permission from our previous work (Tahara et al. 2020). Copyright (2020) American Chemical Society

10.6 Summary

Transdermal vaccination is a safe and easy administration route and has been globally prioritized as an effective transdermal delivery method for antigens. Biocompatible ILs such as choline-geranate and choline-oleate were reported as efficient skin penetration enhancers and solvent mediators. These are important candidates for further study as transdermal delivery additive agents and skin vaccination enhancers during the next generation of biopharmaceutical research.

Acknowledgements We thank Renee Mosi, PhD, from Edanz Group (https://en-author-services. edanz.com/ac) for editing a draft of this manuscript.

References

Adawiyah N et al (2016) Ionic liquids as a potential tool for drug delivery systems. MedChemComm 7:1881–1897
Araki S et al (2015) Ionic liquid-mediated transcutaneous protein delivery with solid-in-oil nanodispersions. MedChemComm 6:2124–2128
Bal SM et al (2010) Advances in transcutaneous vaccine delivery: do all ways lead to Rome? J Control Release 148:266–282
Banerjee A et al (2017) Transdermal protein delivery using choline and geranate (CAGE) deep eutectic solvent. Adv Healthc Mater 6:1601411
Banerjee A et al (2018) Ionic liquids for oral insulin delivery. PNAS 115:7296–7301
Dobler D et al (2013) Ionic liquids as ingredients in topical drug delivery systems. Int J Pharm 441:620–627
Glenn GM et al (1998) Skin immunization made possible by cholera toxin. Nature 391:851–852

Hough WL et al (2007) The third evolution of ionic liquids: active pharmaceutical ingredients. New J Chem 31:1429–1436

Ishii Y et al (2008) A transcutaneous vaccination system using a hydrogel patch for viral and bacterial infection. J Control Release 131:113–120

Kitaoka M et al (2014) Sucrose laurate-enhanced transcutaneous immunization with a solid-in-oil nanodispersion. MedChemComm 5:20–24

Kitaoka M et al (2015) Transcutaneous peptide immunotherapy of japanese cedar pollinosis using solid-in-oil nanodispersion technology. AAPS PharmSciTech 16:1418–1424

Madison KC (2003) Barrier function of the skin: "La Raison d'Être" of the epidermis. J Invest Dermatol 121:231–241

Md Moshikur R et al (2020) Biocompatible ionic liquids and their applications in pharmaceutics. Green Chem 22:8116–8139

Moniruzzaman M et al (2010a) Ionic liquid-assisted transdermal delivery of sparingly soluble drugs. Chem Commun 46:1452–1454

Moniruzzaman M et al (2010b) Ionic liquid-in-oil microemulsion as a potential carrier of sparingly soluble drug: characterization and cytotoxicity evaluation. Int J Pharm 400:243–250

Paul A et al (1995) Transdermal immunization with large proteins by means of ultradeformable drug carriers. Eur J Immunol 25:3521–3524

Piao H et al (2008) A novel solid-in-oil nanosuspension for transdermal delivery of diclofenac sodium. Pharm Res 25:896–901

Roberts LK et al (2005) Clinical safety and efficacy of a powdered Hepatitis B nucleic acid vaccine delivered to the epidermis by a commercial prototype device. Vaccine 23:4867–4878

Silberberg I et al (1976) The role of langerhans cells in allergic contact hypersensitivity. a review of findings in man and guinea pigs. J Invest Dermatol 66:210–217

Sullivan SP et al (2010) Dissolving polymer microneedle patches for influenza vaccination. Nat Med 16:915–920

Tahara Y et al (2008) A solid-in-oil nanodispersion for transcutaneous protein delivery. J Control Release 131:14–18

Tahara Y et al (2012) Solid-in-oil dispersion: a novel core technology for drug delivery systems. Int J Pharm 438:249–257

Tahara Y et al (2020) Biocompatible ionic liquid enhances transdermal antigen peptide delivery and preventive vaccination effect. Mol Pharm 17:3845–3856

Tanner EEL et al (2018) Transdermal insulin delivery using choline-based ionic liquids (CAGE). J Control Release 286:137–144

Tezel A et al (2005) Low-frequency ultrasound as a transcutaneous immunization adjuvant. Vaccine 23:3800–3807

Wakabayashi R et al (2018) Solid-in-oil peptide nanocarriers for transcutaneous cancer vaccine delivery against melanoma. Mol Pharm 15:955–961

Wang C et al (2018) Ionic liquid—microemulsions assisting in the transdermal delivery of Dencichine: preparation, in-vitro and in-vivo evaluations, and investigation of the permeation mechanism. Int J Pharm 535:120–131

Wasserscheid P, Keim W (2000) Ionic liquids—new "solutions" for transition metal catalysis. Angew Chem Int Ed 39:3772–3789

Wilkes JS, Zaworotko MJ (1992) Air and water stable 1-ethyl-3-methylimidazolium based ionic liquids. J Chem Soc Chem Commun 965–967. https://doi.org/10.1039/C39920000965

Yoshiura H et al (2013) Ionic liquid-in-oil microemulsions as potential carriers for the transdermal delivery of methotrexate. J Chem Eng Jpn 46:794–796

Zaffaroni A (1981) ALZA: an enterprise in biomedical innovation. Technovation 1:135–146

Zakrewsky M et al (2014) Ionic liquids as a class of materials for transdermal delivery and pathogen neutralization. Proc Natl Acad Sci 111:13313

Zakrewsky M, Mitragotri S (2016) Therapeutic RNAi robed with ionic liquid moieties as a simple, scalable prodrug platform for treating skin disease. J Control Release 242:80–88

Chapter 11
Ionic Liquids-Based Antibiotics for Resistant Microbial Strains and Drug Polymorphism

Amal A. M. Elgharbawy, Muhammad Moniruzzaman, Normi Ismail, and Shiva Rezaei Motlagh

Abstract Ionic liquids (ILs) are a large group of so-called green solvents with extensive uses in medicine, drug delivery, stabilization of proteins and enzymes, and industrial applications. Active Pharmaceutical Ingredients ionic liquids (API-ILs) are the third generation of ILs. Because of their role precision, biodegradability, and biocompatibility, this class of ILs is gaining interest. This chapter, therefore, aims to recognize the antimicrobial action of non-toxic and biocompatible antibiotics targeting API-ILs that could respond to resistant strains. The World Health Organization (WHO) has published a priority list of pathogens for antibiotic research and development. Among the pathogens are *Enterobacteriaceae* (*Escherichia coli, Klebsiella pneumonia, Serratia* spp., *Enterobacter* spp., *Proteus* spp., *Providencia* spp, and *Morganella* spp), carbapenem-resistant, 3rd generation cephalosporin-resistant, *Pseudomonas aeruginosa* (carbapenem-resistant), and *Acinetobacter baumannii* (carbapenem-resistant). The application of API-ILs could assist in overcoming the challenges correlated to the pharmaceutical industry including drug solubility and polymorphism. At present, biocidal characteristic of large cations, such as imidazolium and benzalkonium species are mostly utilized to hinder growth of yeast or bacteria. In this context, this chapter introduces a background on the applications of ILs as antimicrobial agents for microbial resistance. Their corresponding mechanism of ILs and their effect as an antimicrobial agent was investigated, as well. Since the selection of the organic cations and anions in the ILs formulation can tube

A. A. M. Elgharbawy (✉) · N. Ismail
International Institute for Halal Research and Training (INHART), International Islamic University Malaysia (IIUM), P.O. Box 10, 50278 Kuala Lumpur, Malaysia
e-mail: amalgh@iium.edu.my

A. A. M. Elgharbawy
Department of Biotechnology Engineering, Faculty of Engineering, Bioenvironmental Engineering Research Centre (BERC), IIUM, Selangor, Malaysia

M. Moniruzzaman
Centre of Research in Ionic Liquids (CORIL), Universiti Teknologi PETRONAS (UTP), Seri Iskandar, Malaysia

S. R. Motlagh
Department of Chemical Engineering, Faculty of Engineering, Chulalongkorn University, 10330 Bangkok, Thailand

the biological properties of the ILs, we aim to unveil some of the recent studies that focused on the antimicrobial activities of ILs toward some resistant strains.

Keyword Ionic liquid · Antimicrobial · Antibiotics resistance · Active ingredients · Mechanism

Abbreviations

WHO	The World Health Organization
APIs	Active pharmaceutical ingredients
ILs	Ionic liquids
API-ILs	Active pharmaceutical ingredients Ionic liquids
ICU	Intensive Care Unit
NSAIDs	Nonsteroidal anti-inflammatory drugs
SILs	Switchable ionic liquids
RILs	Reversible ionic liquids
PIL	Poly (ionic liquid)
MDR	Multidrug-resistant
AMX	Amoxycillin
Pen	Penicillin
OSILs	Organic salt ionic liquids
AMP	Ampicillin
[TMAm]	Trimethylammonium
[BMPyr], [HMPyr], [OMPyr]	Butyl, hexyl, octyl pyridinium
[C_nQuin]	1-Alkyl-quinoline
[CDIM]	1-Decyl-3-cinnamylimidazolium
[Br], [Cl]	Bromide, Chloride
[C_nPyr]	1-Alkylpyridine
[HDPyr]	Cetyl pyridinium
[HDPyr][Amp]	Cetyl pyridinium based on ampicillin
[Na]	Novel active
[NTf$_2$]	Bis (trifluoro-methanesulfonyl) imide
[EMIM]	1-Ethyl-3-methylimidazoloium
[BMIM]	1-Butyl-3-methylimidazolium
[OMIM]	1-Octyl-3-methylimidazolium
[DMIM]	1-Decyl-3-methylimidazolium
[EOHDMIM]	1-(2-Hydroxyethyl)-2,3-dimethylimidazolium
[MMF]	Monomethyl Fumarate
[MBF]	Monobutyl Fumarate
[MOF]	Monooctyl Fumarate
[MDF]	Monododecyl Fumarate
[Ch]	Choline
[Met]	Methionine

[Gln]	Glutamine
[TBO]	2-Mercaptobenzoxazole
[TBT]	2-Mercaptobenzothiazole
[TBI]	2-Mercaptobenzimidazole
[DODMA]	Dioctyldimethylammonium
[HDPyr]	Cetyl pyridinium
[PyroC$_x$]	Pyrrolidinium
[P$_{6,6,6,14}$]	Trihexyltetradecylphosphonium
[TMBA]	Trimethyl butylammonium
[TMOA]	Trioctyl methylammonium
[TMDA]	Trimethyl decylammonium
[TMDOA]	Trimethyl dodecylammonium
[TMHDA]	Trimethyl hexadecylammonium
[TBMA]	Tributyl methylammonium
[TOMA]	Trioctyl methylammonium
[DODMA]	Dioctyl dimethylammonium
[EMMor]	4-Etyl-4-methylmorpholinium
[TMPnA]	Trimethyl propanammonium
[EMpip]	1-Ethyl-1-methylpiperidinium
[TBMP]	Tributyl methylphosphonium

11.1   Introduction

The World Health Organization (WHO) created a Global Priority Pathogen List (GPL) of bacteria resisting antibiotics to help prioritize new and efficient antibiotic drugs for research and development (R&D). To date, small and large pharmaceutical companies have primarily been driven by the selection of pathogens for R&D activities based on a set of metrics, such as investor pressure, apparent patient needs, market size, availability of specific technologies, and scientific innovation potential. In line with that, to contribute to solving the problem, efforts are required to develop new antibiotics (Goossens et al. 2006; Goossens 2009; CDC 2019). Bacterial antibiotic resistance is a global public health concern, and in 2017, the World Health Organization (WHO) has drawn up a list of 12 genera, which were prioritized in the critical, high, and medium categories for the creation of alternative antimicrobial drugs.

At the turn of the twenty-first century, the emergence of ILs represents levels of biological activities. This category includes examples of ILs that, for example, are used as active pharmaceutical precursors or ingredients in medicine (APIs). Therefore, the pharmaceutical industry is also engaged in exploring alternative medicines, such as solvents, active liquid forms, co-crystals, or solvents, to prevent the risk of complications of polymorphism (Flieger and Flieger 2020). Since the harmful impact of ILs on the growth of bacteria and fungi has been demonstrated by some studies, this can eliminate the resistance of pathogenic microbes toward antibiotics and other

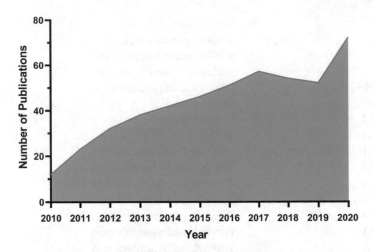

Fig. 11.1 The number of publications in the period between 2010 and 2020 based on Scopus search of the keywords "ionic liquid" and "antimicrobial", as of 26 February 2021

drugs. In medicine, there are several avenues in ILs that can be used, including detergents or dressing substance additives. They are also constituents of the so-called retarded action drugs (Flieger and Flieger 2020).

Even though there are limited reports on ionic liquids' antimicrobial activities, there is a clear growing interest based on the data obtained from Scopus search in the period between 2010 and 2020 (Fig. 11.1).

11.2 Antimicrobial Resistance

In the last two decades, antibiotics have significantly decreased mortality and morbidity and from opportunistic and classical infections (Johnson 2015). Accelerated global use of antibiotics and lapses in the prevention and management of infections has escalated the incidence and extent of antibiotic resistance (AR) relative to the advances achieved. In the United States, an average of two million individuals are infected, and 23,000 died per year with antibiotic-resistant bacteria. In the European Union, AR leads to 25,000 deaths and a further 2.5 million healthcare days at hospitals a year (CDC 2019). The highest prevalence of AR is reported by the Mediterranean countries. Comparably, the average consumption of antibiotics (regular doses per 1000 individuals) in the Mediterranean, eastern, and southern countries appears to be greater than in northern European countries (Goossens et al. 2006; Goossens 2009).

Germs such as bacteria and fungi cause antibiotic resistance to the antibiotics intended to inhibit them. This implies that the antibiotics could not destroy the germs, and they continue growing. The cure of infections caused by antibiotic-resistant germs is complex and sometimes difficult. In most cases, the antibiotic-resistant infection needs long-term hospital admission, additional supervision, follow-up tests, and toxic and extensive therapy. Antibiotic tolerance does not mean the human body gets resistant to antibiotics; it is that the microbe becomes tolerant to the treatment intended to inhibit or kill it (CDC 2019). Bacterial that have been immune to antibiotics are described as 'susceptible' to 'intermediate' or 'resistant' as a susceptibility pattern alteration or from 'intermediate' to 'resistant'. The key drivers of AR are overuse and misuse of antibiotics which result in presumptive treatment and mishandling of antibiotics with excessive doses. Other drivers include the spread of resistant bacteria from patient-to-patient, medical personnel to patients and vice-versa, prophylactic application in animal husbandry, and lack of antibiotics regulation and policy. The World Bank report warned that the global economy could encounter antimicrobial resistance that causes as much damage as the 2008 financial crisis (Asokan et al. 2019).

The resistance of bacteria to antibiotics is a worldwide health challenge; hence, WHO developed 12 families of microbes in 2017 that have been categorized in the critical, high, and medium classes for the production of alternate antimicrobial agents (Havenga et al. 2019). One of the major "priority 1: critical group" species is *Acinetobacter baumannii* (*A. baumannii*) since it is the predominant opportunistic pathogen involved in nosocomial infections. Several strains of *P. aeruginosa* were also sampled from hospital facilities and are involved in nosocomial infections. Table 11.1 shows some common bacteria with high antibiotic resistance.

11.3 Ionic Liquids as an Antimicrobial Agent for Resistant Microbes

ILs are typically classified as salts below 100 °C with melting points (Flieger and Flieger 2020). The twenty-first century is the advent of third-generation ionic liquids that integrate attractive biological characteristics with physicochemical characteristics. This category includes ILs that, for instance, are utilized as active pharmaceutical ingredients or precursors in medicine (APIs). Cations with proven low toxicity and beneficial action (local anesthetic, anticholinergic, antibacterial, antifungal) and anions with certain properties (antibiotics, emollients, NSAIDs, vitamins) have been utilized to obtain a variety of the essential biological properties of ILs. The category of compounds referred to as ILs is currently divided into a variety of sub-categories. The switchable IL group (SILs), also called reversible ILs (RILs), is an important group of ILs. Exposure to external factors, such as oil refining, sulfur dioxide, carbon dioxide captures, and pine hemicellulose extraction, could change the constituents' state from neutral to ionic (Flieger and Flieger 2020). Various studies have demonstrated that in species ranging from bacteria to higher animals, ILs display various levels of toxicity, which largely depends on the cations and their side-chain length. Moreover, anions were proven to relate to toxicity. Compared to traditional organic

Table 11.1 List of some common microbes with high antibiotic resistance with critical threats (CDC 2019, 2020)

Pathogen	Antibiotic resistance	Type of infection	Resistance notes
Acinetobacter	Multi-drug resistant Carbapenem	Pneumonia and wound, bloodstream and urinary tract infections Appears mostly with patients in the ICU	It can spread from hospitals and ICU Can be transmitted by direct contact and can be found on the skin, in the dirt, water or food *Acinetobacter baumannii* is extremely infectious and responsible for 80% or so of infections with*Acinetobacter*
Candida auris	Multi-drug resistant	*C. Auris* has induced serious illnesses in inpatients. *C. Auris* causes wound infections, ear infections and infections in the bloodstream. It has also been isolated from urinary and respiratory samples	It is immune to multiple antifungal agents widely used to treat infections of Candida. With traditional laboratory procedures, it is difficult to differentiate and can be misidentified in laboratories without specific technology
Enterobacteriaceae • *Klebsiella pneumoniae* and other species • *Escherichia coli (E. coli)* species	Carbapenem	Lung infections, pneumonia (pulmonary diseases), infections of the upper respiratory tract, urinary tract (UTI), stomach, febrile neutropenia, meningitis, infections of surgical wounds	*Enterobacteriaceae*, which may become immune to a class of antibiotics known as carbapenems, are a natural, typically harmless part of our digestive tract bacteria
Neisseria gonorrhoeae	Fluoroquinolone antibiotics like ciprofloxacin	It causes gonorrhoea, a sexually transmitted disease. The lack of successful drugs to cure at least 95% of the gonorrhoea population could contribute to major public health issues	Resistance to most drugs used for its treatment. These bacteria are now immune to the antibiotic fluoroquinolone

solvents such as methanol and dichloromethane, most imidazole-based ILs demonstrate higher toxicity (Stasiewicz et al. 2008). The capacity to inhibit several enzymes was also reported. The effect of inhibiting acetylcholinesterase, which results in certain neuronal mechanisms being distorted, tends to be especially harmful. Pyridinium and imidazolium-based ILs have been shown to be responsible for inhibiting the activity of enzymes (Thuy Pham et al. 2010).

As they engage in the nitrogen and carbon cycles, microorganisms are indispensable to the ecosystem. Many studies about the toxicity of ILs of bacteria can be found in the literature. For instance, the evaluation of the toxicity of ILs to *Escherichia coli, Vibro fischeri, Pichia pastoris, Bacillus cereus* and lactic acid-producing bacteria, and others are described in several reports (Flieger and Flieger 2020) (Table 11.2). In contrast to traditional solvents such as phenol, toluene, benzene, the ILs toxicity is more harmful to bacteria and varies according to the organisms. Gouveia et al. (2014) have proposed that the same rules developed for higher organisms do not regulate toxicity against bacteria. The authors found that soil bacteria's susceptibility, namely, *B. subtilis* and *E. coli,* varies significantly with ILs originating from imidazolium, pyridinium, and amino acid-choline. It seemed that ILs were less lethal to *E. Coli* (Gram-negative), tested for inhibition by [Ch][Met] and [Ch][Gln], as opposed to *B. Subtilis* (Gram-positive). More studies have also verified that, compared to Gram-negative, ILs are more lethal to Gram-positive bacteria because of various interactions of ILs with lipid components and peptidoglycan of the cell wall (Saadeh et al. 2009; Gouveia et al. 2014). Existing studies have shown that three major elements should be considered to assess various microorganisms' toxicity, involving bacteria. Those elements are the head group, the head group substituents, and the anion type (Flieger and Flieger 2020). With increasing the alkyl chain length of the imidazolium cations, the bacterial activity decreases. This theory, however, is not substantiated by other reports. In 2014, a number of benzoazole ILs were tested by Czekański et al. (2014), and their antimicrobial activity was evaluated. Among the ILs synthesized, only a few exhibited inhibition of *B. subtillis* and *P. aeruginosa* (Gram-positive), specifically, [EMIM][TBI] and [HMIM][TBT], [BMIM][TBO]. It is important to mention that the ILs head group alteration or the alkyl chain length that is bound to the head group had no effect on the antibacterial behavior with respect to the bacteria examined. About 50% of drugs are being used as salts. It is why the requisite drug properties can be precisely tuned by pairing suitable ions. Any essential drug characteristics, such as melting temperature and solubility, are important in drug manufacturing processes and pharmacodynamics (Black et al. 2007). However, crystal polymorphism in medicinal compounds, which may alter the active ingredient's solubility in a lethal dosage or ineffectual dose, continues to be the most important issue for the pharmaceutical industry. Additional drawbacks related to the solid-state of the drug were identified, as were low solubility and limited bioavailability, besides polymorphic conversions. Consequently, the pharmaceutical industry is still keen to discover alternative types of pharmaceuticals to prevent polymorphism problems (Flieger and Flieger 2020). The therapeutic effectiveness, bioaccessibility, and low solubility of drugs via semi-permeable membranes are generally limited. This will

Table 11.2 Studies on antimicrobial activity of ILs against some microorganisms

	ILs	Main Findings	Type of Microorganism			References
			Gram +ve	Gram −ve	Others	
1	Fumarate Anion (acid anions): 4 series: [BMIM][MMF] [BMIM][MBF] [BMIM][MOF] [BMIM][MDF]	Longer alkyl chain → better antimicrobial activity against *B. subtilis* All ILs are air stable under ambient conditions	*B. subtilis*	*E. coli*	*S. cerevisiae (yeast)*	He et al. (2013)
2	Dicationic pyridinium hydrazone-phenoxy ILs	↑ alkyl chain → ↑ lethal effect	*Clostridium difficile, S.aureus (MRSA)*	*Neisseria gonorrhea, E. coli*	*C.albicans*	Rezki et al. (2019)
3	[Ch]-based ILs: [Ch] phenylacetate, tricinoleate and trichloroacetate	Antimicrobial activity similar to Streptomycin	*S.epidermidis (S273), S.aureus (S276), B.subtilis (B29)*	*E. coli (E266), P.aeruginosa (ATCC 15,442)*	*C. tropicalis (A3), C.albicans (9002)*	Ahmad et al. (2019)
4	Amino acid-based ILs and poly(ionic liquid) (PIL)	Potential for antimicrobial wound dressing	*MRSA*	–	–	Guo et al. (2019)
5	Novel pyrrolidinium-based ILs [PyroC$_x$]Br (x = 4, 6, 8, 10, 12)	Activity of ILs depends on the length of alky chain (C4 < C6 < C8 < C10 < C12) Melittin + [PyroC$_x$]Br → better antibacterial activity	*S. aureus*	*E. coli*	–	Saraswat et al. (2020)

(continued)

Table 11.2 (continued)

	ILs	Main Findings	Type of Microorganism			References
			Gram +ve	Gram −ve	Others	
6	Pyrithione-based ILs ([EMIM], [OMIM], [TMBA], [TMOA], [TMDA], [TMDOA], [TMHDA], [TBMA], [TOMA], [DODMA], [EMMor], [TMPnA], [EMPip], [BMPyr], and [TBMPI]) Pyrithione	Antibacterial: Incorporation of pyrithione into ILs enhanced agar pyrithione diffusion, improved bactericidal action Antiviral: ILs with more than one lengthened side chain	*S. aureus* ATCC 6538, *Geobacillus stearothermophilus* ATCC 7953, *Lactococcus lactis* ATCC 19,435, *Lactobacillus sakei* ATCC 15,521, *Enterococcus faecalis* ATCC 19,433, *Listeria monocytogenes* ATCC 13,932, *Geobacillus stearothermophilus* ATCC 7953, *Bacillus cereus* ATCC 11,778	*Salmonella enterica* ser, *Escherichia coli* ATCC 25,922, *Citrobacter freundii* ATCC 43,864, *Proteus mirabilis* ATCC 29,906, *Pseudomonas aeruginosa* ATCC 27,853, *Pseudomonas aeruginosa* ATCC 27,853, *Typhimurium* ATCC 14,028	Yeasts: *Candida tropicalis* ATCC 750, *Saccharomyces cerevisiae* ATCC 4,000,850 viruses: Phi6, MS2 and P100 (bacteriophages)	Bromberger et al. (2020)
7	Amoxycillin OSILs and Penicillin OSILs ILs + organic salts Pen-OSILs AMX-OSILs	Hydrolysed beta lactam antibiotics (like OSIL) are stronger antimicrobial agents Penicillin-OSILs:100-fold better antimicrobial activity. (Pen-OSILs/AMX-OSILs) + [Ch], [P6,6,6,14] and [HDPy] → growth inhibition	*S. aureus*, MRSA	*E.coli* ATCC 25,922, *E. coli* CTX M2 and CTX M9	*Staphylococcus aureus* (MRSA ATCC 43,300), *Staphylococcus aureus* ATCC 25,923	Ferraz et al. (2020)

(continued)

Table 11.2 (continued)

ILs	Main Findings	Type of Microorganism			References
		Gram +ve	Gram -ve	Others	
8	12 ciprofloxacin and norfloxacin OSILs by combination with 6 organic cations: ammonium, pyridinium and N-methylimidazolium	47- and 1416-fold more soluble in water [HDPyr] + ciprofloxacin + norfloxacin → selective inhibited	*S. aureus* *B. subtilis*	*K. pneumoniae*	-	Santos et al. (2020)
9	MeIm-3.1-PP4 and its parent peptide, 3.1-PP4: IL + AMP	Improved stability Improved antimicrobial activity	*S. aureus, Enterococcus faecalis*	*K. Pneumoniae, K. Pneumoniae, P. aeruginosa, E. coli* (MDR)	–	Gomes et al. (2020)
10	levofloxacin (LVF) + imidazolium-based ILs; [BMIM]Cl and [DMIM]Cl	LVF + [DMIM]Cl has remarkable antimicrobial activity	*S. aureus*	*E. coli*	–	Siddiquee et al. (2021)

result in administering higher doses of the drug to achieve medicinal benefits (Myles et al. 2010; Egorova et al. 2017).

Therefore, APIILs offer an option for the enhancement of solubility and stability in conventional drug delivery systems. Recent studies have shown that many ILs boost protein shelf-life in a solution, preserving the perfect folded structure. It is why the solvents for protein treatment seem to have promising stability. ILs are capable of replacing water for intravenous injection for costly lyophilized drugs in protein-based pharmaceutical formulations. Suitable components would offer an appropriate viscosity, polarity, conductivity, and potential for hydrogen bonding (Myles et al. 2010; Egorova et al. 2017). Antibacterial and antifungal processes have been tested in ILs. Most of them displayed activity against Gram-positive and Gram-negative bacteria, paving the way for the use of these ILs as chemotherapeutics (Lim et al. 2014). As some researchers have demonstrated IL's negative effect on bacterial and fungal growth, we can use these as the solution to eliminate antibiotic-resistant microbes (Flieger and Flieger 2020).

11.4 Mechanism of Action of Ionic Liquids as Antimicrobial Agent

11.4.1 Cations and Anions and ILs Structure

In a study by Madaan and Tyagi (2008), pyridinium ILs exhibits no odor or toxicity when utilized in proper concentrations. They are stable and contain no contaminants such as active chlorine, iodine, phenol, mercury, copper, or other heavy metals. The growth of bacteria, viruses, fungi, protozoa, and other microorganisms is limited by their activities. Therefore, they do not accomplish the necessities of a "universal" antimicrobial agents. Moreover, they do not always provide the whole annihilate of different microorganisms under certain conditions. The past literature presented that the pyridinium-based IL activities are controlled by different mechanisms. They could be adsorbed on the microorganisms' cell membranes and improve different functions in their cell structure. Changes may include the alteration in cell structure and restraint in glycolysis and respiration of microorganisms, principally hindrance of the oxidation system. Generally, these ILs may be adsorbed on the cell wall and bind to one of the biochemical components. Sufficiently binding causes disruption of cell structure and halt the growth of the bacteria, which leads cells to death. It has also been stated that there could be leakage of free glycine, nitrogen, and phosphorus due to the presence of the ILs on the microorganisms (Pernak and Branicka 2003).

A study by Pernak and Branicka (2004) examined the antimicrobial activity of alkylpyridinium IL. They found that an enhance in the length of cation alky chain could improve the antimicrobial activity. On the other hand, nature of the cations (aromatic/non-aromatic) and their position at which they are replaced with the pyridine head group. Additionally, the assessment of both electrostatic impact and antimicrobial activity showed that ILs containing dimethylamino group at 3-position presented greater inhibition activity toward microbes and high electrostatic impact in contrast with the ILs possessing a hydroxy group at 3-position. The head group substituted with pyridinium amphiphiles exposed the significant effect on the antimicrobial activity and increment of the number of head-groups played a role in enhancing the activity. Moreover, a study by LaDow et al. (2011) examined the antimicrobial features of two novel cationic amphiphiles with triple pyridinium and trimethylammonium headgroups. It was stated that to destroy both Gram-positive and Gram-negative bacteria, the amphiphiles ILs holding triple head-groups were most efficient. On the other hand, ILs with triple head-groups could kill and destroy them in shorter duration time in comparison with the single headed ILs. In another study N,N-hexamethylene bis(4-carbamoyl-1-decylpyridinium bromides) and their derivatives exhibited in vitro antimalarial activity toward the *Plasmodium falciparum* FCR-3. The antimalarial activity of amphiphilic ILs were obtained at around 3.5–10 μM. The finding demonstrated that antimalarial activity relies upon the length the cation's alkyl chain. The cation with 8C presented was the most suitable for the application (Fujimoto et al. 2006). Higher hydrophobicity could be achieved with longer alkyl chains, leading to higher cytotoxicity and higher antimicrobial activity of quaternary ammonium compounds and ILs (Gindri et al. 2016).

The antimicrobial activity of C_4, C_6, and C_8 of [Pyr][Br] based ILs toward pure cultures of *P. fluorescens*, *S. cerevisiae*, *B. subtilis*, *E. coli*, and *S. aureus* was detected by Docherty and Kulpa (2005). Conversely, the ILs containing shorter alkyl chain, namely butyl pyridinium bromide, showed less antimicrobial resistant as compared with longer alkyl chain cations among the microbes tested. Dissimilarly, degree of oligomerization could improve the antibacterial and even anti-cholinesterase activities (Mancini et al. 2004). Moreover, the antibacterial activity behavior towards the Gram-positive, *Cocaceae,* is comparable to other salts from imidazolium-based IL. Furthermore, a study by Busetti et al. (2010) described ILs antibacterial activities, namely 1-alkyl-quinoline bromide ([C_nQuin][Br]), with chain length of C_8, C_{10}, C_{12}, C_{14}, C_{16}, and C_{18}. They found that the antimicrobial activities were enhanced with the increase in the chain length. They stated the values of minimal inhibitory concentration (MIC) around 121.2 μ M against *S. aureus* for IL containing alkyl carbon chain (C_{10}). Meanwhile, the value of MIC was around 31.25 μ M using [CDIM][Cl] with the same bacteria. Cornellas et al. (2011) achieved the inhibition at 940 μ M for the Gram-positive bacteria, *S. epidermidis* utilizing 1-alkylpyridine bromide ([C_nPyr][Br]). Additionally, they obtained the values of MIC against *S. epidermidis* at 250 μ M using the short chain imidazolium-based IL, [CMIM][Cl]. This shows that the length of alkyl chains of cation based ILs is not the mere factor that can affect the antibacterial activity. Concerning this, a study by Łuczak et al. (2010) evaluated the ILs with the long carbon chain (C_{10} or, $>C_{10}$) to promote a less threshold of human cells. This is directly link to the surface tension produced by the IL which could impact its viability through the membrane shock (Cornellas et al. 2011). On the other hand, the presence of a hydroxyethyl group (the C_{14} to C_{16} linker) in choline-based ILs improved their antimicrobial action and lowered the cytotoxic effects of this category of ILs (Siopa et al. 2016). The group reported antimicrobial effect toward *Mycobacterium smegmatis, B.subtilis, E. coli, C. albicans* and many others.

11.4.2 Solubility of Active Ingredients

Another study investigated the antibacterial activity of pyridinium-based ILs on *S. epidermidis* and *S. aureus*. They have shown that when the ILs are in combination with antibiotics, the values of MICs could reduce. For instance, the values of MICs for *S. aureus* were around 0.5 and 0.005 μM for [HDPyr][Cl] and [HDPyr][Amp] (ampicillin-coupled IL), respectively. These findings are in agreement with numerous examinations, wherever ILs were utilized as proteins carriers or another molecule transporter in plasma (Cybulski et al. 2008). The physicochemical characteristics of ILs showed that it could increase the solubility of hydrophobic antibiotics due to the occurrence of the hydrocarbon chains which produce the lipophilia feature (Neves et al. 2020). Furthermore, in another study ampicillin ILs were able to reverse the resistance in some medical strains that were earlier isolated and examine as persistence. For instance, *E. coli* TEM CTX M9, CTX M2, and AmpC MOX are good examples of this. The ampicillin ILs exhibited bactericidal effect and growth inhibition on several delicate bacteria, particularly some Gram-negative resistant bacteria as in comparison with the [Na][Amp] and the initial [Br]⁻ and [Cl]⁻ (Ferraz et al. 2014).

A study by He et al. (2013) highlighted the effect of anions, fumarate in this case. They showed that fumarate alkyl group tuned the ILs lipophilicity when altering the chain length. On the other hand, fumarate carboxylic terminal promoted the ILs hydrophilicity. Therefore, it is concluded that the number of alkyl chains relates with the ILs lipophilicity, while water solubility is enhanced by the carboxylate ion. It is also noted that balancing these two drivers may have a significant effect on activities toward microbes (He et al. 2013).

With the suitable choice of organic cation, significant biological alternatives could be attuned with antibacterial effects. Also, this study discloses the influence of ion-pair on the mechanism of ampicillin. The hydrophobic ampicillin counter-ions are needed, as well. The maximum enhancement of activity is attained versus Gram-negative resistant bacteria. These findings are undoubtedly encouraging and appoint to the useful influence of drug delivery aided by IL via exterior membrane of Gram-negative bacteria, creating potentials for novel similar purposes and investigations particularly in reversing drug resistance.

11.4.3 Integration of ILs with Antibiotics

ILs formation by combining ionizable APIs with biocompatible organic counter-ions (API-ILs) has proved its value as it improves the biological availability of poorly water and lipid-soluble substances, decreases, or removes the polymorphism which creates more efficiently developed commercial medicines. In this context, for instance, improved antimicrobial activity was achieved using hydrolysed Amoxicillin (AMX) and Penicillin (Pen) in the form of OSILs over the neat ILs or the antibiotics alone. The combined formula (Pen-OSIL) was 100-fold better against *S.aureus* and *E.coli*. Moreover, integrating the hydrolysed form of the antibiotics (AMX-OSIL and Pen-OSIL) with the cation [HDPyr], [P$_{6,6,6,14}$], and [Ch] effectively constrained the growth of resistant MRSA and *E. coli* strains (Ferraz et al. 2020). Additionally, Santos et al. (2020) reported the possibility to develop highly bioavailable modified formulation of fluoroquinolones with low toxicity against healthy cells and commensal bacteria while possessing antimicrobial activity. They found that [HDPyr]-based OSILs containing norfloxacin and ciprofloxacin were selective toward distinct bacterial strains, *S. aureus* (11-fold) and *K. pneumoniae* (20-fold), respectively at subtoxic concentrations. They have also found that combining ciprofloxacin with [EOHDMIM] or [EMIM] improved the antimicrobial activity. Another study by Siddiquee et al. (2021) has combined levofloxacin and imidazolium-based ILs such as [BMIM][Cl] and [DMIM][Cl] and found that the effect is preferable with longer alkyl chain ILs. Moreover, they reported greater efficiency of bactericidal activity with IL with longer alkyl chains. A low cytotoxicity against HEK-293 (Human embryonic kidney cells) was observed with the application of levofloxacin combined with ILs, while the activity against both *E. coli* and *S.* aureus was enhanced. Similar findings were reported by Yang et al. (2021) were they have used [C$_x$MIM]Cl in combination with tetracycline against *S. cerevisiae, E. coli, P. aeruginosa,* and *S. aureus*. This shows that synergistic interaction between ILs and antibiotics could enhance their antimicrobial activities.

11.4.4   Cell Destructions and Enzymatic Activity

The corrosion inhibition of pyridinium ILs was also investigated in a study by (Wang and Free 2004). Pyridinium-based ILs as cathodic inhibitors could associate increment in cell resistance, polarization, and reduction in common pathways. Adsorption and constitution of insoluble films are usually associated with inhibition. It is through the interaction between the atoms of the metal and salts on the surfaces. Quaternized pyridinium salts are considered as valuable restraint due to the combination with different cationic surfactants in an acidic medium of ferrous surfaces, which could hinder their corrosion by creating a bilayer. In a study presented by (Rezki et al. 2019) the innovative dicationic ILs (DiILs) containing pyridinium hydrazone and phenoxy side chain hybrids were assessed for their antimicrobial activity versus various kinds of vast persistent strains. It includes two fungal strain, two Gram-negative bacteria and two Gram-positive bacteria. The antibacterial screening was accomplished through the method of microdilution. The growth inhibition was also detected by MIC values. The tested DiILs displayed bactericidal effect toward methicillin resistant *E.coli* and *S. aureus* (MRSA). The result showed that DiILs have some influence on the

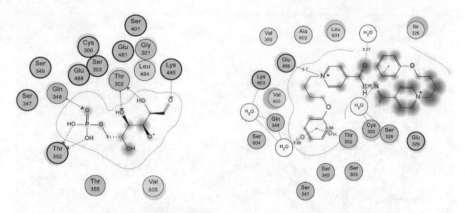

Fig. 11.2 Posing of molecular docking of the active DiIL, [4-Pyridyl][BF₄], in the binding pocket of *E. coli* glucosamine-6-P synthase with two-dimension scheme (Reprinted with permission from Elsevier (Rezki et al. 2019). DiIL: dicationic IL

Gram-positive bacteria as the increase in chain of phenoxy linker resulted in higher MIC values. This could be due to the increment in the hydrophobicity of the DiILs. Figure 11.2 presents the simulation of molecular dynamic in two-dimension via the interaction with surface membrane as a general mechanism of ILs. Also, the chemical structure and binding of ILs as inhibitors might be included in the obstruction of the cell membrane via this mechanism. Therefore, it is validating the docking protocol by means of the high resolution of *E. coli* glucosamine-6-P synthase as the targeted protein (Rezki et al. 2019).

Ineffective sanitation usually results in higher colonization of *Streptococcus* mutants. Biofilms are syntrophic group of microbial cells attach to each and may also attach to a surface. ILs have been utilized to decrease the formation of biofilm and to inhibit demineralization at vulnerable sites in orthodontic equipment (Martini Garcia et al. 2019). Using [BMIM] [NTf₂], the biofilm formation decreased without influencing the cytotoxicity and the physicochemical properties of the orthodontic resin. The research group suggested that on the one hand, the positive charge of imidazole group is accountable for an electrostatic interaction with the negative components of the bacterial cell wall, for instance, N-acetylmuramic and lipoteichoic acids. While on the other hand, the hydrophobic components of the ILs, [NTf₂] and alkyl chain, could result in disorganization of the membrane. See Fig. 11.3. In addition, some studies explained that membranes of prokaryotic cells such as bacteria are composed of negative charges, which facilitates the interaction with IL. Meanwhile, cell membranes of mammals constitute of sterols, sphingolipids, and lecithins leading to neutral net charge. Therefore, the electrostatic attraction between positive imidazole group, for instance, and mammalian is obstructed, which leads to better biocompatibility (Yeaman and Yount 2003; Pendleton and Gilmore 2015).

Moreover, ILs may precipitate and coagulate components of the cytol, including nucleic acids, and result in enzyme denaturation (Fig. 11.4). Hence, pathways that are

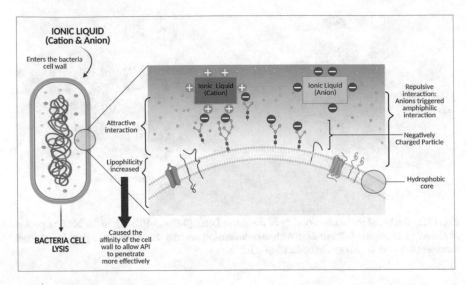

Fig. 11.3 Mechanism of ionic liquids as antimicrobial agents based on electrostatic interactions. Created by Biorender.com

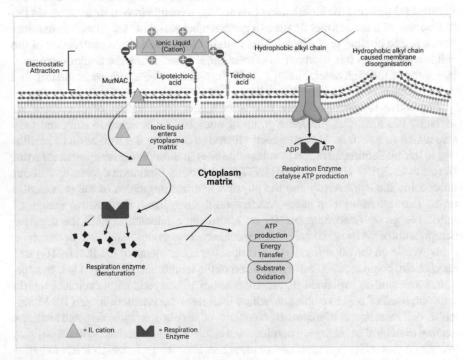

Fig. 11.4 Mechanism of ionic liquids as antimicrobial agents based on the effect on enzymes of some pathways. Created by Biorender.com

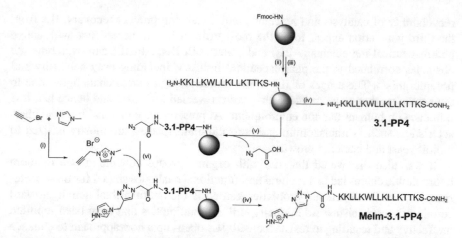

Scheme 11.1 "Synthesis route toward MeIm-3.1-PP4: (i) 1.1 molar equivalents (eq) of MeIm, 1.0 eq of propargyl bromide (80% in toluene), 40 °C, 24 h; (ii) 5 eq of Fmoc-protected amino acid, 10 eq of *N*-ethyl-*N,N*-diisopropylamine (DIEA), and 5 eq of *O*-(benzotriazol-1-yl)-*N,N,N′,N′*-tetramethyluronium hexafluorophosphate (HBTU) in *N,N*-dimethylformamide (DMF), 1 h, room temperature (r.t.); (iii) 20% piperidine in DMF, 15 min, r.t.; (iv) trifluoroacetic acid (TFA)/triisopropylsilane (TIS)/distilled water (95:2.5:2.5 *v/v/v*), 2 h, r.t.; (v) 5 eq of azido acetic acid, 10 eq of DIEA, and 5 eq of HBTU in DMF, 1 h, r.t.; (vi) 1 eq sodium L- ascorbate, 10 eq of DIEA, 10 eq of 2,6-lutidine, 1 eq of Pr-MeIm, and 1 eq of copper(I) bromide in DMF:acetonitrile (MeCN) (3:1 *v/v*). Reprinted with permission from (Gomes et al. 2020), *International Journal of Molecular Science*; published by MDPI, 2020

depending on enzymes such as glycolysis, substrate oxidation, and ATP synthesis and will be hindered, leading to negative impact on cellular respiration (Denyer 1995).

In this context, coupling of antimicrobial methylimidazolium ILs to the N-terminus of peptide was investigated by Gomes et al. (2020) to study the stability, antibiofilm, and antibacterial activities on the resistant strain, *K. pneumoniae*. See Scheme 11.1.

The group highlighted the importance of some enzymes such as tyrosinase in the wound site. Tyrosinase is unstable in an infected wound. However, the addition of MeIm-3.1-PP4 demonstrating a significant increase in stability achieved by coupling IL to the *N*-terminal of the peptide, which prevented the enzyme degradation (Gomes et al. 2020).

11.5 Conclusions and Future Perspectives

ILs are tremendously utilized in a broad diversity of applications as they are considered tuneable and designer solvents. Various ILs are extended for employment as solvents in the chemical industry, and their applications offer plenty of benefits over utilizing conventional solvents. Their benefit goes beyond greater efficiency, recyclability of catalysts and reactants, and enhancing product recovery. ILs from the third generation appear to be the most utilized ILs, in accordance with active pharmaceutical ingredients (APIs) and called API-ILs. API-ILs can overcome the obstacles correlated to the pharmaceutical industry, including drug solubility and polymorphism. The usages of ILs in the biotechnologies were challenging due to their toxicity, though more lately, they were revealed less toxic and hence had less influence on humans and the environment. At present, large cations' biocidal characteristics, such as imidazolium and benzalkonium species, are mostly utilized to inhibit yeast and bacterial growth.

It was also discovered that ILs with organic groups containing one or more hydrophobic chains led to structural and functional analogies shared for these materials. One of the functional properties gained is the antimicrobial activity toward certain microorganisms. Additionally, utilizing antibiotics that have been reported ineffective and resulting in resistance with ILs opens up a new approach to enhance the antibiotic's effectiveness. It is also concluded that ILs would enhance the antibiotics' solubility, yielding more bioavailability of the formulation. Therefore, future prospects of this research include:

1. Although ILs have been studied and proved to be successful in several applications, there are still gaps to be explored and reported, such as microbial resistance studies.
2. Even though some recent studies have reported the antimicrobial activities of some API-ILs, the mechanism of activity is still complex and uncertain and based mostly on speculations and theories. Hence, it is necessary to conduct studies to unveil the mechanisms of antimicrobial activities toward resistant strains.
3. More research on the combination of API-ILs with antibiotics is highly encouraged.
4. Most experiments on API-ILs are conducted at lab-scale considering the high cost of constitutes. However, as API-ILs demonstrated prominent activity toward highly resistant microbes, additional large-scale studies are required to evaluate those substances' practicality as disinfectants of hospitals, for instance.
5. The past literature focused on investigating the effect of cations of ILs as antimicrobial agents; hence, the effect of anions as antimicrobial agents may be assessed further to enhance the effect against resistant microbial strains.
6. It is highly recommended to consider the effect on human cells, thermal stability, and shelf-life of API-ILs to improve the effect against resistant microbial strains.
7. To date, there is still uncertainty regarding the potential risks and the environmental impacts of ILs categories. Therefore, future studies should consider the prominence on the waste of chemical and pharmaceutical industries, treatment implication, and inspecting the ecotoxicology to focus on underlying issues and better utilization of ILs in medicine.

Acknowledgements This study is funded the Research Management Centre, IIUM [RMCG20-021-0021] (Elgharbawy, A.).

References

Ahmad H, Zaharudin NS, Majid NNA et al (2019) Synthesis and characterization of new choline-based ionic liquids and their antimicrobial properties. J Adv Res Fluid Mech Therm Sci 54:124–132

Asokan GV, Ramadhan T, Ahmed E, Sanad H (2019) WHO global priority pathogens list: a bibliometric analysis of medline-pubmed for knowledge mobilization to infection prevention and control practices in Bahrain. Oman Med J 34:184–193. https://doi.org/10.5001/omj.2019.37

Black SN, Collier EA, Davey RJ, Roberts RJ (2007) Structure, solubility, screening, and synthesis of molecular salts. J Pharm Sci 96:1053–1068

Bromberger B, Sommer J, Robben C et al (2020) Evaluation of the antimicrobial activity of pyrithione-based ionic liquids. Sep Purif Technol 251:117309. https://doi.org/10.1016/j.seppur.2020.117309

Busetti A, Crawford DE, Earle MJ et al (2010) Antimicrobial and antibiofilm activities of 1-alkylquinolinium bromide ionic liquids. Green Chem 12:420–425

CDC (2019) Antibiotic resistance threats in the United States. U S Dep Heal Hum Serv. https://www.cdc.gov/drugresistance/pdf/threats-report/2019-ar-threats-report-508.pdf

CDC (2020) MOST DANGEROUS ANTIBIOTIC-RESISTANT BACTERIA Antibiotic/Antimicrobial Resistance (AR/AMR). In: Drug Resist. https://www.cdc.gov/drugresistance/biggest-threats.html. Accessed 13 Apr 2020

Cornellas A, Perez L, Comelles F et al (2011) Self-aggregation and antimicrobial activity of imidazolium and pyridinium based ionic liquids in aqueous solution. J Colloid Interface Sci 355:164–171. https://doi.org/10.1016/j.jcis.2010.11.063

Cybulski J, Wiśniewska A, Kulig-Adamiak A et al (2008) Long-Alkyl-chain quaternary ammonium lactate based ionic liquids. Chem Eur J 14:9305–9311. https://doi.org/10.1002/chem.200800973

Czekański L, Almeida TSD, Mota JP et al (2014) Synthesis of benzoazole ionic liquids and evaluation of their antimicrobial activity. Biomed Biopharm Res 11:227–235

Denyer SP (1995) Mechanisms of action of antibacterial biocides. Int Biodeter Biodegrad 36:227–245. https://doi.org/10.1016/0964-8305(96)00015-7

Docherty KM, Kulpa CF Jr (2005) Toxicity and antimicrobial activity of imidazolium and pyridinium ionic liquids. Green Chem 7:185–189

Egorova KS, Gordeev EG, Ananikov VP (2017) Biological activity of ionic liquids and their application in pharmaceutics and medicine. Chem Rev 117:7132–7189. https://doi.org/10.1021/acs.chemrev.6b00562

Ferraz R, Silva D, Dias AR et al (2020) Synthesis and antibacterial activity of ionic liquids and organic salts based on penicillin G and amoxicillin hydrolysate derivatives against resistant bacteria. Pharmaceutics 12:221. https://doi.org/10.3390/pharmaceutics12030221

Ferraz R, Teixeira V, Rodrigues D et al (2014) Antibacterial activity of Ionic liquids based on ampicillin against resistant bacteria. Rsc Adv 4:4301–4307

Flieger J, Flieger M (2020) Ionic liquids toxicity-benefits and threats. Int J Mol Sci 21:6267. https://doi.org/10.3390/ijms21176267

Fujimoto K, Morisaki D, Yoshida M et al (2006) Antimalarial effect of bis-pyridinium salts, N, N′-hexamethylenebis (4-carbamoyl-1-alkylpyridinium bromide). Bioorg Med Chem Lett 16:2758–2760

Gindri IM, Palmer KL, Siddiqui DA et al (2016) Evaluation of mammalian and bacterial cell activity on titanium surface coated with dicationic imidazolium-based ionic liquids. RSC Adv 6:36475–36483. https://doi.org/10.1039/C6RA01003B

Gomes A, Bessa LJ, Correia P et al (2020) "Clicking" an ionic liquid to a potent antimicrobial peptide: on the route towards improved stability. Int J Mol Sci 21:6174. https://doi.org/10.3390/ijms21176174

Goossens H (2009) Antibiotic consumption and link to resistance. Clin Microbiol Infect 15:12–15

Goossens H, Guillemot D, Ferech M et al (2006) National campaigns to improve antibiotic use. Eur J Clin Pharmacol 62:373–379

Gouveia W, Jorge TF, Martins S et al (2014) Toxicity of ionic liquids prepared from biomaterials. Chemosphere 104:51–56

Guo J, Qian Y, Sun B et al (2019) Antibacterial amino acid-based poly(ionic liquid) membranes: effects of chirality, chemical bonding type, and application for MRSA skin infections. ACS Appl Bio Mater 2:4418–4426. https://doi.org/10.1021/acsabm.9b00619

Havenga B, Ndlovu T, Clements T et al (2019) Exploring the antimicrobial resistance profiles of WHO critical priority list bacterial strains. BMC Microbiol 19:1–16. https://doi.org/10.1186/s12 866-019-1687-0

He B, Ou G, Zhou C et al (2013) Antimicrobial ionic liquids with fumarate anion. J Chem 473153. https://doi.org/10.1155/2013/473153

Johnson AP (2015) Surveillance of antibiotic resistance. Philos Trans R Soc B Biol Sci 370:20140080

LaDow JE, Warnock DC, Hamill KM et al (2011) Bicephalic amphiphile architecture affects antibacterial activity. Eur J Med Chem 46:4219–4226

Lim GS, Zidar J, Cheong DW et al (2014) Impact of ionic liquids in aqueous solution on bacterial plasma membranes studied with molecular dynamics simulations. J Phys Chem B 118:10444–10459

Łuczak J, Jungnickel C, Łącka I et al (2010) Antimicrobial and surface activity of 1-alkyl-3-methylimidazolium derivatives. Green Chem 12:593–601

Madaan P, Tyagi VK (2008) Quaternary pyridinium salts: a review. J Oleo Sci 57:197–215

Mancini I, Sicurelli A, Guella G et al (2004) Synthesis and bioactivity of linear oligomers related to polymeric alkylpyridinium metabolites from the Mediterranean sponge Reniera sarai. Org Biomol Chem 2:1368–1375

Martini Garcia I, Jung Ferreira C, de Souza VS et al (2019) Ionic liquid as antibacterial agent for an experimental orthodontic adhesive. Dent Mater 35:1155–1165. https://doi.org/10.1016/j.den tal.2019.05.010

Myles L, Gore R, Špulák M et al (2010) Highly recyclable, imidazolium derived ionic liquids of low antimicrobial and antifungal toxicity: a new strategy for acid catalysis. Green Chem 12:1157–1162

Neves YF, Eloi ACL, de Freitas HMM et al (2020) Imidazolium salts as alternative compounds to control diseases caused by plant pathogenic bacteria. J Appl Microbiol 128:1236–1247

Pendleton JN, Gilmore BF (2015) The antimicrobial potential of ionic liquids: a source of chemical diversity for infection and biofilm control. Int J Antimicrob Agents 46:131–139. https://doi.org/10.1016/j.ijantimicag.2015.02.016

Pernak J, Branicka M (2003) The properties of 1-alkoxymethyl-3-hydroxypyridinium and 1-alkoxymethyl-3-dimethylaminopyridinium chlorides. J Surfactants Deterg 6:119–123

Pernak J, Branicka M (2004) Synthesis and aqueous ozonation of some pyridinium salts with alkoxymethyl and alkylthiomethyl hydrophobic groups. Ind Eng Chem Res 43:1966–1974

Rezki N, Al-Sodies SA, Ahmed HEA et al (2019) A novel dicationic ionic liquids encompassing pyridinium hydrazone-phenoxy conjugates as antimicrobial agents targeting diverse high resistant microbial strains. J Mol Liq 284:431–444

Saadeh SM, Yasseen Z, Sharif FA, Shawish HMA (2009) New room temperature ionic liquids with interesting ecotoxicological and antimicrobial properties. Ecotoxicol Environ Saf 72:1805–1809

Santos MM, Alves C, Silva J et al (2020) Antimicrobial activities of highly bioavailable organic salts and ionic liquids from fluoroquinolones. Pharmaceutics 12:694

Saraswat J, Aldahmash B, AlOmar SY et al (2020) Synergistic antimicrobial activity of N-methyl substituted pyrrolidinium-based ionic liquids and melittin against Gram-positive and Gram-negative bacteria. Appl Microbiol Biotechnol 104:10465–10479. https://doi.org/10.1007/s00253-020-10989-y

Siddiquee MA, Saraswat J, Imtiyaz K et al (2021) In-vitro cytotoxicity, synergistic antibacterial activity and interaction studies of imidazolium-based ionic liquids with levofloxacin. J Mol Liq 325:115125. https://doi.org/10.1016/j.molliq.2020.115125

Siopa F, Figueiredo T, Frade RFM et al (2016) Choline-based ionic liquids: improvement of antimicrobial activity. Chem Sel 1:5909–5916. https://doi.org/10.1002/slct.201600864

Stasiewicz M, Mulkiewicz E, Tomczak-Wandzel R et al (2008) Assessing toxicity and biodegradation of novel, environmentally benign ionic liquids (1-alkoxymethyl-3-hydroxypyridinium chloride, saccharinate and acesulfamates) on cellular and molecular level. Ecotoxicol Environ Saf
71:157–165

Thuy Pham TP, Cho CW, Yun YS (2010) Environmental fate and toxicity of ionic liquids: a review.
Water Res 44:352–372. https://doi.org/10.1016/j.watres.2009.09.030

Wang W, Free ML (2004) Prediction and measurement of corrosion inhibition of mild steel using
nonionic surfactants in chloride media. Corros Sci 46:2601–2611

Yang DD, Paterna NJ, Senetra AS et al (2021) Synergistic interactions of ionic liquids and
antimicrobials improve drug efficacy. iScience 24(1):101853. https://doi.org/10.1016/j.isci.2020.
101853

Yeaman MR, Yount NY (2003) Mechanisms of antimicrobial peptide action and resistance.
Pharmacol Rev 55(1):27–55. https://doi.org/10.1124/pr.55.1.2

Chapter 12
Ionic Liquid Assisted Microemulsions for Drug Delivery

Alireza Salabat

Abstract In this chapter after the definition of traditional microemulsions, a new type of microemulsion system, namely, ionic liquid based microemulsion, has been introduced. Ionic liquids comprising microemulsions may be acting as polar or non-polar phases as well as surfactant agent. Then, three types of ionic liquid based microemulsions will be formed as non-aqueous IL microemulsions, aqueous IL microemulsions, and IL/oil/water microemulsions. It has been discussed that the ionic liquid based microemulsions are able to overcome the inabilities of conventional microemulsions to dissolve a number of chemicals which are water insoluble and result in better pharmacological activities of drugs. Most importantly, in the drug delivery field, ionic liquid based microemulsions can serve better bioavailability, permeation, and stability. Additionally, one of the most important applications of ionic liquid based microemulsions, as drug nanocarrier, has been discussed in details. The mechanism of solubility enhancement of an active pharmaceutical ingredient and transdermal drug delivery by using ionic liquid based microemulsion is highlighted. It is also discussed that hydrophobic nontoxic ionic liquids can be used to prepare novel ionic liquid based microemulsions for intravenous administration.

Keywords Ionic liquid · Microemulsion · Nanocarrier · Drug delivery

Abbreviations

IL	Ionic liquid
RTIL	Room temperature ionic liquid
SAIL	Surface active ionic liquid
ME	Microemulsion
MEG	Microemulgel

A. Salabat (✉)
Department of Chemistry, Faculty of Science, Arak University, 38156-8-8349 Arak, Iran
e-mail: a-salabat@araku.ac.ir

Institute of Nanosciences and Nanotechnology, Arak University, 38156-8-8349 Arak, Iran

SFME	Surfactant free microemulsion
O/W	Oil-in-water
W/O	Water-in-oil
IL/W	Ionic liquid-in-water
W/IL	Water-in-ionic liquid
IL/O	Ionic liquid-in-oil
O/IL	Oil-in-ionic liquid
BC	Bicontinuous phase
DLS	Dynamic light scattering
API	Active pharmaceutical ingredient
TDD	Transdermal drug delivery
UV-vis	UV–vis spectroscopy
TEM	Transmission electron microscopy
AFM	Atomic force microscopy
TEWL	Trans-epidermal water loss values
ATR-FTIR	Attenuated total Reflectance Fourier Transform Infrared
MMA	Methyl methacrylate
AOT	Sodium bis(2-ethylhexyl)sulfosuccinate
IPM	Isopropyl myristate
ACV	Acyclovir
MTX	Methotrexate
Den	Dencichine
Ars	Artemisinin
SC	Stratum corneum
Ch	Cholinium
CA	Carboxylic acids
$[C_2mim][BF_4]$	1-ethyl-3-methylimidazolium tetrauoroborate
$[C_4mim][BF_4]$	1-butyl-3-methylimidazolium tetrauoroborate
$[C_8mim][BF_4]$	1-octyl-3-methylimidazolium tetrauoroborate
$[C_4mim][PF_6]$	1-butyl-3-methylimidazolium hexauorophosphate
$[C_8mim][NTf_2]$	1-octyl-3-methylimidazolium bis(triuoromethylsulfonyl)imide
[Ch][Ole]	Choline oleate
[Ch][Lin]	Choline linoleate
[Ch][Eru]	Choline erucate
[Ch][For]	Choline formate
[Ch][Lac]	Choline lactate
[Ch][Pro]	Choline propionate
$[C_1mim][(CH_3O)_2PO_2]$	Dimethylimidazolium dimethylphosphate
[HOEmim][Cl]	1-hydroxyethyl-3-methylimidazolium chloride
$[C_4mim][C_{12}SO_3]$	1-butyl-3-methylimidazolium dodecanesulfate

12.1 Introduction

Active pharmaceutical ingredients (APIs) usually commercialized in several dosage forms in which crystalline forms have been the preferred option. However, about 70% of the drugs present low water solubility, which may compromise the bioavailability and therapeutic efficacy and, thus, fail in the later stages of development. The irregular gastrointestinal absorption of solid forms along with the low therapeutic efficiency and possible toxicity and side effects of polymorphs are major concerns to overcome. On the other hand, the APIs' solubility in aqueous solution, dissolution, and bioavailability is also dependent on particle size and properties. Attempting to improve the drugs solubility in water as well as their bioavailability, several strategies have been investigated (Kalepu and Nekkanti 2015; Savjani et al. 2012). Micro-structured fluids such as microemulsions or nanoemulsions, nanoparticles and vesicles could be a suitable choice for this purpose.

Much research in recent years has focused on the microemulsion method, a system generally comprised of water, oil, and surfactants which are said to be the modern colloidal carriers. The microemulsion system is seen to be exquisite as the biocompatible and stable system with high solubilization capacity for drug molecules. ILs based microemulsions are a newly emerging carriers with great potential in pharmaceutics and drug delivery (Pedro et al. 2020). In recent years, ionic liquids (ILs) or green/designer solvent have attracted much attention in contrast to the conventional solvents (usually organic solvents and water) as environmentally benign media for reaction, separation, and multidisciplinary chemistry (Smiglak et al. 2007; Chaturvedi 2011). ILs are unique class of designable and tunable solvents with wide electrochemical window, essentially zero volatility, high thermal stability, nonflammability, extremely high electrical conductivities, and wide liquid range. ILs usually are 1-1 electrolyte containing a huge organic cation and an inorganic anion, which the ions can be varied by a simple ion-exchange procedure. ILs have melting points below 100 °C, and at below room temperature are known as room temperature ionic liquids (RTILs) (Petkovic et al. 2011).

In drug delivery systems, ILs may be applied as novel pharmaceutical forms of API-ILs. ILs are excellent solvents for a wide range of biopolymers, such as proteins, DNA, and polysaccharides, being used in their processing into films and micro and nanoparticles with potential for drug delivery. Drug delivery systems using ILs can be categorized as follows: Nanocarriers, nanoparticles, nanofibers, ionogels, polymerizable systems, membranes, and vesicles (Pedro et al. 2020). The design of IL-excipient with tunable lipophilicity/hydrophilicity character is advantageous, especially when used as solubility-enhancing agent or as nanocarrier for complex amphiphilic APIs. Ionic liquid based microemulsion droplets formed in continuous oil or aqueous phase using nontoxic surfactants could be efficient carriers for sparingly soluble drugs which are dissolved in ILs phases in microemulsions. Some relevant reviews and book chapters recognizing the advances of ILs in different areas of pharmaceuticals development, spanning from their formulation, biological activity, and application on drug delivery are available (Egorova et al. 2017; Tanner et al. 2019). However, there is no review paper or book on the application of IL based microemulsion systems in drug delivery or pharmaceutical field.

In this chapter, after definition of traditional microemulsion systems, three types of IL based microemulsions namely non-aqueous IL microemulsions, aqueous IL microemulsions, and IL/oil/water microemulsions will be introduced in details. Then, IL based microemulsions as drug nanocarriers will be discussed.

12.2 Ionic Liquid Based Microemulsions

Microemulsions are defined as a system of water, oil, and amphiphile which is a single optically isotropic and thermodynamically stable liquid solution (Eastoe 2005). In practice, the key difference between emulsions and microemulsions are that the former, while they may exhibit excellent kinetic stability, are fundamentally thermodynamically unstable, and will eventually phase separate (Salabat et al. 2008). Then emulsions are cloudy solutions, while microemulsions are clear or translucent. Basically, when surfactants are incorporated into immiscible mixtures of oil and water, the surfactant molecules can locate at the oil/water interface which is thermodynamically very favorable (Kale and Deore 2017).

The free energy of microemulsion formation can be considered to depend on the extent to which surfactant lowers the surface tension of the oil–water interface and the change in entropy of the system such that

$$\Delta G_f = \gamma \Delta A - T \Delta S, \tag{12.1}$$

where ΔG_f is the free energy of formation, γ is the surface tension of the oil–water interface, ΔA is the change in interfacial area on microemulsification, ΔS is the change in entropy of the system which is effectively the dispersion entropy, and T is the temperature. When a microemulsion is formed, the change in ΔA is very large due to the large number of very small droplets formed. On dispersion, the droplet number increases and ΔS is positive. If the surfactant can reduce the interfacial tension to a sufficiently low value, the energy term in Eq. (12.1) ($\gamma \Delta A$) will be relatively small and positive, thus allowing a negative free energy change, that is, spontaneous microemulsification.

A well-known classification of the microemulsions (MEs) is that of Winsor (1948) who identified four general types of phase equilibria. In Winsor type I, the surfactant-rich water phase coexists with the oil phase where surfactant is only present as monomers at small concentration and oil-in-water (o/w) MEs form. Alternatively, in Winsor type II, the surfactant is mainly in the oil phase and water-in-oil (w/o) MEs form. The surfactant-rich oil phase coexists with the surfactant-poor aqueous phase. Type III corresponds to a three-phase system, where a surfactant-rich middle-phase co-exists with both excess water and oil surfactant-poor phases, while Winsor type

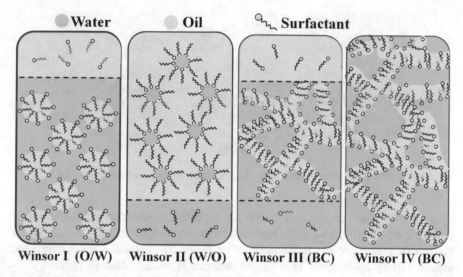

Fig. 12.1 Winsor classification of microemulsions

IV is a single phase isotropic solution which forms upon addition of higher amount of surfactant. Both Winsor III and IV are bicontinuous (BC) MEs (Fig. 12.1).

A literature survey shows that numerous reviews and books have been published summarizing studies focused on the formulation of MEs, as well as their character-ization and their applications in different fields (Gradzielski 2008; Callender et al. 2017; Lawrence and Rees 2012; Stubenrauch 2008). Recently, it has been demon-strated that the polar and non-polar phases of the classical microemulsions, which are water and oil, can be replaced by ILs (Kunz et al. 2011). The use of ILs as dispersed or continuous phases in MEs makes them attractive due to their unique and unusual properties. Greaves et al. in a review paper mentioned that ILs with long alkyl chains behave like amphiphiles, which introduce them suited for the develop-ment of targeted MEs (Greaves and Drummond 2008). Recently, in many researches, it is well identified that ILs, depending on their cation and anion properties, can be used as polar as well as non-polar solvents to form stable MEs. ILs as dispersed or continuous phases form the different types of ionic liquid base MEs, including (a) non-aqueous IL MEs, which water replaced by ILs, (b) aqueous IL MEs, which organic solvent replaced by ILs and (c) IL/oil/water MEs, which ILs act as surface active agent (Hejazifar et al. 2020).

12.2.1 Non-Aqueous IL MEs

Basically, many organic compounds and APIs are only slightly soluble or may be unstable in water phase. Then new ME formulations are demanded to replace water

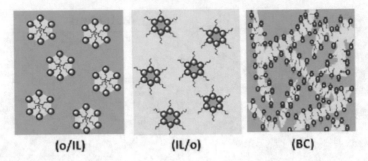

(o/IL)   **(IL/o)**   **(BC)**

Fig. 12.2 Structures of non-aqueous MEs (yellow color for oil phase and green color for IL phase)

by a suitable polar compounds. Some ILs have the required properties to act as polar phases in MEs; hence, the replacement of water by ILs is a currently active application area especially in pharmaceutical sector. Therefore ILs are currently the most intensely investigated type of non-aqueous IL based MEs. In general, non-aqueous ionic liquid based MEs consist of an IL, an organic apolar solvent and an additional surfactant (Eastoe et al. 2005). These compositions can form IL-in-oil (IL/O), oil-in-IL (O/IL), and bicontinuous (BC) typical MEs structures, as shown schematically in Fig. 12.2. Zech et al. presented a book chapter on non-aqueous microemulsions containing ILs (Zech et al. 2011). They reviewed published research works on these types of ME systems and investigated their properties in details and some applications.

The most frequently used IL for the preparation of non-aqueous IL MEs is 1-butyl-3-methylimidazolium tetrafluoroborate ($[C_4mim][BF_4]$), which for the first time is introduced as a polar domain by Gao et al. (2004). Many research groups studied the micropolarity of non-aqueous IL MEs by different methods (Gao et al. 2006a; Li et al. 2007; Xu et al. 2013a; Rojas et al. 2013). The $[C_4mim][BF_4]$-in-toluene ME in the presence of the nonanionic surfactant Triton X-100 has also been investigated by Gao et al. who applied various methods for its characterization, including DLS, conductivity, UV–vis spectroscopy, and TEM (Gao et al. 2006a). The results indicate that the initially formed ME droplets are enlarged by the addition of $[C_4mim][BF_4]$, and successive addition of $[C_4mim][BF_4]$ leads to evolution of large-sized ME droplet clusters in the range of 200–400 nm. NMR spectroscopic analysis reveals that the special structures and properties of $[C_4mim][BF_4]$ and TX-100 combined with the apolar nature of toluene contribute to the formation of such self-assemblies. The researches approved that ionic compounds, e.g., $Ni(NO_3)_2$, $CuCl_2$, $CoCl_2$, or biochemical reagents, could be solubilized into the IL/O ME droplets, suggesting their potential application in drugs formulation, nanomaterial synthesis, biological extractions, or as solvents for enzymatic reactions.

A novel non-aqueous IL ME system containing $[C_4mim][BF_4]$ and methyl methacrylate (MMA) introduced as IL-in-MMA ME, a soft template for synthesis of poly(methyl methacrylate) (PMMA)-TiO_2 nanocomposite (Mirhosseini and Salabat 2015). The photocatalytic activity of the prepared nanocomposite was studied, and the result showed that the IL acts as photo-sensitizer for TiO_2. It is found that IL can extend the photo response of TiO_2 to the visible region and improve the photo-catalytic efficiency of TiO_2 nanoparticles under visible light in degradation of a pollutant model. The photocatalytic-based antibacterial activity, for the destruction of Escherichia coli and Staphylococcus aureus, of the prepared nanocomposite film was also investigated (Salabat and Mirhoseini 2015).

12.2.2 Aqueous IL MEs

Organic solvents, which constitute the oil phase of MEs, are typically volatile, toxic, and flammable; these properties are undesirable from an environmental perspective. Because of the unique and attractive physicochemical properties of ILs, water-immiscible ILs appear to be an ideal replacement of the organic solvents used as the oil phase in MEs. This type of microemulsions may be form IL-in-water (IL/W), water-in-ILs (W/IL), or BC microstructures, as shown schematically in Fig. 12.3.

The most frequently used hydrophobic IL for the preparation of aqueous IL microemulsions is $[C_4mim][PF_6]$ because of their relatively simple preparation and reasonable price (Gao et al. 2006b; Anjum et al. 2008; Sun et al. 2014; Mirhoseini et al. 2018). Gao et al. introduced the first aqueous IL microemulsion comprising $[C_4mim][PF_6]$/TX-100/water (Gao et al. 2005). They identified IL/W, BC, and W/IL phase regions as well as hydrodynamic diameter of ME droplets, by cyclic voltammetry, dynamic light scattering, and UV–vis methods. 1-octyl-3-methylimidazolium bis(triuoromethylsulfonyl)imide ($[C_8mim][NTf_2]$) as a hydrophobic IL was used to form aqueous IL ME containing AOT/1-hexanol as surfactant/co-surfactant by Moniruzzaman et al. (2008). They confirmed W/IL structure by DLS and protein solubilization experiments. Hydrophobic interactions between the AOT tails and the long hydrocarbon chain of the IL combined with the strong tendency of the AOT head group to interact with water molecules result in aqueous droplet formation in the IL. The enhanced green fluorescent protein was used as an indicator of the system's ability to solubilize protein molecules.

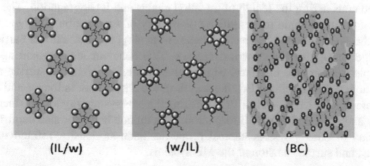

(IL/w)   (w/IL)   (BC)

Fig. 12.3 Structures of aqueous MEs (blue color for aqueous phase and green color for IL phase)

Fig. 12.4 Phase boundaries of microemulsions: I) IPM + S_{mix} + Water, II) [OMIM][PF_6] + S_{mix} + Water, and III) [BMIM][PF_6] + S_{mix} + Water (Parsi and Salabat 2020)

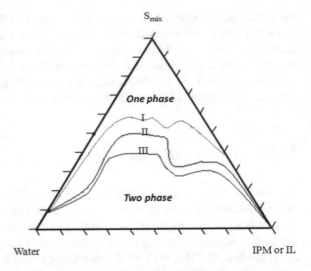

Recently, two hydrophobic ILs of [C_4mim][PF_6] and [C_8mim][PF_6] were used to construct of IL/W microemulsion formulations containing nonionic surfactant/co-surfactant of tween-80/transcutol®P (S_{mix}), as potential carriers for sparingly soluble celecoxib drug (Parsi and Salabat 2020). These IL MEs investigated precisely and compared with a traditional ME system composed tween-80/transcutol®P as surfactant/co-surfactant and isopropyl myristate (IPM) oil phase. Determining the boundaries of the phase diagrams is an essential aspect in microemulsion systems. For this purpose, the pseudo-ternary phase diagrams of the O/W microemulsion and IL/W microemulsions have been obtained by titration method and compared in Fig. 12.4.

As can be seen from Fig. 12.4, the stability region in IL/W MEs is larger than O/W ME. It means that ME formation with ILs is more favorable than IPM. This observation may be related to the role of ILs, both as oil and as surfactant in the ME structures. On the other hand, the stability region in the ME containing [C_8mim][PF_6], interestingly, is smaller than the ME containing [C_4mim][PF_6]. It means that ME formation with [C_4mim][PF_6] is more favorable than [C_8mim][PF_6]. This observation can be analyzed considering the length of the alkyl chains and its aggregation. The longer length of the alkyl chain makes greater possibility of aggregation. The aggregation makes the system more systematic, increases the alkyl chain rotational barrier, and hence the entropy of the system decreases. This behavior leads to sorting inside the droplets and on the interface, which results a slight increase in surface tension as mentioned. In [C_8mim][PF_6], the length of the alkyl chain is longer and then is aggregated more than [C_4mim][PF_6], making the ME system organized, decreasing entropy and unfavorable to ME formation. This unique behavior also confirmed by measuring of other physicochemical properties such as density, refractive index, viscosity, and surface tension of the ME systems.

12.2.3 IL/oil/water MEs

As mentioned in the introduction section, the formation of stable MEs between two immiscible liquids, e.g., oil and water, requires the presence of surfactants to reduce the interfacial tension between two phases. Surfactants may be anionic, cationic, non-ionic, or zwitterionic in nature. Despite the variety and wide application range of these surfactants, they cause some notable drawbacks when used in ME formation. Stabilization of the ME system requires elevated concentrations of both surfactant and co-surfactant. Monocationic alkyl imidazolium based ILs, as active surface agent, mostly have been studied by many authors to understand their aggregation behavior (Blesic et al. 2007, 2008; Sarac et al. 2017). A research also showed that different counterions, namely, Cl^-, PF_6^-, and NTf_2^-, have different effect on the aggregation behavior of 1-dodecycl-3-methylimidazolium based ionic liquids in water. Formation of aggregates in aqueous solution was observed only when chloride anions were employed as counterions, whereas phase separation was observed prior to the formation of aggregates when PF_6^- and NTf_2^- were employed (Blesic et al. 2008). Surface active ionic liquids (SAILs) have a number of advantages compared to conventionally employed surfactants and hence suitable for the formation of IL-oil–water microemulsions.

It has been reported that the SAILs possessed better surface-active properties, enhanced antimicrobial and skin permeation activity, and high temperature stability and temperature insensitivity, and they even considered as environmentally friendly surfactants (amino acids and choline containing SAIL) as compared to conventional surfactants (Kaur et al. 2018). The SAILs also possess combined properties of ILs and surfactants. Zech et al. formulated and characterized a ME containing SAIL of 1-hexadecyl-3-methylimidazolium chloride where 1-butyl-3-methylimidazolium tetrafluoroborate and ethylammonium nitrate were used as polar components and dodecane as the nonpolar component (Zech et al. 2009). However, the use of volatile organic solvents as the nonpolar phase in these types of MEs is unsuitable because of their high toxicity and nonbiodegradable nature. Safavi et al. have investigated 3-octyl-1-methylimidazolium chloride as a SAIL-containing ME where a hydrophobic IL, 1-butyl-3-methylimidazolium phosphate, was used as a substitute for the organic solvent and water as the polar component (Safavi et al. 2010). Unfortunately, the above-mentioned SAILs are not suitable for applications in biological science because of existence toxic cations such as ammonium, imidazolium, pyrrolidinium, and morpholinium. Recently, Goto et al. in their researches introduced some biocompatible cholinium fatty acid IL surfactants as potential alternatives to conventional surfactants (Ali et al. 2019; Korban Ali et al. 2020). They prepared three halogen-free and biocompatible choline fatty acid-based ILs with different chain lengths of choline oleate ([Ch][Ole]), choline linoleate ([Ch][Lin]), and choline erucate ([Ch][Eru]), and their micellar properties in aqueous solutions have been investigated (Ali et al. 2019). The surface-active properties and thermodynamic properties of micellization (ΔG°_m, ΔH°_m, and ΔS°_m) of the SAILs were investigated by tensiometry, conductometry, and dynamic light scattering measurements. The critical micelle concentration of the SAILs was found to be 2–4 times lower than those of conventional surfactants. The cytotoxicity evaluation approved that [Ch][Ole] has lower toxicity than the analogous ILs surfactants, and it is almost similar to that of the conventional surfactant Tween 80. Their results clearly suggest that these environmentally friendly SAILs can be used as a potential alternative to conventional ILs for various purposes, including biological applications.

SAILs based MEs systems may be also called surfactant free (detergent-less) microemulsion (SFME) contained no traditional surfactants. SFMEs have subsequently attracted much attention, and their phase and interfacial tension behavior have been found to be similar to those of surfactant-based microemulsions. A nonaqueous IL ME containing [C$_4$mim][BF$_4$], toluene and ethanol, which referred as SFME system by distinct microregions of IL-in-toluene, bicontinuous and toluene-in-IL introduced by Xu et al. (2013a). Single-phase microemulsion and multiphase regions were also observed and confirmed with different methods in the ternary phase diagram of [C$_4$mim][PF$_6$], N,N-dimethylformamide, and water SFME system, which containing no surfactant (Xu et al. 2013b). Such SFMEs are expected to have specific applications, e.g., in nanocomposite synthesis, reaction engineering, drug delivery, and separation, because of their IL-containing and surfactant-free natures.

12.3 Ionic Liquid Based Microemulsions as Drug Nanocarriers

Drugability and bioavailability of APIs are extremely dependent to the solubility and permeability factors. Dissolution of an API in the gastrointestinal tract is a prerequisite for its absorption. The dissolved drug molecules then should be absorbed into the blood circulation to perform their therapeutical effects, which requires permeation of drug molecules across various biomembranes, such as the epithelium for oral or pulmonary absorption, the stratum corneum for transdermal absorption, and the cornea for ocular absorption. It is found that the outstanding properties of microemulsions include high solubilization capacity, thermodynamic stability, and straightforward preparation highlighted their great potential as modern colloidal carriers for drug delivery. Nevertheless, conventional microemulsions are yet to fully attain their true potential in this field because of low solubility or permeability of some drugs in water and oil. Taking into account the tunable properties of ionic liquids (ILs), notably hydrophobicity, viscosity, and solubility, studies postulated that ILs could be a good prospect in the preparation of useful microemulsions in drug formulations. In many research works, IL based microemulsions have shown potential for drug delivery by improving both the solubility and permeability of APIs (Moniruzzaman and Goto 2011). This is because with the correct structural selection of cations and anions of ILs, it is possible to optimise the microemulsion formulation for specific application.

12.3.1 Ionic Liquid Based Microemulsions for Enhancing Drug Solubility

In the pharmaceutical industry, it is well known that many promising drugs which are sparingly soluble never enter a formulation stage due to their poor solubility as well as difficulties in delivery. Some methods have been employed to enhance the solubility or permeation rate of such drugs, but many researches indicate that sparingly soluble drugs need to be formulated with superior solvent systems and a novel delivery technology. As green and tailorable solvents, ILs have shown potential for drug delivery by improving both the solubility and permeability of drug substances. The solvation capacity of ILs for various drugs, such as amphotericin B, albendazole, danazol, itraconazole, coumarin, erythromycin, and paclitaxel, are tens of thousand-fold more than water and conventional organic solvents.

The solubilizing mechanisms in ILs are not yet fully understood because the results are incompatible with the dictum 'like dissolves like'. The solvation capability of ILs has been considered to be related to the anions instead of the cations within ILs. Hydrophilic ILs containing an acetate or dimethylphosphate anion were found to have superior solvation capability compared with ILs containing noncoordinating anions, such as tetrafluoroborate, hexafluorophosphate, and bis(trifluoromethylsulfonyl)imide. The most important factor is whether IL anions can form hydrogen bonds with the drug molecules. The hydrogen-bonding ability and delocalization of charges between oxygen are attributed as the reasons for the different solvation capability of ILs. Then, the solvation capability of ILs can be tailored by tuning the cation and anion combination (Huang et al. 2020). On the other hand, the use of surfactants above the critical micelle concentration can be also an appealing alternative to increase API's solubility. By varying the cation type and alkyl chain length and the nature and size of the counterion, it is possible to change the ILs' hydrophilic-hydrophobic balance. ILs with surfactant behavior are displaying high potential to increase the solubility of pharmaceutical agents in aqueous media (Faria et al. 2017).

An IL based microemulsion (ME) have both the advantages of IL and ME, which can overcome the inability of conventional ME to dissolve poorly water soluble drugs, even for those poorly soluble in organic solvents. Moniruzzaman et al., for the first time, reported that hydrophilic ILs having coordinating anions are very effective in dissolution of acyclovir (ACV), which is practically insoluble in water, and its solubility in pharmaceutical grade solvents is very limited (Moniruzzaman et al. 2010a). As an interesting result, they suggested that although ACV-IL formulation was not suitable for drug administration, IL-in-oil microemulsions (IL/O MEs) could be used for ACV delivery. In a later study, the influence of ILs type on the formation IL/O MEs, formation mechanism and drugs encapsulation efficiency were studied (Moniruzzaman et al. 2010b). They constructed the nanometer-sized ionic liquid droplets in isopropyl myristate were formed with a blend of nonionic surfactants, polyoxyethylene sorbitan monooleate (Tween-80), and sorbitan laurate (Span-20). Then, a set of ILs was tested as a dispersed phase. It was observed that ILs

Table 12.1 Solubility of drugs molecules in various systems at 25 °C (Moniruzzaman et al. 2010b)

Systems	Drug solubility (mg mL^{-1})^a		
	ACV	MTX	Dantrolene sodium
IL/O microemulsion^b	4.98	6.90	3.65
IL free micelles^c	0.10	0.85	0.03
W/O microemulsion^d	0.30	1.20	0.18
Water	0.50	0.12	0.10
IPM	0.036	0.02	ND

^{a}The data represents the average of the three experiments
^{b}Microemulsion compositions are as follows: [surfactant] = 20 wt.%, where the weight ratio of Tween-80 to Span-20 was 2:3. [IL] = 5 wt.%. [IPM] = 75 wt.%
^{c}Micelle contains 20 wt.% surfactants (Tween-80:Span-20 = 2:3 (w:w)) in IPM
^{d}Microemulsion prepared with [surfactant] = 20 wt.%, where the weight ratio of Tween-80 to Span-20 was 2:3. [Water] = 5 wt.%. ND: not detectable

possessing coordinating anions with strong hydrogen bond acceptor were most effective in forming microemulsion droplets. The solubility of three drug molecules of ACV, methotrexate (MTX) and 1-[(5-(p-nitrophenyl) furfurylidene) amino] hydantoin sodium (dantrolene sodium), that are insoluble or sparingly soluble in most of the conventional solvents was investigated. It was observed that such MEs are able to solubilize higher amount of the drugs than their individual components and water. The solubility data for various systems collected in Table 12.1.

As discussed before, the successful dissolution of such sparingly soluble drug molecules in the IL/o microemulsion may be attributed to the formation of hydrogen bonds between the IL anions and the polar groups of drug molecules like as cellulose compounds. To check this phenomenon, they examined the solubility of ACV in [C$_2$mim][BF$_4$]/tween-80/span-20/IPM microemulsions and found that the solubility was essentially zero. In fact, BF$_4^-$ anion is a weakly coordinating anion and consequently, it has very poor ability to form hydrogen bonding with other molecules. The effect of IL contents on the solubility of ACV in the microemulsions was also investigated. The results showed that the solubility increases with the increase of IL content. It is also approved that the solubility of ACV in IL/o microemulsions by varying the overall surfactants concentrations at fixed IL content and no significant difference in ACV solubilization was observed. These results once again support that IL plays the main role in solubilization of drugs in IL-based MEs. Since the solubility of all drugs in IPM is very low, it can be assumed that the drugs molecules exist in the hydrophilic IL core, which is stabilized by the surfactants mixtures creating an interfacial film between IPM and IL. Figure 12.5 shows a schematic representation of IL/O ME containing drug molecules.

Recently, solubility of celecoxib, as API with poor solubility in water (0.0033 mg·mL^{-1} at 298.15 K), investigated in IL/W MEs containing [C$_4$mim][PF$_6$] and [C$_8$mim][PF$_6$] and compared with a traditional O/W ME system (Parsi and

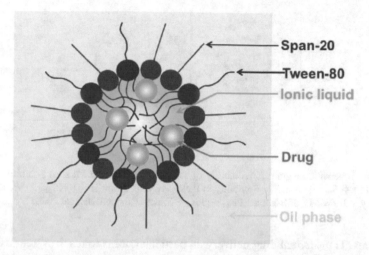

Fig. 12.5 Schematic representation of ionic liquid-in-oil (IL/O) microemulsions containing drug molecules (Moniruzzaman et al. 2010b)

Salabat 2020). The results showed that at the same composition the solubility in the IL/W MEs is more than the solubility in the O/W ME, which can be related to the interaction of surfactant/co-surfactant and ionic liquids (or oil) molecules with celecoxib. This more interaction between ILs and celecoxib may be related to the π orbitals of imidazolium ring in the structure of the ionic liquids with the π orbitals of the celecoxib molecules. Comparison of the celecoxib solubility in two types of IL/W MEs shows that ME containing $[C_8mim][PF_6]$ dissolve celecoxib less than ME containing $[C_4mim][PF_6]$. This interesting and unexpected result might be again related to the higher organized structure of the $[C_8mim][PF_6]$ compared to $[C_4mim][PF_6]$, as confirmed from density, refractive index, viscosity, and conductivity data. On the other word the hydrophobicity of the ILs does not appear to be a predominant factor on the celecoxib solubility, as a hydrophobic active ingredient, in IL/W systems. As a result, the special structures of the IL/W MEs, as approved by physicochemical properties, can remarkably affect celecoxib solubility. This concept is shown schematically in Fig. 12.6. The results confirmed that physicochemical properties trend is in agreement with the solubility data. The remarkable results of this research hint that the structure and then properties of the ILs molecules in a normal solution or pure state may be extremely different to that in a ME template.

12.3.2   Ionic Liquid Based Microemulsions for Transdermal Delivery

When considering the topical application of APIs, the drug delivery system must target one or more different skin layers and underlying tissues or skin associated

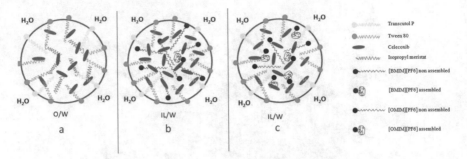

Fig. 12.6 Schematic diagram for solubility of celecoxib in different MEs: a) Solubility in O/W ME of IPM + S_{mix} + Water, b) Solubility in IL/W ME of $[C_4mim][PF_6]$ + S_{mix} + Water, and c) Solubility in IL/W ME of $[C_8mim][PF_6]$ + S_{mix} + Water (Parsi and Salabat 2020)

structures. Transdermal drug delivery, in particular, aims to reach systemic circulation, representing an alternative to parenteral and oral routes, while avoiding pre-systemic metabolism. In the last two decades, ILs have been studied as promising pharmaceutical agents or formulation components in order to tackle the challenges in topical and transdermal delivery systems. In this respect, IL based MEs have been investigated for delivery of poorly water-soluble drugs at topical and trans-dermal level. The first IL-based microemulsion reported for transdermal delivery by Moniruzzaman et al. aimed to improve membrane transport of a sparingly soluble API, namely, the antiviral drug acyclovir (Moniruzzaman et al. 2010c). In this work, a blend of two nontoxic surfactants, Tween-80 and Span-20, was used in combination with imidazolium-based ILs to form stabilized IL droplets. As schematically shown in Fig. 12.5, the external phase (oil phase) of the ME is constituted by isopropyl myristate. It is found that, IL based ME containing dimethylimidazolium dimethyl phosphate ($[C_1mim][(CH_3O)_2PO_2]$) presented superior ability to dissolve the selected API and form stable droplets in the formulation. This improvement was justified by the hydrogen bonding interactions between the polar groups of acyclovir and the IL anions. The in vitro study across Yucatan micropig porcine skin (performed on Franz diffusion cells) allowed for verifying an increase in acyclovir's skin permeability of several orders of magnitude, as well as the API's transdermal permeation when using the IL/O system as drug carrier.

The obtained permeation results of ACV from IL/O MEs and others formulations considered as control experiments (e.g., IL, IPM, surfactants + IPM, W/O MEs, and commercially available ACV cream) have been reported in Table 12.2. Note that ACV was suspended in IPM, surfactants + IPM, and W/O ME formulations, whereas ACV was solubilized in IL and IL/o MEs. As can be seen from Table 12.2, the skin permeability of ACV increased by several orders of magnitude, when IL/O ME was used. More significantly, application of IL/O system induced significant transdermal permeation of ACV whereas other formulations showed essentially zero (below the detection level).

Table 12.2 Acyclovir penetration into and across the skin using various formulations at 32.4°C[a]; All data are presented as mean ± SD from three experiments (Moniruzzaman et al. 2010c)

Formulations	Topical delivery ($\mu g/cm^3$)	Transdermal delivery ($\mu g/cm^2$)
IL	ND	ND
IPM[b]	0.18 ± 0.34	ND
Surfactants + IPM[c]	3.86 ± 1.65	ND
W/O ME[b,c,d]	4.57 ± 3.17	ND
IL/O ME[c,e]	10.78 ± 3.12	1.95 ± 1.0
ACV cream[f]	1.21 ± 0.10	ND

[a]All formulations contained 5 mg/mL ACV and penetration time 24 h
[b]Almost all ACV remained as suspended powders in formulations
[c]IL/O ME2 as mention in Table 12.1 in which the weight fraction of surfactants was 20 wt% (Tween-80: Span-20 = 3:2 (w/w)
[d]w/o emulsion contained 4 wt% water
[e]The system containing 4 wt% IL
[f]ACV herpecia topical cream (equivalent to 5 mg ACV) used for penetration study

In fact, drugs are able to enter the skin only in a dissolved state and IL/O MEs can solubilize a large ACV amount in the IL droplets which play the role of a drug reservoir. Besides, lipophilic components of IL/O MEs can easily disrupt the SC lipid structure to move nanometer size IL droplets containing solubilized drug into the skin. The drug can easily be diffused from the IL droplets into external phase when absorption (in vitro) takes place. In case of IL + ACV formulation, although ACV was fully solubilized in the formulations, the diffusion of ACV into the skin was hindered by the hydrophobic barrier function of the skin, provided by the highly organized structure of the SC due to the highly hydrophilic nature of IL. Besides, the formulations containing a large amount of IPM a well-known enhancer to overcome the barrier of SC, the presence of a large amount of suspended ACV probably obstructed the skin. Taking these results together, it is possible to conclude that the excellent skin permeation rate was obtained using IL/O MEs due to the following dual effects. Firstly, a large amount of ACV was totally dissolved in nanometer-sized IL droplets, which led to larger concentration gradient toward the skin. Secondly, IPM and surfactants were used to overcome the barrier of SC lipid structure.

Ex-vivo diffusion of celecoxib loaded in IL-based MEs and microemulgels (MEGs) through the rat skin by using Franz diffusion cell was investigated and compared with traditional MEs by Salabat and Parsi (2021). The investigated systems were composed of the same nonionic surfactant/co-surfactant of tween-80/transcutol®P, and different oil phases of isopropyl myristate, $[C_4mim][PF_6]$ and $[C_8mim][PF_6]$, for construction of O/W and ILs/W MEs, named as ME1, ME2, and ME3 respectively. In order to prepare MEGs samples, the drug-loaded MEs were mixed with carbopol gel solution at the ratio of 1:1 (v/v). The selected compositions, from the phase diagrams, obtained by this research group (Parsi and Salabat 2020), were containing 5% of the oil or ILs and 55% surfactant/co-surfactant in all systems. The permeation profiles of celecoxib from the ME formulations, as cumulative drug release ($\mu g/cm^2$) versus time, is shown in Fig. 12.7. As can be seen, drug releases from IL/W MEs formulations are remarkably more than that of O/W ME formulation. It means, as like as drug solubility that discussed in the previous section, MEs containing ILs favor the transdermal delivery of celecoxib. On the other hand, drug release from ME containing [C_4mim][PF_6] is more than the ME containing [C_8mim][PF_6]. This interesting trend is exactly compatible with celecoxib solubility in the ME systems and can be concluded that the thermodynamic activity of the formulations should be arranged as follows: ME2 > ME3 > ME1. These differences in thermodynamic activity among the various MEs may affect the drug permeation through the skin because of the presence of imidazole group in the ILs which has been reported to have penetration enhancing effect (Wang et al. 2018). It should be noted that droplets of the microemulsions are in nanometer size and then diffusion process is not in control of droplets size. It means that there is the same chance for the number of vesicles that can interact with a fixed area of stratum corneum.

The mean cumulative flux values were also calculated for all ME and MEG formulations (Table 12.3). The improved flux values for ME2 and ME3 as compared to ME1 are again gave interesting results of this research. The cumulative flux for microemulsion containing [C_4mim][PF_6] is extraordinary more than microemulsion containing [C_8mim][PF_6] and then better penetration enhancer. The mean cumulative flux trend from ILs/W MEG formulations is same as ILs/W ME formulations and remarkably more than that of O/W MEG formulation. Another important *ex-vivo* permeation parameter is lag time (the first of drug detection), which is calculated and reported in Table 12.3. These values indicating the drug permeation characters through rat skin was significantly affected by the composition of microemulsion formulation. As can

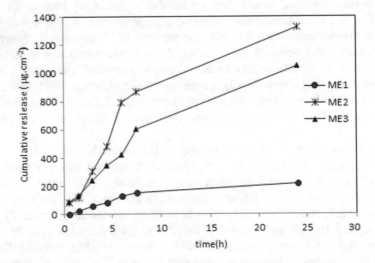

Fig. 12.7 Release of celecoxib from microemulsion formulations (Parsi and Salabat 2021)

Table 12.3 *Ex-vivo* release parameters of average cumulative flux (J), lag time (LT), and release percent for different microemulsion and microemulgel formulations (Salabat and Parsi 2021)

Formulations	ME1	ME2	ME3	MEG1	MEG2	MEG3
J (μg.cm^{-2}.h^{-1})	22.1	122.9	71.0	17.3	71.7	59.1
LT (h)	1.5	0.5	0.5	1.5	3	3
Release % after 7.5 h	14.0	80.4	56.1	11.1	41.8	35.1
Release % after 24 h	20.0	98.1	97.1	15.2	77.3	61.1

be seen the lag time of the ME1 is three times higher than the lag times of ME2 and ME3. This result approves again the previous trend for permeation from ILs based microemulsions.

An IL based ME containing 1-hydroxyethyl-3-methylimidazolium chloride ([HOEmim][Cl]) was developed for topical delivery of Dencichine by Wang et al. (2018). Dencichine (β-N-oxalyl-L-α, β-diaminopropionic acid, Den) was found to exert its hemostatic activities in a paracrine fashion, by increasing intracellular calcium, reducting cAMP, and releasing TXA2 after binding to the AMPA receptors, which resulted in platelet aggregation. This compound showed potential for clinical usage for diseases characterized by hemorrhage. However, further applications were limited by the low bioavailability, along with some side effects. Therefore, trans-dermal administration could be an alternative way for Den to improve its bioavail-ability, and also reduce the side effects. Due to the hydrophilic nature of Den, it is exhibited a poor permeability in the lipophilic membrane of the stratum corneum (SC), which led to a low efficiency in transdermal delivery (Li et al. 2015). Therefore, acceleration and promotion methods, such as ME method, for the Den skin perme-ation were needed to produce desired pharmacological effects. In this research, the proposed IL based ME (O/W) was composed of 50% water/[HOEmim][Cl] mix (1:1) as water phase, 20% tween 80/[C$_4$mim][C$_{12}$SO$_3$] mix (1:1) as surfactant, 10% propylene glycol as co-surfactant, and 20% IPM as oil phase. In vitro skin perme-ation study carried out in a Franz diffusion cell through the skin of mice and skin permeation parameters of J (steady permeation flux) and Q$_{12h}$ (cumulative amount of Den which had permeated through the mice skin at 12 h) were calculated.

The formulations of IL-ME (ME1) and ME without IL (ME3) containing 1% of Den and also a Den hydrogel and a drug aqueous solution, as the comparisons, have been evaluated for skin permeation enhancement and the results appeared in Fig. 12.8. Aqueous solution of the drug produced a permeation rate of 3.1×10^{-2} mg/cm^2/h in the initial penetration stage. However, the flux was found to be reduced sharply over time, and at the end of the permeation, it was only 0.241×10^{-2} mg/cm^2/h. These results were attributed to the rapid consumption of drug in the donor solution and the barrier function of the SC. Once the Den reached saturated concentrations in skin membranes, the diffusion process were retarded during the later stage (12–24 h). The Q$_{24h}$ was determined to be only 0.27 mg/cm^2. In Carbopol gel, after a poor permeation at the initial stage, the drug was subjected to a sustained and accel-erated permeation profile. Carbopol is one of the most common thickening agents

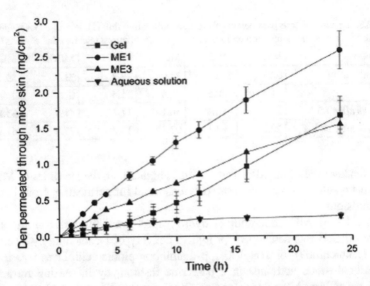

Fig. 12.8 Skin permeation profiles of Den from different formulations (Wang et al. 2018)

for hydrogel. The cross-linking structures provided by the acrylic acid backbone are responsible for water absorption, swelling and drug control-release and finally the skin permeability significantly enhanced with the Q_{24h} of 1.66 mg/cm^2. In contrast, as seen in Fig. 12.8, due to the low viscosity and molecular interference from the vehicle, the microemulsion formulations (ME1 and ME3) produced the relatively higher permeation rates during the entire penetration process. Following the rapid drug permeation rate during the initial stage, the flux stayed in the stable ranges of $11.4 \times 10^{-2} - 9.2 \times 10^{-2}$ mg/cm^2/h for the ME1, and $6.12 \times 10^{-2} - 5.72 \times 10^{-2}$ mg/cm^2/h for the ME3. Additionally, the Q_{24h} were approximately 2.57 mg/cm^2 for the ME1 and 1.54 mg/cm^2 for the ME3, respectively, which showed significant improvements when compared to the aqueous solution.

In this research, two ILs, [HOEmim][Cl] and [C$_4$mim][C$_{12}$SO$_3$], demonstrated their enhancements on skin permeation and were incorporated into the aqueous and surfactant phases respectively. Therefore, the Q_{24h} of the Den from the ME1 was significantly higher than that of the ME3, which suggested an enhancement of the skin permeability. In transdermal methods IL based ME systems have been reported as carriers for poorly soluble drugs. Due to the excellent solubilization abilities, the vehicle greatly increased concentration gradients, leading to the intensifying of driving force for drug diffusion. However, regarding the hydrophilic nature of Den, the solubilization effects of ILs on the compound did not account. The enhancement of skin permeability might be attributed to the moderation of the SC barrier properties. The influence of the ILs-ME on skin permeability was approved by evaluation of trans-epidermal water loss values (TEWL). The results showed that after administration of normal saline, the TEWL value of the mice showed little changes with time. However, the TEWL values of ME2 and ME4 groups were increased sharply at the initial stage, indicating the improvements in the skin permeability. DSC and ATR-FTIR methods also confirmed the ability of IL based MEs to reduce the skin barrier properties by disrupting the regular and compact arrangements of corneocytes, and moderating the surface properties of the SC.

Zhang et al. proposed a ME system containing a transdermal enhancer IL of 1-hydroxyethyl-3-methylimidazolium chloride ([HOEmim][Cl]) to improve the transdermal delivery of artemisinin (Ars), as a poorly water soluble drug (Asbill and Michniak 2000). In the studied IL based ME, deep eutectic lidocaine ibuprofen (Lid-Ibu) was selected as the oil phase, tween-80/span-20 as surfactant mixture, and ethanol as co-surfactant. The ingredients of ME were selected based on the solvation ability for Ars or their permeation capability.

In-vitro transdermal assay of the IL based ME showed a remarkable enhancement of Ars transport through the skin, with the permeation flux being threefold of the value for isopropyl myristate system in 6 h. The high permeation of IL based ME could be due to the fact that [HOEmim][Cl] as an enhancer could permeate into the SC layer quickly to disturb the regular arrangement of the skin, which led to a massive permeation of the surfactant phase to further damage the tight arrangement of SC. On the contrary, the systems without [HOEmim][Cl] possessed relatively lower average permeation fluxes. The impact of IL-based ME on stratum corneum (SC) was also investigated by DSC, ATR-FTIR, and AFM, which unveiled that this ME type possesses the ability of reducing the SC barrier by disrupting the regular arrangement of keratin, resulting in enhancement of transdermal delivery of Ars. The obtained results suggested that this IL based ME, with low toxicity, has an excellent capability to promote the transdermal delivery of Ars, which might also be a promising vehicle for the skin delivery of other hydrophobic natural drugs.

Recently, by considering the two important factors, biocompatibility and toxicity, cholinium (Ch)-based ILs comprising anions that derived from carboxylic acids (CA) were used in IL/O microemulsions to increase the transdermal delivery of acyclovir (Islam et al. 2020). Hydrophilic [Ch][CA] ILs (cholinium formate, cholinium lactate, and cholinium propionate) were used as the non-aqueous polar phase and a surface-active IL [Ch][Ole] (cholinium oleate) as the surfactant in combination with a co-surfactant, Span 20, in a continuous oil phase. Based on the phase diagram, different formulations of IL based MEs were prepared (Table 12.4).

The maximum solubility of ACV in the ILs was obtained as 203, 208, and 278 mg/mL for [Ch][For], [Ch][Lac], and [Ch][Pro], respectively, which was significantly higher compared with Milli-Q (0.41 mg/mL) and IPM (0.03 mg/mL). As the ILs were hydrophilic in nature with strong H-bond accepting anions, ACV may be dissolved in these ILs through the formation of H-bonds, van der Waals forces, or π-π interactions between the polar groups of the drug and the IL anions. On the other hand, though the ACV-loading capacity of the IL-free S/Co$_{mix}$/IPM system was very low (0.15 mg/mL for S/Co$_{mix}$ = 2:1, 15 wt.%), the capacity increased dramatically by incorporating IL into this system (7.7 mg/mL). The ACV-loading capacity was decreased with decreasing [Ch][Ole] content in the S/Co, and it was significantly decreased when [Ch][Ole] < Span-20 in the S/Co mixture. The ACV-loading capacity also depended on the type of IL. It was found that ME containing [Ch][Pro]

Table 12.4 Contents of the microemulsion (ME) formulations[a] (Islam et al. 2020)

Formulations	ILs	Surfactant: co-surfactant (Weight Ratio)		
		Surfactant co-surfactant		
		[Ch][Ole]	Tween 80	Span 20
ME1	[Ch][Pro]	2	–	1
ME2	[Ch][Pro]	3	–	2
ME3	[Ch][Pro]	1	–	1
ME4	[Ch][Pro]	2	–	3
ME5	[Ch][Pro]	1		3
ME6	[Ch][Pro]	1	1	1
ME7	[Ch][For]	2	–	1
ME8	[Ch][Lac]	2	–	1
ME9[b]	Milli-Q	–	2	1

[a]MEs were prepared with overall 15 wt.% S/Co$_{mix}$ and 3 wt.% [Ch][CA] ionic liquid (IL) in isopropyl myristate (IPM). [b]ME9 was set as control where 3 wt.% Milli-Q was used as a replacement for [Ch][CA] IL

had a significantly higher loading capacity, owing to the higher ACV solubilizing capacity. It has also been reported that the drug loading capacity of IL/O MEs highly depends on the categories of IL (Kandasamy et al. 2018).

In vitro drug permeation studies were performed using Yucatan micro pig (YMP) skin. The topical and transdermal delivery of ACV from IL/O MEs has been investigated and compared with other formulations (Fig. 12.9a). The topical delivery of ACV from IPM, S/Comix/IPM, and W/O MEs was very low, while the transdermal delivery was below the detection limit. Interestingly, compared with the other formulations, the IL/O ME demonstrated significantly enhanced topical and transdermal delivery with values of 36.47 and 45.05 μg/cm^2, respectively. This dramatically enhanced permeation using the IL/O MEs was found because of their high drug solubilizing capacity and promising drug conveyances technique. Generally, drugs are administrated into the skin in a solubilized state. A large amount of drug was loaded into the core of the IL/O ME solubilized by IL, which could act as a drug reservoir and provide a greater concentration gradient to the skin. Whereas, IPM as a potential enhancer disrupts the barrier function of the skin, which facilitated to enter the nano-sized drug-loaded IL droplets into the skin. Nonetheless, ACV was solubilized state in IL, but as it is hydrophilic in nature, IL alone could not deliver ACV due to of the strong hydrophobic barrier functions of the skin. On the other hand, though IPM, S/Comix/IPM, and W/O MEs disrupt the barrier function as they contain IPM, ACV could not permeate across the skin from these formulations because ACV was suspended in these systems, which probably obstructed the access of ACV to the skin (Moniruzzaman et al. 2010c).

It has been reported that the molar of individual surfactants can influence drug delivery by controlling the physicochemical properties of the MEs (Huang et al.

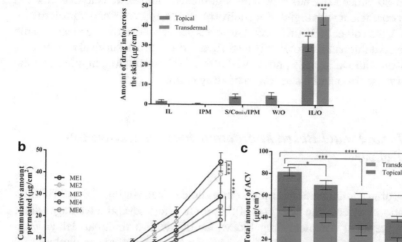

Fig. 12.9 **a** Topical and transdermal delivery of ACV from various drug carriers after 48 h, where S/Co$_{mix}$/IPM: 15 wt.% S/Co$_{mix}$ at a 2:1 ratio in IPM. W/O: ME9. IL/O: ME1; **b** transdermal permeation profile of ACV from various IL/O MEs with S/Co ratios; **c** the total (topical + transdermal) delivery of ACV from various IL/O MEs with varying S/Co ratios after 48 h. All the drug carriers contained 5 mg/mL ACV (Islam et al. 2020)

2008). Therefore, the delivery of ACV from various MEs with varying S/Co ratios was studied. From the cumulative permeation profiles (Fig. 12.9b), it can be seen that ME1 enhanced the transdermal delivery significantly compared with the other MEs. Other permeation parameters, including transdermal flux, permeability coefficient, diffusion coefficient, and skin partition coefficient were determined from the cumulative permeation profile. It was found that all these parameters were increased with increasing [Ch][Ole] content in the MEs. In fact, the transdermal delivery of drug mainly depends on the transdermal flux and permeation coefficient. The highest transdermal flux (1.43 μg/cm^2/h) and permeation coefficient (2.86 × 10^{-4} cm/h) were both found for ME1 because of the higher skin partition and diffusion coefficients (indicating better solvent distribution ability into the deeper layers of the skin) resulting in ME1 having the highest transdermal delivery of ACV.

In addition, the topical delivery of ACV was investigated. The total (topical and transdermal) delivery after 48 h is presented in Fig. 12.9C. It was observed that, as for transdermal delivery, topical delivery was also favored using ME1, having a higher [Ch][Ole] content, resulting in the highest drug delivery. This result could be explained based on the higher interfacial area and stability of ME1. It has been reported that a larger stable interfacial area of ME droplets favors transdermal and topical delivery. Comparing the transdermal delivery between ME1 and ME6, it was revealed that [Ch][Ole] significantly enhanced the permeability compared with Tween-80. Since ME1 has shown the significantly higher transdermal flux and permeation coefficient with higher drug delivery results, topically and transdermally (Fig. 12.9 C), than other MEs, this formulation has been proposed as the most suitable nano-carrier. Additionally, cytotoxicity tests revealed a high cell survival rate (>90%) in comparison with Dulbecco's phosphate-buffered saline solution, highlighting the potential of these formulations as low toxic drug carriers.

12.3.3   Ionic Liquid Based Microemulsions for Intravenous Delivery

Intravenous drug administration is a preferable choice when aiming to bypass biological absorption barriers. The selection of this route offers several advantages since it provides the most complete drug bioavailability with a minimal delay. Foreseeing intravenous drug delivery applications, ILs have recently been explored for the development of polymer nanocomposites such as conjugation of SAILs with chitosan (Bharmoria et al. 2013). In another approach, polydopamine nanoparticles that were loaded with doxorubicin and the IL [C4mim][PF6] were recently developed for cancer treatment (Tang et al. 2016). The IL was employed as a microwave sensitizer to prepare these novel nanoplatforms for combined chemotherapy and microwave thermal therapy by intravenous administration. The antitumor efficacy of doxorubicin-loaded IL-polydopamine nanoparticles was demonstrated in in vitro and in vivo experiments in the treatment of tumors in mice, after intravenous injection via tail vein. The referred nanoparticles exhibited high inhibition effect when combined with the microwave thermal irradiation, acting in the tumor ablation without inducing significant tissue toxicity.

The possibility to manipulate and design the IL structure allows for the use of IL-based MEs as novel carriers of low-water soluble APIs. Accordingly, hydrophobic nontoxic ILs were used to prepare novel IL/W MEs for intravenous administration of amphotericin B (Esson et al. 2020). Amphotericin B is an antifungal agent that, due to its low water solubility (<1.0 μg mL^{-1}) and self-aggregation in aqueous media, presents undesired side effects, thus being its intravenous drug delivery a challenge. In a preliminary study, high contents (>5.0 mg mL^{-1}) of the API were solubilized in a new hydrophobic dicholinium-based IL with the bis(trifluoromethanesulfonyl)imide ([NTf2]$^{-}$) anion. The mixture of this hydrophobic IL with a hydrophilic cholinium-based IL resulted in the solubilization of the drug, preventing the concentration-dependent aggregation with controlled release of the API. Despite the maintenance of the antifungal activity of the API, and the low toxicity toward embryo-larval zebrafish models, further studies are required in order consider this formulation adequate for intravenous administration.

12.4   Conclusions and Outlook

Microemulsion is a kind of single optically isotropic and thermodynamically stable dispersion containing oil, surfactant, and aqueous phases. MEs have many advantages, such as enhanced drug transdermal ability, good solvation ability for active pharmaceutical ingredients, excellent thermodynamic stability, and easy fabrication, which make MEs widely used in TDD, extraction, cosmetic production, and so on. However, solubility of APIs in traditional solvents usually used as the ingredients of MEs is quite limited. In the past decade, ILs based MEs have been developed and tested as potential nanocarriers for drug solubility enhancer, TDD, and intravenous delivery.

In this chapter, the different types of IL based ME systems have been introduced and compared with traditional ME systems. ILs comprising MEs exhibit flexibility acting as polar or non-polar phases. Apart from that, the amphiphilic properties of some ILs allow them to function as additives or surfactants that stabilize the formed ME systems. The combination of all these properties, renders IL based MEs much more versatile than conventional MEs or even pure ILs. The IL based MEs are able to overcome the inabilities of conventional MEs to dissolve a number of chemicals which are water insoluble and result in better pharmacological activities of the drugs. Most importantly, in the drug delivery field, ILs can serve as permeation enhancers.

It is discussed that the hydrogen-bonding ability and delocalization of charges between oxygen attributed as the reasons for the different solvation capability of ILs in the ME composition. Then, the solvation capability of ILs can be tailored by tuning the cation and anion combination. On the other hand, by varying the cation type and alkyl chain length and the nature and size of the counterion, it is possible to change the ILs' hydrophilic-hydrophobic balance. ILs with surfactant behavior are displaying high potential to increase the solubility of pharmaceutical agents in aqueous media. Many researches on the transdermal drug delivery by using IL based MEs have been showed remarkable enhancement of APIs transport through the skin. The high permeation of IL based ME could be due to the fact that IL as an enhancer could permeate into the SC layer quickly to disturb the regular arrangement of the skin, which led to a massive permeation of the surfactant phase to further damage the tight arrangement of SC. Additionally, it is also discussed that hydrophobic nontoxic ILs can be used to prepare novel IL based MEs for intravenous administration of APIs.

Some advanced researches introduced new IL based MEs by considering the two important factors, biocompatibility and toxicity. These new nontoxic systems, called cholinium based ILs MEs, comprising anions that derived from carboxylic acids and used to increase the transdermal delivery of APIs with little irritation in the skin. Overall, IL based MEs have potential to overcome solubility, bioavailability, permeation, and stability concerns that are associated to traditional pharmaceuticals. Moreover, the pharmaceutical application of IL based MEs primarily focuses on transdermal drug delivery. Thus, a huge development space is offered for oral, pulmonary, ocular, and parenteral drug delivery in this field.

References

Ali MK, Moshikur RM, Wakabayashi R, Tahara Y, Moniruzzaman M, Kamiya N, Goto M (2019) Synthesis and characterization of choline-fatty-acid-based ionic liquids: a new biocompatible surfactant. J Colloid Interface Sci 551:72–80

Anjum N, Guedeau-Boudeville MA, Stubenrauch C, Mourchid A (2008) Phase behavior and microstructure of microemulsions containing the hydrophobic ionic liquid 1-butyl-3-methylimidazolium hexauorophosphate. J Phys Chem B 113:239–244

Asbill CS, Michniak BB (2000) Percutaneous penetration enhancers: local versus transdermal activity. Pharm Sci Technol Today 3:36–41

Bharmoria P, Singh T, Kumar A (2013) Complexation of chitosan with surfactant like ionic liquids: molecular interactions and preparation of chitosan nanoparticles. J Colloid Interface Sci 407:361–369

Blesic M, Marques MH, Plechkova NV, Seddon KR, Rebelo LPN, Lopes A (2007) Self-aggregation of ionic liquids: micelle formation in aqueous solution. Green Chem 9:481–490

Blesic M, Lopes A, Melo E, Petrovski Z, Plechkova NV, Lopes JNC, Seddon KR, Rebelo LPN (2008) On the self-aggregation and uorescence quenching aptitude of surfactant ionic liquids. J Phys Chem B 112:8645–8650

Callender SP, Mathews JA, Kobernyk K, Wettig SD (2017) Microemulsion utility in pharmaceuticals: implications for multi-drug delivery. Int J Pharm 526:425–442

Chaturvedi D (2011) Ionic liquids: a class of versatile green reaction media for the syntheses of nitrogen heterocycles. Curr Org Chem 8:438–471

De Faria ELP, Shabudin SV, Claúdio AFM, Válega M, Domingues FMJ, Freire CSR, Silvestre AJD, Freire MG (2017) Aqueous solutions of surface-active ionic liquids: remarkable alternative solvents to improve the solubility of triterpenic acids and their extraction from biomass. ACS Sustain Chem Eng 5:7344–7351

Eastoe J (2005) Microemulsions. In: Cosgrove T (ed) Colloid science: principles, methods and applications. 1st. Wiley, pp 48–84

Eastoe J, Gold S, Rogers SE, Paul A, Welton T, Heenan RK, Grillo I (2005) Ionic liquid-in-oil microemulsions. J Am Chem Soc 127:7302–7303

Egorova KS, Gordeev EG, Ananikov VP (2017) Biological activity of ionic liquids and their application in pharmaceutics and medicine. Chem Rev 117:7132–7189

Esson MM, Mecozzi S, Mecozzi S (2020) Preparation, characterization, and formulation optimization of ionic-liquid-in-water nanoemulsions toward systemic delivery of amphotericin B. Mol Pharm 17:2221–2226

Gao H, Li J, Han B, Chen W, Zhang J, Zhang R, Yan D (2004) Microemulsions with ionic liquid polar domains. Phys Chem Chem Phys 6:2914–2916

Gao Y, Han S, Han B, Li G, Shen D, Li Z, Du J, Hou W, Zhang G (2005) TX-100/water/1-butyl-3-methylimidazolium hexauorophosphate microemulsions. Langmuir 21:5681–5684

Gao Y, Wang S, Zheng L, Han S, Zhang X, Lu D, Yu L, Ji Y, Zhang G (2006a) Microregion detection of ionic liquid microemulsions. J Colloid Interface Sci 301:612–616

Gao Y, Li N, Zheng L, Zhao X, Zhang S, Han B, Hou W, Li G (2006b) A cyclic voltammetric technique for the detection of micro-regions of bmimPF$_6$/tween 20/H$_2$O microemulsions and their performance characterization by UV-Vis spectroscopy. Green Chem 8:43–49

Gradzielski M (2008) Recent developments in the characterisation of microemulsions. Curr Opin Colloid Interface Sci 13:263–269

Greaves TL, Drummond J (2008) Ionic liquids as amphiphile self-assembly media. Chem Soc Rev 37:1709–1726

Hejazifar M, Lanaridi O, Schroder KB (2020) Ionic liquid based microemulsions: a review. J Mol Liq 303:112264

Huang YR, Lin YH, Lu TM, Wang RJ, Tsai YH, Wu PC (2008) Transdermal delivery of capsaicin derivative-sodium nonivamide acetate using microemulsions as vehicles. Int J Pharm 349:206–211

Huang W, Wu X, Qi J, Zhu Q, Wu W, Lu Y, Chen Z (2020) Ionic liquids: green and tailor-made solvents in drug delivery. Drug Discov Today 25:901–908

Islam MR, Chowdhury MR, Wakabayash R, Kamiya N, Moniruzzaman M, Goto M (2020) Ionic liquid-in-oil microemulsions prepared with biocompatible choline carboxylic acids for improving the transdermal delivery of a sparingly soluble drug. Pharmaceutics 12:392

Kale SN, Deore SL (2017) Emulsion microemulsion and nanoemulsion: a review. Sys Rev Pharm 8:39–47

Kalepu S, Nekkanti V (2015) Insoluble drug delivery strategies: review of recent advances and business prospects. Acta Pharm Sin B 5:442–453

Kandasamy S, Moniruzzaman M, Sivapragasam M, Rashid M, Ibrahim M, Mutalib A (2018) Separation and purification technology formulation and characterization of acetate based ionic liquid in oil microemulsion as a carrier for acyclovir and methotrexate. Sep Purif Technol 196:149–156

Kaur M, Singh G, Kumar S, Navnidhi KTS (2018) Thermally stable microemulsions comprising imidazolium based surface active ionic liquids, non-polar ionic liquid and ethylene glycol as polar phase. J Colloid Interface Sci 511:344–354

Korban Ali M, Moshikur RM, Wakabayashi R, Moniruzzaman M, Kamiya N, Goto M (2020) Biocompatible ionic liquid surfactant-based microemulsion as a potential carrier for sparingly soluble drugs. ACS Sustain Chem Eng 8:6263–6272

Kunz W, Maurer E, Klein R, Touraud D, Rengstl D, Harrar A, Dengler S, Zech O (2011) Low toxic ionic liquids, liquid catanionics, and ionic liquid microemulsions. J Dispers Sci Technol 32:1694–1699

Lawrence MJ, Rees GD (2012) Microemulsion-based media as novel drug delivery systems. Adv Drug Del Rev 64:175–193

Li N, Gao Y, Zheng L, Zhang J, Yu L, Li X (2007) Studies on the micropolarities of bmimBF$_4$/TX-100/Toluene ionic liquid microemulsions and their behaviors characterized by UV-Visible spectroscopy. Langmuir 23:1091–1097

Li L, Yang XY, Yang Y, Cui XM, Wang CX (2015) Percutaneous permeation characteristic research of dencichine. Chin Herb Med 46:2563–2567

Mirhosseini F, Salabat A (2015) Ionic liquid based microemulsion method for fabrication of poly(methyl methacrylate)-TiO$_2$ nanocomposite as highly efficient visible light photocatalyst. RSC Adv 5:12536–12545

Mirhoseini F, Salabat A (2018) Photocatalytic filter. US Patent 0104678 A1

Moniruzzaman M, Goto M (2011) Ionic liquids: future solvents and reagents for pharmaceuticals. J Chem Eng Jpn 44:370–381

Moniruzzaman M, Kamiya N, Nakashima K, Goto M (2008) Formation of reverse micelles in a room-temperature ionic liquid. ChemPhysChem 9:689–692

Moniruzzaman M, Tahara Y, Tamura M, Kamiya N, Goto M (2010a) Ionic liquid assisted transdermal delivery of sparingly soluble drugs. Chem Commun 47:1452–1454

Moniruzzaman M, Kamiya N, Goto M (2010b) Ionic liquid based microemulsion with pharmaceutically accepted components: formulation and potential applications. J Colloid Interface Sci 352:136–142

Moniruzzaman M, Tamura M, Tahara Y, Kamiya N, Goto M (2010c) Ionic liquid-in-oil microemulsion as a potential carrier of sparingly soluble drug: characterization and cytotoxicity evaluation. Int J Pharm 400:243–250

Parsi E, Salabat A (2020) Comparison of O/W and IL/W microemulsion systems as potential carriers of sparingly soluble celecoxib drug. J Solution Chem 49:68–82

Pedro SN, Freire CSR, Silvestre AJD, Freire MG (2020) The role of ionic liquids in the pharmaceutical field: an overview of relevant applications. Int J Mol Sci 21:8298

Petkovic M, Seddon KR, Rebelo LPN, Pereira CS, Cristina S (2011) Ionic liquids: a pathway to environmental acceptability. Chem Soc Rev 40:1383–1403

Rojas O, Tiersch B, Rabe C, Stehle R, Hoell A, Arlt B, Koetz J (2013) Nonaqueous microemulsions based on N, N-alkylimidazolium alkylsulfate ionic liquids. Langmuir 29:6833–6839

Safavi A, Maleki N, Farjami F (2010) Phase behavior and characterization of ionic liquids based microemulsions. Colloids Surf A 355:61–66

Salabat A, Mirhoseini F (2015) Applications of a new type of poly(methyl methacrylate)/TiO$_2$ nanocomposite as an antibacterial agent and reducing photocatalyst. Photochem Photobiol Sci 14:1637–1643

Salabat A, Parsi E (2021) Ex vivo evaluation of celecoxib release from ionic liquid-based microemulsions and microemulgels for topical applications. J Iranian Chem Soc. https://doi.org/10.1007/s13738-020-02122-7

Salabat A, Eastoe J, Mutch KJ, Tabor RF (2008) Tuning aggregation of microemulsion droplets and silica nanoparticles using solvent mixtures. J Colloid Interface Sci 318:244–251

Sarac B, Medos Z, Cognigni A, Bica K, Chen LJ, Bester-Rogac M (2017) Thermodynamic study for micellization of imidazolium based surface active ionic liquids in water: effect of alkyl chain length and anions. Colloids Surf A: Physicochem Eng Asp 532:609–617

Savjani KT, Gajjar AK, Savjani JK (2012) Drug Solubility: importance and enhancement techniques. ISRN Pharm 2012:195727. https://doi.org/10.5402/2012/195727

Smiglak M, Metlen A, Rogers RD (2007) The second evolution of ionic liquids: from solvents and separations to advanced materials-energetic examples from the ionic liquid cookbook. Acc Chem Res 40:1182–1192

Stubenrauch C (2008) Microemulsions: background, new concepts, applications, perspectives, 1st edn. Wiley-Blackwell, United Kingdom

Sun Y, Yan K, Huang X (2014) Formation, characterization and enzyme activity in water-in-hydrophobic ionic liquid microemulsion stabilized by mixed cationic/nonionic surfactants. Colloids Surf B: Biointerfaces 122:66–71

Tang W, Liu B, Wang S, Liu T, Fu C, Ren X, Tan L, Duan W, Meng X (2016) Doxorubicin-loaded Ionic liquid-polydopamine nanoparticles for combined chemotherapy and microwave thermal therapy of cancer. RSC Adv 6:32434–32440

Tanner EEL, Curreri AM, Balkaran JPR, Selig-wober NC, Yang AB, Kendig C, Fluhr MP, Kim N, Mitragotri S (2019) Design Principles of ionic liquids for transdermal drug delivery. Adv Mater 31:1901103

Wang C, Zhu J, Zhang D, Yang Y, Zheng L, Qu Y, Yang X, Cui X (2018) Ionic liquid-microemulsions assisting in the transdermal delivery of dencichine: preparation, in-vitro and in-vivo evaluations, and investigation of the permeation mechanism. Int J Pharm 535:120–131

Winsor P (1948) Hydrotropy, solubilisation and related emulsification processes. Trans Faraday Soc 44:376–398

Xu J, Zhang L, Yin A, Hou W, Yang Y (2013a) Nonaqueous ionic liquid microemulsions of 1-butyl-3-methylimidazolium tetrauoroborate, toluene and ethanol. Soft Matter 9:6497–6504

Xu J, Zhang L, Li C, Zhan T, Hou W (2013b) Ionic liquid microemulsions of 1-butyl-3-methylimidazolium hexafluorophosphate, N, Ndimethylformamide, and water. RSC Adv 3:21494–22500

Zech O, Thomaier S, Bauduin P, Rück T, Touraud D, Kunz W (2009) Microemulsions with an ionic liquid surfactant and room temperature ionic liquids as polar pseudo-phase. J Phys Chem B 113:465–473

Zech O, Harrar A, Kunz W (2011) Nonaqueous microemulsions containing ionic liquids: properties and applications. In: Kokorin A (ed) Ionic liquids: theory, properties, new approaches, 1st. Chapter, vol 11. InTechOpen, pp 245–270

Printed in the United States
by Baker & Taylor Publisher Services